D1379774

GEOMORPHOLOGY

PRESENT PROBLEMS AND
FUTURE PROSPECTS

GEOMORPHOLOGY

PRESENT PROBLEMS AND FUTURE PROSPECTS

EDITED FOR
THE BRITISH GEOMORPHOLOGICAL RESEARCH GROUP
BY

C. EMBLETON

D. BRUNSDEN

AND

D. K. C. JONES

1978

OXFORD UNIVERSITY PRESS

Oxford University Press, Walton Street, Oxford OX2 6DP

OXFORD LONDON GLASGOW NEW YORK
TORONTO MELBOURNE WELLINGTON CAPE TOWN
IBADAN NAIROBI DAR ES SALAAM LUSAKA ADDIS ABABA
KUALA LUMPUR SINGAPORE JAKARTA HONG KONG TOKYO
DELHI BOMBAY CALCUTTA MADRAS KARACHI

British Library Cataloguing in Publication Data

British Geomorphological Research Group. Conference, London, 1976
Geomorphology.
1. Geomorphology – Congresses
I. Title II. Embleton, Clifford
551.4 GB400 77-30542

ISBN 0-19-874078-6

Typeset by Hope Services, Wantage
Printed in Great Britain
by Fletcher & Son Ltd., Norwich

PREFACE

This volume contains sixteen papers presented at a conference of the British Geomorphological Research Group held in London, 5-9 April 1976. Seven speakers were invited from overseas to deliver lectures, and other papers were presented by British contributors. The conference was organised by D. Brunsden and C. Embleton of King's College, and D. K. C. Jones and J. B. Thornes of the London School of Economics. Financial support was received from several sources, including the Royal Society and the British Council; the United States Army, Office of European Research, contributed generously and made it possible to invite participants from Europe and North America.

C. EMBLETON

CONTENTS

Introduction — ix
J. B. THORNES

1. Bases for theory in geomorphology — 1
R. J. CHORLEY

2. The character and problems of theory in contemporary geomorphology — 14
J. B. THORNES

3. El Asunto del Arroyo — 25
L. B. LEOPOLD

4. Fluvial processes in British basins — 40
K. J. GREGORY

5. Slopes: 1970–1975 — 73
A. YOUNG

6. Some recent progress and problems in the study of mass movement in Britain — 84
D. B. PRIOR

7. Glacial geomorphology: present problems and future prospects — 107
J. GJESSING

8. British glacial geomorphology: present problems and future prospects — 132
R. J. PRICE

9. Periglacial geomorphology: present problems and future prospects — 139
J. DEMEK

10. Periglacial geomorphology in Britain — 154
R. S. WATERS

11. Tropical geomorphology: present problems and future prospects — 162
I. DOUGLAS

12. Denudation in the tropics and the interpretation of the tropical legacy in higher latitudes – a view of the British experience 185
M. F. THOMAS

13. Research in coastal geomorphology: basic and applied 203
H. J. WALKER

14. Coastal geomorphology in the United Kingdom 224
C. A. M. KING

15. Applied geomorphology: a British view 251
D. BRUNSDEN, J. C. DOORNKAMP, and D. K. C. JONES

16. The future of geomorphology 263
G. H. DURY

Index 275

INTRODUCTION

J. B. THORNES
(*London School of Economics*)

This book comprises a series of essays arising from a conference held in London in the spring of 1976. The conference, organised by the British Geomorphological Research Group, sought to establish the present status of the subject as seen by practitioners from Britain and abroad. Speakers of international standing were invited to discuss contemporary and likely future developments in their own particular fields. Not all fields have been covered; for example, there are no particular papers on Quaternary studies or on structural geomorphology but, in addition, several British geomorphologists were invited to examine the contribution of British work in various aspects of geomorphology. Taken together, the papers present a comprehensive and well-informed picture of the 'State of the Art' in the 1970s.

A true perspective of this collection of papers can only be gained if it is realised at the outset that geomorphology and associated scientific disciplines are undergoing rapid and accelerated development throughout the world. Because of the resultant enormous rate of production of material, it has become increasingly difficult for workers to be acquainted with all the literature even in their own research areas, let alone that published in other parts of the discipline. Consequently the return period for previously published ideas has become shorter. Such a situation makes the occasional appraisal of current research on a subject-wide basis both valuable to individual workers and extremely important for the well-being of the subject as a whole.

The nature and scope of geomorphology has changed radically over the last three decades. Although a great deal of credit must go to two American engineers, E. W. Lane and R. E. Horton, for their seminal works in the 1930s, the major flux in geomorphology at the turn of the century is to be attributed to the work of J. H. Mackin and A. N. Strahler. Mackin established a change in ideas from the cyclical models prevalent in the first half of the twentieth century towards the more generally acceptable concept of geomorphological systems in dynamic equilibrium. Strahler, in his paper on the 'Equilibrium theory of slopes', not only grasped the concept but also initiated studies that tested the concept in the more general framework of systems theory. The over-all benefits resulting from both of these contributions are only now beginning to be fully realised.

The progressive acceptance of the need to quantify landforms, both in simple and multivariate situations, was largely due to Strahler's active pursuit of terrain studies through grants to Columbia University from the United States Office of Naval Research. Strahler's work, together with other notable contributions, led to M. Melton's important investigation of the correlation structure of variables in a drainage basin. The recurrent use of these techniques in succeeding years and the development of more complex descriptive measures has meant that the problems and procedures involved in the use of such techniques are now widely disseminated.

In a somewhat similar vein, the measurement and description of dynamic change, as opposed to description of the static state vectors, played an important role in the shift towards a more scientifically-based geomorphology. Among these studies, S. A. Schumm's research into the rapidly developing slopes and channels at Perth Amboy seems highly significant. Form description, if pursued diligently and over time, was expected to reveal the nature of the underlying process. In the following 15 years this theme was expanded into the widespread attempt to examine the rates of operation of geomorphological processes. It is a remarkable fact, signalled by several of the contributors to this volume, notably A. Young, K. J. Gregory and I. Douglas, that even now we are still basically ignorant of the rates of operation of many of the processes that are regarded as fundamental in terms of the creation of landforms.

It is difficult to avoid the impression from the contributions gathered together in this volume that there has been a distinct shift away from the description of landforms, except in the context of mapping exercises undertaken as part of applied work. In its place there appear to have been three major new developments that appear likely to play an increasingly important role in the next 10 years. The first of these is the direct application of mechanical principles to form development. Strahler's 'Dynamic basis of geomorphology' (1952) and the slope stability studies of K. Terzhagi and A. W. Skempton immediately come to mind. The latter provide an example of stimulus to pure research from an applied viewpoint. These works have influenced especially the study of mass-movement phenomena, and D. B. Prior traces the increased interest and activity in this area by British geomorphologists. He points out that there is here a large overlap in research with engineers and geologists, which leads him to conclude that geomorphologists will have to acquaint themselves more and more with the fundamental concepts and methods of soil and rock mechanics.

Fluvial geomorphologists, by contrast, appear to have made remarkably small inroads into mechanics. Gregory illustrates graphically the intimate mingling of recent fluvial work with the rapidly developing field of hydrology. However, there appears to be little conflict here, for only a few of the geomorphologists he cites are especially concerned with process–form relationships. The majority have contemporary processes, particularly their rates of operation and problems of measurement, as the centre of concern. Despite the weight often attached to fluid properties in textbooks and teaching programmes, there appears to have been a failure to come to grips with these properties, except in a few instances in theoretical work, some of which are noted by J. B. Thornes in his paper. In part this may reflect a failure to recognise the fundamental questions. It is refreshing to find L. B. Leopold seeking a rephrasing of such questions, in the field, as a direct result of field observation. In particular he suggests that the slopes of arroyos are the key variable of dynamic interest and one whose control on form is only poorly understood after many years of observation. On the basis of hydraulics alone, he argues, quasi-equilibrium and unsteady behaviour cannot be separated. R. J. Chorley is also concerned that we should be careful not to lose sight of the larger-scale forms and responses in what may loosely be called the 'earth-surface processes syndrome'.

Certainly the process theme is given prominence in all the papers presented. Though this is in part owing to the selection of papers by process area, within any one of these areas it is clear that the study of processes is a major preoccupation. A major part of coastal studies, as reflected by H. J. Walker's paper, is the investigation of geomorphological agents, a trend he sees as continuing for some time to come. Here too, the reliance of theoretical developments on measurement techniques is spelt out, but the mechanics of wave motion and the transformation and dissipation of wave energy in the surf zone are regarded as of utmost importance. One particular area of increasing importance and interest concerns the recent studies of bio-erosion. This interface between rock and organism is also assuming an increasing role in slope studies as revealed by Young's paper.

As organisers of the conference, we attached importance to the rather special nature of mechanics and processes in the tropical, glacial and periglacial environments. In the tropics, however, Douglas argues that the geomorphology is not of a special kind. Similarly, in his review of work in glacial environments, J. Gjessing argues that the alternation of glacial and subaerial processes and antecedent conditions are important in determining the resulting form. Both Douglas and Gjessing argue for further comparative studies of the range of tropical and glacial environments respectively in order to reach a better understanding of form and process. In both of these extreme environments, generalisations in the past have relied on rather particular and spatially isolated case studies and tended to concentrate on spectacular, unusual or clearly defined landforms.

Current research in glacial geomorphology is now concentrating on studies of erosional and depositional processes, in the field and the laboratory; on developing models of erosion, transport and sedimentation, and on morphometric studies. The latter, compared with fluvial morphometric studies, are still in their infancy but, as Gjessing shows, accurate description, analysis and classification of phenomena such as plastic scouring forms, *roches moutonnées*, and the whole range of glacial macroforms, is essential to attempts to understand their modes of origin.

Investigations by J. Demek and R. S. Waters in the third 'special' environment considered, the periglacial environment, are also seen to be process-based. It is therefore perhaps surprising that rather less progress into process-rate measurement and the theory of cryogenic processes has been made. However, studies of the genesis, composition and structure of frozen ground are accelerating, the distribution of permafrost is becoming known in detail in many areas, and the distinctive landforms are being classified and analysed. Landforms such as pingos, ice-wedge polygons and thermokarst features are being used as diagnostic indicators of the types and conditions of frozen ground.

The second common area of interest, touched upon by almost all the authors, is the role of theory in geomorphology. It can be fairly said that, after the shift in emphasis towards the study of process and mechanics, the development of theoretical studies over the last 15 years represents the major shift in geomorphology. In the middle 1960s, the highly important work of A. E. Scheidegger, W. B. Langbein and W. E. H. Culling led to a significant reappraisal, not by any means completely worked through, of approaches to the subject. Chorley examines the development of theory-types and, in particular, expresses serious reservations about present-day theory development on the grounds that it carries the subject into direct competition with other traditional and well-established disciplines. Thornes outlines the development of the steady-state school and concludes that current and future theoretical work, particularly in fluvial and work studies, will above all be related to unsteady behaviour. Young sees the complementary development of numerical iterative three-dimensional solutions of the type investigated by F. Ahnert and A. Armstrong on the one hand, and the closed-form equation approach exemplified in Britain by the work of M. J. Kirkby. Young and several other contributors remark on the few attempts to provide data for the direct comparison of deduced forms and actual landforms. In this respect the dichotomy between the long-term responses and the short-term process studies, noted by C. A. M. King in her chapter on coastal studies in Britain, is an important aspect of the evaluation of theoretical models. It seems likely that a period of challenge and refutation will be engendered by the encouraging leaps in theory over the past few years.

The third and final theme which is almost all-pervading is the interface between geomorphological processes and human activity. The theme has gained great momentum since E. H. Brown's inaugural address, at University College London, on 'Man as a geomorphological agent'. The eagerness with which some geomorphologists espoused the subject of hydrology may in part reflect this trend, for there can be little doubt that much of the work described by Gregory has direct application. Yet in the tropics, too, Douglas argues forcibly and very persuasively that future geomorphological investigations must pay particular attention to 'the problems of most importance to the people of tropical countries and endeavour to give them priority'. Demek emphasises the applications of periglacial geomorphology, insisting that the economic utilisation of the periglacial zone requires careful planning because of the highly sensitive systems and the catastrophic responses that may result from interference. In coastal studies, applied research is aimed at solving specific problems, usually those of pressing economic significance. Walker's view is worth reiterating here; 'such an emphasis will almost certainly lead to a reduction in the production of fundamental discoveries, discoveries that actually make applied research meaningful'. Chorley also expresses some reservation about 'conventionalist theory'. D. Brunsden, J. C. Doornkamp and D. K. C. Jones adopt a more aggressive view towards applied work and incline towards a higher level of technical training and professional conduct, stimulated by increasing contact with professionally organised geologists and civil engineers and others working at an interdisciplinary level. Whatever the relative merits of these cases, it is difficult to avoid the conclusion that geomorphologists are destined, in the next few years, to play a larger role in solving problems of direct relevance to the prevailing social and economic climate. As Walker states, 'the relative proportions of basic and applied research appear to be very closely tied to the state of the economy'.

On the last day of the conference the general discussion involved these three areas: the role and development of applied research, the growth of new theory arising out of newly acquired mathematical skills, and the problems of the modelling of mechanical processes. G. H. Dury then gazed into his crystal ball! Faced with the unenviable task of predicting future developments in the subject, he provided an entertaining introduction to a lively discussion. This debate was expectably unstructured and occasionally repetitive, but it was possible to detect several themes that slowly emerged and began to crystallise into a consensus of opinion. The role of applied studies, process modelling and theory were again heavily stressed but several delegates saw interdisciplinary studies, the application of concepts and methods from other disciplines, and a need to re-examine the basic principles of geomorphology itself, as the most fruitful approaches.

A strong plea was made to use the ecosystem as the basic unit of study with ecology, geomorphology, pedology, climatology and geology forming the most desirable subject grouping for degree-level training and for applied research. Many delegates were concerned that palaeo-studies and a return to the study of the evolution of landforms should play a central role. This approach was recommended by several engineers in the audience who suggested that we should concentrate on extending short-term process mechanics as developed by such subjects as hydraulics and soil mechanics to the longer time-scales required for an understanding of landforms.

This provoked a lively exchange concerning the scales at which work should be carried out with, predictably, those of the 'present process school' joining the pedologists in arguing for small-scale, short temporal studies, and the denudation chronologists arguing for large-scale studies, accurate dating of events, and theoretical models for the extrapolation of the results of short-term work to the long term. All agreed, however, that the major task of immediate importance was the need to understand and obtain information on the operation of processes and the evolution of form in the graded (10^3-10^4 years) time-span.

Finally, some delegates pleaded for improvements in technical proficiency and in the standards of equipment and testing procedures. The relative merits of mapping, measurement, counting and analysis were outlined and there was general agreement that we were moving toward an era of professionalism and exciting discovery. As one reviewer put it (*Geophemera*, no.9, 1976), 'Geomorphology is alive and well, with good prospects'.

REFERENCES

Ahnert, F. (1973), 'COSLOP 2–a comprehensive model program for simulating slope profile development', *Geocom Programs* **8**, 24 pp.

––, (1976), 'Darstellung des Struktureinflusses auf die Oberflächenformen im theoretischen Modell', *Z. Geomorph., Suppl. Bd* **24**, 11–22.

Armstrong, A. (1976), 'A three-dimensional simulation of slope forms', *Z. Geomorph., Suppl. Bd.* **25**, 20–8.

Brown, E. H. (1970), 'Man shapes the Earth', *Geogr. J.* **136**, 74–85.

Culling, W. E. H. (1963), 'Soil creep and the development of hillside slopes', *J. Geol.* **71**, 127–61.

Horton, R. E. (1932), 'Drainage basin characteristics', *Trans. Am. geophys. Un.* **13**, 350–61.

–– (1933), 'The role of infiltration in the hydrologic cycle', *Trans. Am. geophys. Un.* **14**, 446–60.

Kirkby, M. J. (1971), 'Hillslope process-response models based on the continuity equation', *Inst. Br. Geogr. Spec. Publ.* **3**, 15–30.

–– (1975), 'Deterministic continuous slope models', *Univ. of Leeds, Dept. of Geography, Working Pap.* **130**, 28 pp.

Lane, E. W. (1937), 'Stable channels in erodible materials', *Trans. Am. Soc. civ.*

Engnrs. **102**, 123–94.
Langbein, W. B.: see Scheidegger and Langbein (below).
Mackin, J. H. (1948), 'Concept of the graded river', *Bull. geol. Soc. Am.* **59**, 463–512.
Melton, M. A. (1958), 'Correlation structure of morphometric properties of drainage systems and their controlling agents', *J. Geol.* **66**, 442–60.
Scheidegger, A. E. & Langbein, W. B. (1966), 'Probability concepts in geomorphology', *U.S. geol. Surv. Prof. Pap.* **500C**, 14 pp.
Schumm, S. A. (1956), 'The evolution of drainage systems and slopes in badlands at Perth Amboy, New Jersey', *Bull. geol. Soc. Am.* **67**, 597–646.
Skempton, A. W. (1945), 'Earth pressure and the stability of slopes' in *The principles and application of soil mechanics* (Instn. of Civil Engineers, London), 50–3.
–– (1964), 'The long-term stability of clay slopes', *Géotechnique* **14**, 75–102.
Strahler, A. N. (1950), 'Equilibrium theory of erosional slopes approached by frequency distribution analysis', *Am. J. Sci.* **248**, 673–96 and 800–14.
–– (1952), 'Dynamic basis of geomorphology', *Bull. geol. Soc. Am.* **63**, 923–38.
Terzaghi, K. (1943), *Theoretical soil mechanics* (New York), 510 pp.
–– (1950), 'Mechanism of landslides', *Bull. geol. Soc. Am.*, Berkey Volume, 83–122.

I

BASES FOR THEORY IN GEOMORPHOLOGY

RICHARD J. CHORLEY
(*University of Cambridge*)

Whenever anyone mentions theory to a geomorphologist, he instinctively reaches for his soil auger. A recent questionnaire from the British Geomorphological Research Group on the teaching of geomorphology in University Geography Departments in the United Kingdom requested comments regarding the 'methodological component' of each course–in itself an interesting phrase. The responses were unenthusiastic. Two of the most common attitudes were either that there is no need to distinguish methodology from techniques, or that the scientific method is obvious and therefore needs no discussion. I do not agree with these views and, to adapt an aphorism, believe that the only true prisoners of theory are those who are unaware of it. It is indeed disturbing that attitudes to theory are so negative at a time when the bases of our theories are changing so rapidly and drastically with, as I believe, profound implications for the future of geomorphology.

Geomorphology is that science which has for its *objects* of study the geometrical features of the earth's terrain, an understanding of which has been attempted in the past within clearly definable, but not always clearly defined, spatial and temporal scales and in terms of the processes which produced, sustain and transform them within those scales. Each generation of geomorphologists has redefined both these scales and processes in manners which have appeared to accord best with the contemporary *aims* of the science. The changing theoretical bases of geomorphology derive from these changing aims, which determine the character that geomorphological theories assume. The problem is that these changing aims have often been so intimately bound up with general contemporary currents of thought and, in an essentially small scholarly discipline, identified with the personal preferences and ambitions of a few influential practitioners, that it has not always been possible to recognise clearly that conscious choices are possible regarding aims and theory, or even that these aims exist at all. Occasionally, however, light is shed on the interactions between scale and process, on the one hand, and geomorphological aims, on the other. Such a light derived from the paper published more than a decade ago by S. A. Schumm and R. W. Lichty (1965), and it is instructive to recognise, in a period of rapid and considerable changes in not only the aims but also, perhaps for the first time, in the objects of our study, that this is the most recent paper of theoretical consequence in geomorphology. Theory always preconditions practice, but always implicitly, and in times of rapid change the game is commonly well advanced before its rules are understood, much less its aims explicitly formalised. In systems terms, practising geomorphologists react to theoretical inputs with a high noise level and long relaxation time and only equilibriate happily with theory during long periods of stable, uncomplicated input.

It now seems clear that the implicit aims which have preconditioned theory in geomorphology during the past two centuries show evidence of having been associated with a number of recognisable phases which may conveniently, if not unambiguously, be entitled as follows:

Teleological. This concerned studies of terrain aimed at relating geomorphological observations to current views of the overall design of nature and to concepts involving final causes. Such an approach has existed in a wide variety of disciplines and was designed to facilitate general views of the real world which satisfy both the dictates of the intellect as well as existing orthodoxy.

Immanent. This subsumes theories which concentrate on explaining the characteristics of landforms in terms of their indwelling or inherent nature.

Historical. The explanation of landforms in terms of a narrative reconstruction of a series of events assumed to have led up to the present. Such palaeogeographical speculations are often found to rely on allegorical or metaphysical underpinnings.

Taxonomic. The explanation of landforms by their grouping into spatially associated classes. Such an approach is invariably connected with an information explosion and its occurence in time is to this extent scientifically fortuitous. Its development in geomorphology took place during the growth of the ecological concept in natural science and it thus acquired overtones of both ecology and functionalism which help it to maintain some of its contemporary attraction.

Functional. The basis of theory relying on the mainstream logical positivist thesis that real world phenomena can be explained by showing them to be instances of repeated and predictable regularities in which form and function can be assumed to be related. Theory of this type derives from the view that science is empirically based, rational, objective and aimed at providing explanation and prediction on the basis of observed regular relationships.

Realist. This is a philosophical extension of the functional approach to logical positivism, which accepts many of the tenets of the latter. It is based on the view that explanation involves something more than prediction based upon observed regularities. It thus seeks to penetrate 'behind' the external appearances of phenomena to the essences of mechanisms which necessitate them as the result of chains of causal connection. The realist is concerned with the identification and investigation of detailed causal mechanisms and of the underlying structures of which the external forms are the artifacts (R. Keat and J. Urry, 1975).

Conventionalist. These theories depart from logical positivist attitudes in the belief that no useful distinctions can be drawn between observation and theory, and that the development of the former precedes the latter. Moreover, it is postulated that external reality cannot be usefully held to exist independently of our theoretical beliefs and concepts. Thus the adoption of theory is largely a matter of convention in which moral, aesthetic and utilitarian values play an essential part (Keat and Urry, 1975).

TELEOLOGICAL AND IMMANENT BASES OF THEORY

Teleological explanations of landforms were fundamental to the development of the science of geomorphology for many hundreds of years, their origins reaching back to instinctive desires to employ observations of the present world to justify the ways of God (or of gods) to man. In some senses it might be argued that until the latter part of the eighteenth century the true object of geomorphological study was not the landform itself but the mind of the Almighty, of which the landform was held to be an outward and visible manifestation. Thus the work of Ray, Buffon and Cuvier was linked with that of the Middle Ages by a conscious desire to understand the terrestrial economy within the guidelines of existing orthodoxy. In terms of theory building, this had the advantage of permitting one to evoke and to dismiss at will arbitrary past causes, providing that their magnitude and frequency accorded with orthodox views. It is instructive to recall that the most compelling work in geomorphology grounded in teleological principles was published by Dean Buckland in 1820, no more than some 150 years ago.

The decline of the old teleology was due to a breakdown in confidence regarding the magnitude and frequency of events which it presupposed. It was natural that it should be replaced by a causal basis of theory founded upon events of smaller magnitude both in space and time. These immanent theories relied upon the assumed inherent features of both exogenetic processes and endogenetic characteristics in order to permit theoretical explanations to be erected on spatial and temporal scales which encouraged human observation and testing. The geomorphological work of J. Hutton and J. Playfair was based upon the assumption that spatial patterns of erosion and sedimentation were autocorrelated and that temporal patterns were parts of non-stationary, but nevertheless continuous, series. This approach which leans, at base, so heavily upon seventeenth and eighteenth century advances in the calculus found expression in concepts embodying a belief in the 'nice adjustment' of landforms in space which was expressed so succinctly in *Playfair's Law* (1802) and in the temporal complement of erosion and deposition upon which Hutton's theory rested. Although this work vastly expanded the time-scale within which theory could operate, it is interesting that it fostered a view of an inherent association between erosion and deposition, uplift and subsidence, and between form and process which, in terms of geomorphology, found expression in the almost organic view of drainage systems adopted by Playfair.

The other manifestation of immanence was the nineteenth-century development of ideas relating landforms to geology. This came about both because of the close association between 'rocks and relief' in the scientifically advanced areas of western Europe and north-eastern North America and because of the symbiotic relationship which came to exist between the interpretation of landforms and the mapping of bedrock geology. This relation, so manifest in the work of W. Smith, J. P. Lesley and J. W. Powell, produced such a strikingly successful level of prediction that its reverberations still dominate some of our textbooks today. Indeed, so successful a geomorphological explicator did lithology and structure become that the relationship was assumed to be so close as to preclude the need to establish causal links between rocks and relief in terms of detailed studies of the manner by which certain differences in rock types support the recurring differences observed to exist in terrain. It is only in very recent years that detailed work has been attempted on such geomorphological topics as the significance of rock strength (E. Yatsu, 1966) and the manner by which jointing becomes topographically exploited.

Notwithstanding all their apparent philosophical and practical strength, the immanent bases of geomorphological theory which were present in the first half of the nineteenth century were soon to be supplanted. Once again it was a shift in scale emphasis which occasioned the change. Only on a large spatial scale could one be impressed either by the overriding spatial integration of drainage systems and slopes or by the simple relationships of rocks and relief. As detailed topographical and geological mapping proceeded during the second half of the last century, particularly in Great Britain and the United States, attention became focused upon the spatially inconsequential but naggingly recurrent examples of terrain that departed from what might be expected on the grounds of spatial patterns of either erosion or geological control. It was these anomalies which heralded the protracted 'classical' period of theory directed towards historical explanation in geomorphology.

HISTORICAL THEORY

The goal of historical explanation is not prediction but retrodiction, the derivation and testing of singular descriptive statements about a sequence of real, individual but interrelated past events (G. G. Simpson, 1963, p. 25; D. B. Kitts, 1976, p. 38), constituting a chronology of specific dates and places at which events occurred (Kitts, 1963). Its basic question is not the 'How?' of science, nor the 'Why?' of philosophy, nor the 'What for?' of humanism, but 'How come?' (Simpson, 1963, p. 28). History is not susceptible to the application of laws in the logical positivist sense of that term, because, whereas scientific laws are abstractions divorced from individual cases, history deals with unique events and non-repeatable processes (K. R. Popper, 1957). For this reason neither the 'laws of evolution' nor even 'evolutionary trends' are laws in an accepted causal sense (Kitts, 1974, p. 12), despite the application of notions of indeterminacy–much less the so-called 'laws of historical succession' (Kitts, 1963). It is the association of geomorphology with geological theory which explains the essentially historical theoretical basis for the main body of classical geomorphology, as exemplified by the model of the cycle of erosion, by denudation chronology and by tectonic geomorphology.

Comte suggested that the development of theory construction commonly follows three stages, proceeding from the theological to the metaphysical and thence to the positivist. If Buckland is a geomorphological practitioner of the first, and G. K. Gilbert of the last, then W. M. Davis's cycle can be shown to be an example of the second stage. The cycle of erosion was the result of the first successful attempt to develop in geomorphology a theoretical model of any power, and it was aimed at the understanding and genetic classification of landforms on regional spatial scales and within geological time-scales in a combined metaphysical, organic and allegorical manner. Much has been written elsewhere regarding this theoretical basis for geomorphology and it is sufficient here to point to its decline in terms of its inevitable failure to provide a law of historical succession which could be individually applied without leaving too much outstanding unexplained noise.

At the same time as the cyclical model was being developed, denudation chronology came into being, initially quite separately. Denudation chronology was based on the notion of historical succession, just as the cycle was founded on the evolutionary concept. However, the cycle soon provided the temporal yardstick against which morphological evidence for cycles and parts of cycles (i.e. the structurally and erosionally anomalous water gaps, wind gaps, bevelled ridges, breaks of slope, accordances, terraces, peneplains and the like) were measured. Apart from remaining a device for teaching and landform classification, the cycle was soon subsumed into the simple historical pattern of denudation chronology. The aim of the latter was to use landforms to illuminate the more recent stages of earth history, especially those bearing on tectonic and eustatic matters, so that although landforms remained the objects of study, the aim of study was broadly geological. Great efforts were made to justify such work on the basis of scientific theory, not least in terms of the so-called method of multiple-working hypotheses (Gilbert, 1886 and 1896; T. C. Chamberlin, 1897). Davis's study of the rivers and valleys of Pennsylvania (1889) was held to demonstrate his pure and secure use of this analytical method (D. W. Johnson, 1940) but if it was not apparent in his earlier work it was made abundantly clear by his study of coral reefs (Chorley, R. P. Beckinsale and A. J. Dunn, 1973) that Davis's theoretical techniques involved the embracing of an initial conclusion, largely from studies of maps, and then in justifying this conclusion by argument and a minimum of carefully selected field observations. Davis's skill at sequential-type puzzles, together with the immense lengths of unsignposted geological time within which he could manoeuvre, allowed him great freedom for hypothesising and he was able to give strength to his retrodiction by his skill at adducing multiple inferences to support belief in his proposed events. In his classic paper on Pennsylvania he retreated to the assumed secure origins of the Permian orogeny (wrongly, as it turned out) and then used Occam's Razor to prune out the most likely sequence of subsequent events which might have resulted in the existing features. His method was based on logical likelihood operating in the absence of fact. It was an act of faith based on the assumption of recurring patterns of historical development. It is a measure of Davis's mental dexterity that, beginning with an assumption made eventually unnecessary by the island-arc theory of marginal deformation, he could have faced the problem of the regional reversal of drainage and accomplished this by piecemeal means apparently to everyone's satisfaction except his own. It is also illuminating, regarding the nature of theory in denudation chronology, that Johnson's supplanting of Davis's theory of Appalachian drainage came about not because new evidence was adduced but because the theory of the former was more internally logical and economical than that of the latter.

The tectonic theories of geomorphology, of which that of W. Penck is the best known, were theoretically similar to those of denudation chronology but commanded less popularity because of language, political and personal considerations on the one hand, and less technical assumptions on the other. Among the latter the major difficulty was the manner in which the topographical discontinuities so requisite to the provisions of the necessary temporal hooks upon which historical theory might be suspended could be produced by the mutual interaction of continuous exogenetic and endogenetic processes.

The decline of the historical theories of geomorphology, dating from about the Second World War, came about because of another collapse of scale. They broke down because their time scales were so large and unsignposted that they became the playground for unbridled and untestable speculation. The field became dominated by the spinners of ingenious historical sagas, following themes that were traditional both in development and outcome. Like the Navajo, many historical geomorphologists either died out or retreated into a better and more ordered world. Those who did not pursued temporally-restricted goals based on stratigraphy and were content to compose accurate phrases rather than visionary sagas. Only new techniques, particularly of dating, are in my view capable of broadening the geomorphological horizons of the latter and of giving reality to the dreams of the former.

TAXONOMIC BASES OF THEORY

If theory is the child of necessity, and if historical geomorphology developed in response to geological imperatives, then it is clear that the growth of regional taxonomic studies in geomorphology stemmed from its geographical affiliations. The emphasis on the identification of morphological regions, commonly possessing climatic identity, which commenced in earnest about 1890 and is still going on today, developed with the initial aim of using landforms to provide a regional taxonomic framework for, as well as a generalisation regarding, the wide-ranging and rapidly-accumulating information accruing from the colonial consolidation by Britain, France, Germany, Russia and the United States (D. R. Stoddart, 1969). Like all taxonomic theories, those in geomorphology developed as the result of attempts to digest unusually rapid increases in the generation of data. However, the theoretical banality of taxonomy has caused it to assume the gloss of more challenging theory and thus in geomorphology we find historical/cyclic, functional/climatic and interactive/ecological developments of regional taxonomy, not to mention the social/utilitarian ones upon which present land classifications rest. It is no accident of history that Davis's formalisation of the arid cycle, A. Penck's examination of the climatic features of the land and A. Demangeon's *Picardie* all appeared in the same year (1905). Nor is it remarkable that the spatial functionalism of climatic geomorphology was also the product of the period which gave us Vidal de la Blache's regional geography and Clements' formalisation of the concept of ecology. In geomorphology these theoretical bases crystallised during the early part of this century around two centres: climatic geomorphology, with its cyclic and geographical implications, and morphological geomorphology linked by geography to the broader concept of 'natural regions'. Both theoretical bases are still with us today, although in a much modified form—a situation attributable to their amorphous theoretical bases, their flexible aims and the continually rising flood of topographic data at all spatial scales. To a large extent climatic geomorphology has been shorn of its cyclic implications mainly because of the local structural and lithological idiosyncracies that caused A. Hettner (1928, see 1972, p. 136) to resist the theory of the 'normal' cycle. The insupportable cyclic assumptions regarding geological structure which were most obviously exposed in the cases of the arid and glacial cycles (Davis, 1905, 1906) have now been removed. Moreover, the increasingly ideographic approach to the treatment of landforms within climatic subdivisions stems from the recognition both that the spatial scales originally adopted were too large to damp down local structural and lithological noise and that the temporal scales were too large to avoid the deep polygenetic confusions associated with complicated and persistent climatic changes. To a considerable extent, morphological geomorphology was associated with the Russian regional landscape science as developed by Dokuchayev in the last decade or so of the nineteenth century and continued by his students and followers, notably by Berg (A. G. Isachenko, 1973). The theoretical basis of this approach was much more direct and pragmatic than that of climatic geomorphology, or even than that of its Western counterpart of natural geographical regions. For this reason, and as a result of considerable experimentation both with its flexible spatial and temporal scales and with its aims, it has been able to adapt to modern conventionalist theory.

FUNCTIONAL THEORIES

If not at first the most apparently profound, the most rapid change in mainstream geomorphology occurred just after the Second World War with the advent of what at the time was variously termed 'new' geomorphology, scientific geomorphology or quantitative geomorphology. This approach purported to relate geomorphological forms to their governing processes, but in reality linked form with a more nebulous notion of 'function'. This theoretical basis had been previously propounded by several geologists, geographers and engineers isolated from the main current of teaching in earth science, notably by Gilbert, and was also present to some degree in the ecological overtones of climatic geomorphology. Although, as we have seen, the science has already been prepared by the shortcomings of the historical and taxonomic approaches to turn its attention to the study of landforms of smaller spatial scale and to changes within shorter time scales, it was the posthumous publication by R. E. Horton (1945) which was the immediate instrument of change. Horton's geomorphological work is especially interesting because it contained the same two themes as were exemplified by the work of Gilbert. On the one hand it treated the relationships between erosional forms, or between forms and gross hydrological transfers, and on the other it looked at detailed erosional processes, sometimes on the microscale. It is most significant that Horton did not succeed in uniting these two themes, a fact best demonstrated by the failure of his theory of micro-rilling and cross-grading to provide a secure genetic model for the development of large-scale drainage networks. Horton's work was embraced by A. N. Strahler to whom fell the responsibility of formulating these ideas in a broader geomorphological setting, presenting them to a wider public, and placing them in their historical perspective against Gilbert's receding bulk. What emerged in the 1950s was a classic functional science with mesoscale landforms as the objects of study, but with either form (e.g. Strahler's (1950) channel slope) or mass transfer events (e.g. L. B. Leopold's (1953 *et seq.*) and G. H. Dury's (1962) use of discharge) employed as surrogates for the processes of which the forms were assumed to be the function. The supposed functional associations were secured by a rapidly expanding repertoire of statistical correlation techniques, deriving as a logical necessity, contrary to much popular belief, from the theoretical basis of functionalism and not *vice versa*. The functional approach, especially involving considerations of process or its surrogates, is most clearly adaptable to morphological features of medium to small spatial scale and particularly of relatively rapid temporal change. It was thus no accident that the most important work of the period derived from the Perth Amboy and other badlands (Schumm, 1956a, 1956b), the hydraulic geometry of alluvial channels (Leopold and T. Maddock, 1953), beach forms (W. C. Krumbein and H. A. Slack, 1956), the Verdugo Hills (Strahler, 1950) and low-order basins in semi-arid environments (M. A. Melton, 1957). It is of note that the rapidity of change necessary to support the observed functional correlations often appeared implicitly rather than as an explicit result of these studies. Where detailed studies of process were pursued they tended to lead the researcher into the study of forms which seemed on too small a scale to be geomorphologically viable at the time, and it is more than coincidence that Strahler's least influential paper of this period expounded his dynamic basis of geomorphology (1952a). Functional theories in geomorphology were thus based upon statistical entitation and not upon the establishment of causal process links. Statistical entitation later achieved its most exaggerated manifestation in studies of stream networks after the middle 1960s.

It should not be assumed that functional geomorphology developed without concern for historical theory. Although direct measurements of the rate and character of form changes were restricted to a narrow range of specialised environments, surrogates for time were explored, such as distance (R. A. G. Savigear, 1952), stream order and hypsometric integral (Strahler, 1952b). Despite more recent misgivings over the employment of such ergodic devices, these studies were not without their rationale at the time. More direct links with classical work in denudation chronology were provided by the functional studies of J. T. Hack (1960) and R. F. Flint (1963) in the Appalachians and, more recently, by B. C. Worssam (1973) in the Weald. It was, however, at the outset of the establishment of the functional theory that J. H. Mackin (1948) made the most basic contribution to such unification of theory in geomorphology as took place during the next two decades. He suggested how physical theory provided a template from which a linkage of timebound (historical) and timeless (functional) concepts could be architected, as was later attempted by Schumm and Lichty (1965), and he introduced the idea of systems analysis to the science which was later developed (Strahler, 1950; Melton, 1958; Chorley, 1962; A. D. Howard, 1965) in such a manner as to predestine the recent theoretical shift of the discipline to a more realist basis. During the 1960s it became clear that the majority of mesoscale landform assemblages represent a palimpsest of superimposed and interlocking process-response systems of highly varying relaxation times. It also emerged that regularity of form does not automatically imply regularity of process, nor do similar forms imply similar antecedent processes. It therefore became apparent that the ability to relate form to process in any real sense hinges on an appreciation of rates of change (M. G. Wolman and J. P. Miller, 1960). With this understanding came recognition of the key theoretical roles played in geomorphology by works dealing with measurement of rates of operation (e.g. A. Geikie, 1868; Schumm, 1963). In the absence of such ordered information, a wide variety of theoretical approaches are possible with little hope of testing their relative validity. More basically perhaps, the strictly functional theories in geomorphology have proved inadequate because it became clear that one was treating the artifacts resulting from the operation of geomorphological systems, rather than the systems themselves. It is thus clear why more recent work has tended to concentrate on detailed investigations of the transfer of mass and energy, of which geomorphological forms are often the outward and visible manifestation.

REALIST THEORY

It has been seen how there has been a growing conviction among geomorphologists over the past 15 years or so that they must penetrate behind the appearances of external form to the essential physical and chemical mechanisms which are assumed to sustain and transform them. Although the systems approach has assisted in this development (Chorley and B. A. Kennedy, 1971; J. R. L. Allen, 1974), the seeds of process realism have been present among students of landforms during the whole of the 200-year period with which this paper has been concerned. They were explicit in the later work of Gilbert (1909; 1914), in that of the Swedish school of geomorphology (F. Hjulström, 1935; A. Sundborg, 1956) as well as of Horton (1945), Leopold (1953 *et seq.*) and Schumm (1956 *et seq.*). The realist view was expounded at length by A. E. Scheidegger (1961) and Dury (1972) has given a penetrating analysis of the shift towards realism in the 1960s. Nevertheless, it is only in recent years that such work has fused within the geomorphological profession into a guiding school of thought. In my view, considerable irony is to be found in the reflection that despite the apparently smooth and inevitable development of realist theory in geomorphology, its effect may well be much more profound than any which have been identified in this paper so far. Implicit in such a development are changes in attitude, not only to the spatial and temporal scales of study, but to the aims and very objects of geomorphological investigation. Nor has the demand for realist explanations in geomorphology been entirely satisfied by the interesting and inevitable application of probability models (Leopold and W. B. Langbein, 1962; Scheidegger and Langbein, 1966), any more than the elaborators of T. Hägerstrand's pioneer work have been able to satisfy similar demands in geography.

The most obvious result of the rise of realist theory in geomorphology has been to accelerate the existing tendencies towards studying landforms on more restricted scales of space and time. This has had the effect of disarticulating the science and of causing the goal established by Mackin (1948) and Schumm and Lichty (1965) to recede. A much more fundamental change has been that the objects of basic geomorphological interest, perhaps for the first time, have changed from mesoscale forms to microscale processes. Despite a short-lived euphoria in the late 1960s, with the important reiteration by Langbein and Leopold (1966 and 1968) and others (going back to Wolman (1955) and before) that the gross geometry of an alluvial channel is itself a manifestation of hydraulic roughness, geomorphology has steadily merged with an earth-surface processes syndrome. Some may regard this as no bad thing, particularly, as we shall see, as it conveniently paves the way for the adoption of conventionalist theory. Others may find it disturbing that a recent working party of the Natural Environment Research Council on the water-produced landforms, itself an example of a system with short memory and long relaxation time, was composed of twice as many engineers as geomorphologists. A decade ago hydrology, the science dealing with the distribution of terrestrial water in space and time, had objects and aims of study which were reasonably distinct from those of geomorphology; today geomorphologists have made this distinction far less secure. If the implications for our discipline were not so serious, a cynic might derive some amusement from the fact that attempts by geomorphologists to gain financial support for their work by shifting their theoretical base towards that of alleged utility have coincided with a time of economic recession which has brought their redefined discipline into direct competition with others on grounds where geomorphology is most vulnerable. At best the unbridled pursuit of process has profound implications for changes of scale in geomorphological research, at worst it sufficiently changes the objects of study as to bring the traditional integrity of the discipline into question. It is interesting that in their attitude to weathering, geomorphologists have long had an example of the paralysis of theory which accompanies a preoccupation with micro-processes. Instead of boldly concentrating on the immanent characteristics of the products of weathering and their implications for slope development, or on observed or inferred rates of weathering, or on the geometrical features and geomorphological implications of the weathered mass, geomorphologists have too often assumed the role of inept biochemists, and become preoccupied with process in a field so complex that no competent biochemist would have the temerity to attack it at that scale. The belief that a complete understanding of microscale processes is an essential prerequisite for the study of mesoscale forms would debar Professor Colin Buchanan from his calling as an expert on traffic flow because he is not a trained motor mechanic. The point I am making is that geomorphology can only continue to make a unique contribution to the earth sciences if, in the study of process, physical truth is sufficiently coarsened in both space and time as to accord with the scales on which it is profitable to study geomorphologically-viable landform objects. Furthermore, it should not be overlooked that the contraction of scale towards human dimensions in realist geomorphology, combined with utilitarian considerations, is already paving the way for the inevitable application of conventionalist theory to our science.

CONVENTIONALIST THEORIES

Thus we find the present theoretical basis of geomorphology to be a complex pastiche of largely residual attitudes. Some branches of our discipline have embarked on the uncharted sea of realism, others are dipping their toes into the rising tide of functionalism, while still others are immersed in historicism. It is no part of my task to look into the future, still less to suggest that there is any evolutionary inevitability regarding a possible succession of bases for theory. Nevertheless I may be permitted to end with some speculations prompted both by recent tendencies in our subject and by possible analogous happenings in other disciplines. First, it is clear that, for reasons of syllabus, popularity (professional, student and lay), finance, 'relevance' and the like, geomorphology is becoming increasingly committed, if not wholly then in significant part, to a role involving its relationship to human well-being and aspirations. Secondly, in disciplines further advanced along this road than ours there is an increasing gulf between the theoretical bases dictated by the scientific methods of logical positivism and those prompted by utilitarian and social doctrinaire theories (Keat and Urry, 1975). Thirdly, this gulf is one which may truly divide radically different environmental aims.

It is now generally recognised, even by the most ardent logical positivists, that generation of hypothesis precedes accumulation of data, at least in part. If data are not generated in a theoretical vacuum then it is clear that the growth of theory must derive in some measure from the existing intellectual climate. Therefore, not only the types of scientific work in which it is thought proper to engage, but also their theoretical underpinnings, are vulnerable to the pressures of convention. Those of you who have attended this paper prepared to discuss hydrodynamics, the physics of distributed shear, or other empirical relationships under the guise of geomorphological theory (Scheidegger, 1961) may find this suggestion rather fanciful, but perhaps I can elaborate my meaning. The quest for utility not only conditions the objects of our work, but also the manner in which theory may emerge. Utilitarian approaches to geomorphology will either result in large-scale work of which the intellectually-sterile taxonomic morphological mapping is the most depressing precursor, or in a piecemeal concentration on small-scale realist systems. The latter will encourage treatment in a manner which attempts to broaden spatial or temporal scales by processes of aggradation and which will commonly produce, as have attempts in ecology, highly-interlocked complex models whose empirical linkages predestine a catastrophic behaviour outside arbitrary equilibrium limits. The construction of such models in geomorphology would, of course, be very convenient for the syllabus integration of the Geography Departments which produce the majority of our geomorphologists, but it may not be a secure base for geomorphological theory. Fortunately, geomorphology in the United States (not least by virtue of the lead given to the United States Geological Survey by Leopold), the land classification of the CSIRO in Australia and, perhaps ironically even, the development of landscape science in Russia (predating even Engels' *Dialectics of Nature*, 1872–82) have preserved in my view the proper distinction between nature and society which is necessary to set apart objective management from subjective symbiosis. Conventional social attitudes to the environment affect theory in science, not only indirectly by means of their influence on preferred subject matter, but by much more direct and insidious controls over basic attitudes. Perhaps my fears are groundless and our colleagues are right in believing in a simplistic role for theory in geomorphology. On the other hand, we must not ignore the possibility that the wheel of theory may be coming full circle from one teleology to another, from an old religious, to a new social orthodoxy.

REFERENCES

Allen, J. R. L. (1974), 'Reaction, relaxation and lag in natural sedimentary systems', *Earth Sci. Rev.* **10**, 263–342.

Buckland, W. (1820), *Vindiciae Geologicae* (Oxford), 38 pp.

Chamberlin, T. C. (1897), 'The method of multiple working hypotheses', *J. Geol.* **5**, 837–48.

Chorley, R. J. (1962), 'Geomorphology and general systems theory', *U.S. geol. Surv. Prof. Pap.* **500–B**, 10 pp.

— Kennedy, B. A. (1971), *Physical geography: a systems approach* (Prentice-Hall, London), 370 pp.

—, Beckinsale, R. P. and Dunn, A. J. (1973), *The history of the study of landforms: Vol. 2. The life and work of William Morris Davis* (Methuen, London), 874 pp.

Davis, W. M. (1889), 'The rivers and valleys of Pennsylvania', *Natn. geogr. Mag.* **1**, 183–253.

— (1905), 'The geographical cycle in an arid climate', *J. Geol.* **13**, 381–407.

— (1906), 'The sculpture of mountains by glaciers', *Scott. geogr. Mag.* **22**, 76–89.

Demangeon, A. (1905), *La Picardie* (Armand Colin, Paris), 496 pp.

Dury, G. H. (1962), 'Bankfull discharge: an example of its statistical relationships', *Bull. int. Ass. scient. Hydrol.* **6 (3)**, 48–55.

— (1972), 'Some current trends in geomorphology', *Earth Sci. Rev.* **8**, 45–72.

Engels, F. (1872–82), *Dialectics of Nature* (Translated by C. Dutt, Lawrence and Wishart, London, 1940), 383 pp.

Flint, R. F. (1963), 'Altitude, lithology and the Fall Zone in Connecticut', *J. Geol.* **71**, 683–97.

Geikie, A. (1868), 'On denudation now in progress', *Geol. Mag.* **5**, 249–54.

Gilbert, G. K. (1877), *The geology of the Henry Mountains* (U.S. Department of the Interior, Washington), 170 pp.

— (1886), 'The inculcation of the scientific method by example', *Am. J. Sci.*, **Ser.3, 31**, 284–99.

— (1896), 'The origin of hypothesis, illustrated by the discussion of a topographic problem', *Science* **3**, 1–13.

— (1909), 'The convexity of hilltops', *J. Geol.* **17**, 344–50.

— (1914), 'The transportation of debris by running water', *U.S. geol. Surv. Prof. Pap.* **86**, 263 pp.

Hack, J. T. (1960), 'Interpretation of erosional topography in humid temperate regions', *Am. J. Sci.* **258 A**, 80–97.

Hettner, A. (1928), *Die Oberflächenformen des Festlandes* (2nd ed., Teubner, Leipzig and Berlin). Translated by P. Tilley (Macmillan, London, 1972), 193 pp.

Hjulström, F. (1935), 'Studies of the morphological activity of rivers as illustrated by the River Fyris', *Bull. geogr. Inst. Univ. Uppsala* **25**, 221–527.

Horton, R. E. (1945), 'Erosional development of streams and their drainage basins: Hydrophysical approach to quantitative morphology', *Bull. geol. Soc. Am.* **56**, 275–370.

Howard, A. D. (1965), 'Geomorphological systems—Equilibrium and dynamics', *Am. J. Sci.* **263**, 302–12.

Isachenko, A. G. (1973), *Principles of landscape science and physical-geographic regionalization* (Translated by R. J. Zatorski; Melbourne University Press), 311 pp.

Johnson, D. W. (1940), 'Studies in scientific method: the analytical method of presentation', *J. Geomorph.* **3**, 156–62 and 257–62.

Keat, R. and Urry, J. (1975), *Social theory as science* (Routledge and Kegan Paul, London), 278 pp.

Kitts, D. B. (1963), 'Historical explanation in geology', *J. Geol.* **71**, 297–313.

— (1974), 'Physical theory and geological knowledge', *J. Geol.* **82**, 1–23.

— (1976), 'Certainty and uncertainty in geology', *Am. J. Sci.* **276**, 29–46.

Krumbein, W. C. and Slack, H. A. (1956), 'Relative efficiency of beach sampling methods', *Tech. Memo. Beach Eros. Bd U.S.* **90**, 43 pp.

Langbein, W. B. and Leopold, L. B. (1966), 'River meanders—theory of minimum

variance', *U.S. geol. Surv. Prof. Pap.* **422-H**, 15 pp.
—— and —— (1968), 'River channel bars and dunes—theory of kinetic waves', *U.S. geol. Surv. Prof. Pap.* **422-L**, 20 pp.
Leopold, L. B. and Langbein, W. B. (1962), 'The concept of entropy in landscape evolution', *U.S. geol. Surv. Prof. Pap.* **500-A**, 20 pp.
—— and Maddock, T. (1953), 'The hydraulic geometry of stream channels and some physiographic implications', *U.S. geol. Surv. Prof. Pap.* **252**, 57 pp.
Mackin, J. H. (1948), 'Concept of the graded river', *Bull. geol. Soc. Am.* **59**, 463–512.
Melton, M. A. (1957), 'An analysis of the relations among elements of climate, surface properties, and geomorphology', *Office of Naval Research Project NR 389-042, Tech. Rep. 11, Dept. Geol., Columbia Univ., New York,* 102 pp.
—— (1958), 'Correlation structure of morphometric properties of drainage systems and their controlling agents', *J. Geol.* **66**, 442–60.
Penck, A. (1905), 'Climatic features in the land surface', *Am. J. Sci.* **Ser. 4, 19**, 165–74.
Playfair, J. (1802), *Illustrations of the Huttonian theory of the Earth* (Edinburgh), 528 pp.
Popper, K. R. (1957), *The poverty of historicism* (Beacon Press, Boston), 166 pp.
Savigear, R. A. G. (1952), 'Some observations on slope development in South Wales', *Trans. Inst. Br. Geogr.* **18**, 31–51.
Scheidegger, A. E. (1961), *Theoretical geomorphology* (Springer-Verlag, Berlin), 333 pp. 2nd edition (1970), 435 pp.
—— and Langbein, W. B. (1966), 'Probability concepts in geomorphology', *U.S. geol. Surv. Prof. Pap.* **500-C**, 14 pp.
Schumm, S. A. (1956a), 'Evolution of drainage systems and slopes in badlands at Perth Amboy, New Jersey', *Bull. geol. Soc. Am.* **67**, 597–646.
—— (1956b), 'The role of creep and rainwash on the retreat of badland slopes', *Am. J. Sci.* **254**, 693–706.
—— (1963), 'The disparity between present rates of denudation and orogeny', *U.S. geol. Surv. Prof. Pap.* **454-H**, 13 pp.
—— and Lichty, R. W. (1965), 'Time, space and causality in geomorphology', *Am. J. Sci.* **263**, 110–19.
Simpson, G. G. (1963), 'Historical science' in Albritton, C.C. (ed.), *The fabric of geology* (Freeman, California), 24–48.
Stoddart, D. R. (1969), 'Climatic geomorphology: review and re-assessment', *Progr. Geogr.* **1**, 159–222.
Strahler, A. N. (1950), 'Equilibrium theory of erosional slopes, approached by frequency distribution analysis', *Am. J. Sci.* **248**, 673–96 and 800–14.
—— (1952a), 'Dynamic basis of geomorphology', *Bull. geol. Soc. Am.* **63**, 923–38.
—— (1952b), 'Hypsometric (area-altitude) analysis of erosional topography', *Bull. geol. Soc. Am.* **63**, 1117–42.
Sundborg, A. (1956), 'The river Klarälven—a study of fluvial processes', *Geogr. Annlr* **38**, 125–316.
Wolman, M. G. (1955), 'The natural channel of Brandywine Creek, Pennsylvania', *U.S. geol. Surv. Prof. Pap.* **271**, 56 pp.
—— Miller, J. P. (1960), 'Magnitude and frequency of forces in geomorphic processes', *J. Geol.* **68**, 54–74.
Worssam, B. C. (1973), 'A new look at river capture and the denudation history of the Weald', *U.K. Inst. geol. Sci. Rep.* **73/17**, 21 pp.
Yatsu, E. (1966), *Rock control in geomorphology* (Sozosha, Tokyo), 135 pp.

2

THE CHARACTER AND PROBLEMS OF THEORY IN CONTEMPORARY GEOMORPHOLOGY

J. B. THORNES
(*London School of Economics*)

This paper starts from the assumption that the establishment of theory is the ultimate and desirable goal of a scientific discipline; that the character of theories and their creation and evolution up to the middle of the century has been dealt with by R. J. Chorley in the preceeding chapter; and that it is neither necessary nor desirable to review the whole field of geomorphology. In any case, it is now beyond the range of competence of any individual to cast the net so widely.

It is a little over 15 years since the appearance of A. E. Scheidegger's monumental work on *Theoretical geomorphology* and over 10 years since Chorley's paper (1964) on 'Geography and analogue theory' was published. In this time, *Models in geography* (R. J. Chorley & P. Haggett, 1967) and *The mechanics of erosion* (M. A. Carson, 1971b), as well as A. Johnson's little-known *Physical processes in geology* (1970), have appeared, so that theory may now be thought to be well-established. The fact is, however, that the development of theory remains in the hands of remarkably few practitioners and is very unevenly spread across the subject as a whole. This is consistent with the general view that it is much more difficult to create theory than to develop it, but it does suggest that we are on the edge of a massive, new and exciting phase of theoretical development. The second half of this paper will try to speculate on the form which such developments are likely to take.

If we look at British geomorphology over the past 10 years we can pick out several major contributions to theory in geomorphology. Among the most notable are the work of W. E. H. Culling and M. J. Kirkby on slopes, the glacial work of G. S. Boulton and D. J. Drewry and the fluvial work of J. R. L. Allen. This rather conservative list reflects a personal selection and there are others, such as C. D. Curtis's work on weathering and I. Statham's work on scree slopes, but the overall impression is that we are somewhat weak on theory and strong on observation. The list also reflects a highly conservative use of the word *theory* on which I shall elaborate a little.

To describe theory as 'any speculative fantasy' does not carry us very far; even describing it as a 'system of statements—a language discussing the facts the theory is said to explain' leaves too much to the imagination. Rather, I prefer the more rigid view taken by D. Harvey (1969) of 'scientific theory' which is a systematic account of some field of study derived from a set of general propositions. These propositions may be taken as postulates (or axioms) as in geometry, or principles more or less strongly confirmed by experience. From these axiomatic statements, the theorems are derived by a calculus which usually comprises mathematics, and the theorems are translated into empirical experience by a text. Formalised theory is also recognised as having a domain of application.

REVIEW

The character of geomorphological theory

There are few geomorphological statements so fully formalised as Harvey's definition suggests, although *Theoretical geomorphology* comes closer than any other work. More often, one is dealing with partially formalised or incomplete theory. Harvey (1969) following R. S. Rudner, recognises four basic categories of scientific theory:

(i) Deductively complete theory, such as M. F. Dacey's (1968) formulation of the profile of a random stream;
(ii) Theories that are built up with reference to other theories which may be completely or incompletely specified;
(iii) Quasi-deductive theories in which there is incomplete elaboration because of technical difficulties, because the primitives are relative, or because the theories are essentially inductive systematisations which only follow probabilistically;
(iv) Non-formal theories, or verbal statements that may or may not be capable of formalisation.

Category (ii) theories in geomorphology commonly developed from basic laws of physics or chemistry, and the basic propositions comprise the fundamental laws of mechanics, fluid mechanics or statistical mechanics. Virtually all the recent work on river mechanics falls into this class, as exemplified by the work of R. A. Callendar (1969). In a completely different sense, the deductively formal theory of slope development of J. P. Bakker and J. W. N. Le Heux (1950) relies heavily on geometry.

More commonly, geomorphologists have been able to *develop*, as opposed to simply applying, a body of complete or partial theory from another subject. The best examples of this, and among the most successful, have been the theory of diffusion, the related theory of random flights and the theory of kinematic waves.

In category (iii) the most common problems, leading to incomplete formalisation of theory, are the technical difficulties in obtaining solutions to problems. This has led to the adoption of theory for which solutions are already known to exist. The theory of slope development owes much to H. S. Carslaw and J. C. Jaeger (1959) in so far as they provided solutions to a variety of cases of the diffusion of heat through solids. More recently, the appeal of the method of characteristics (M. B. Abbott, 1966) in solving hyperbolic partial differential equations in time and space has directed slope development theory (J. C. Luke, 1976) and the theory of overland flow (D. A. Woolhiser, 1975).

The incompleteness of geomorphological theory also arises from the lack of primitives, but this is rather by choice. W. M. Davis is often accused of making the doubtful primitive assumption that higher places would be subject to more erosion than lower places. Later this was replaced by an assumption that steeper places would be more eroded; we currently assume that steeper places with more rapid flow are subject to more erosion. None of these assertions is yet deductively or formally derived and this is the sense of incompleteness implied by category (iii).

Geomorphology abounds in theories of category (iv). Some of these are capable of formalisation for rigorous development. An excellent example of this is Boulton's (1972) partially formalised theory of sub-glacial erosion according to the thermal régimes of ice sheets. Another example is the adoption of the theory of queues in the modelling of scree slopes subject to constraints on supply and removal (J. B. Thornes, 1971). Both these papers represent partially formalised conceptual structures in which major gaps in process theory occur, in the first as to the actual processes of erosion, and in the second as to the processes of movement within the scree.

Weaker forms of theory are widespread. These include 'reticular' or historical explanation, such as speculations on the denudation chronology of a particular region (D. K. C. Jones, 1974). Another type is that arising from empirical observations, such as W. B. Bull's (1975) recent attempt to grace all processes identified by power-function regressions with the status of theory. There is no shortage of *potential* theory; the time has come for a more rigid formalisation of existing theory.

Theory within the fields of geomorphology

A second way of looking at theory in geomorphology would be to enquire about the major *themes* that have been developed. Three main fields of investigation have been the object of theory development in the scientific sense: glaciological processes, slope modelling and fluvial processes. Of these, slope modelling has received most attention reflecting, in part at least, its early start with the work of O. Fisher (1866). Macroscopic modelling by A. Young (1963) based on geometrical considerations and process laws, and by Kirkby (1971) based on process laws and invariance principles, has been followed by the exciting work of T. R. Smith and F. P. Bretherton (1972) and Luke (1972, 1974). Paralleling this work has been the investigation of slope mechanics from a theoretical and experimental standpoint. It is remarkable that geomorphologists have applied themselves principally to modelling small-scale processes rather than to modelling massive failures, though some verbal theorising on this scale has been carried out by D. Brunsden (1973) and Brunsden and Jones (1977).

In fluvial geomorphology we may identify several developments. The first of these is the attempt to develop coherent theory relating to drainage networks. The work on random topological networks by R. L. Shreve (1966) and nested hexagonal hierarchies of basins by M. J. Woldenburg (1972) are exemplary in terms of theory development. Unfortunately, it seems in retrospect that the misidentification of the relevant fundamental properties (loss of the metric) and the failure to provide an adequate text have meant that the implications of the work for sediment and water routing are either not self-evident or have been overlooked. Recent theoretical (Kirkby, 1976a) and empirical (Thornes, 1976) efforts are being made to overcome these limitations. Another great stimulus to the development of theory in geomorphology has been the effort to solve the problem of indeterminacy in hydraulic geometry. Besides a direct attack on the problem (F. M. Henderson, 1963; P. Ackers, 1964; and R. M. Li *et al.* 1976), it led to a significant introduction of theoretical ideas from statistical mechanics, culminating in the work of C. T. Yang (1971) on entropy modelling. T. R. Smith's (1974) attempt to place this work into a more geomorphological framework, despite the questionable nature of some of the assumptions (for example, linear increase in width), seems bound to encourage a renewal of the search for solutions to this problem.

Finally, reference should be made to the rapid and extensive growth of theory in the field of glaciology and, to a lesser degree, in glacial geomorphology. The early tradition of theory and the important stimulus given by the work of J. F. Nye (1965, 1969, 1971) and J. Weertman (1972) should have led to significant developments in landform geomorphology.

General theory

The third way of considering the contemporary development of theory in geomorphology is to enquire which major theoretical propositions have achieved acceptance *across* the fields of geomorphology. Taking the broadest and most important case first, the last decade has seen the introduction of systems theory notably by Chorley (1962, 1964) and Chorley and B. A. Kennedy (1971). While this has provided an important stimulus we are still awaiting a more formal application of some of the fundamental aspects of this theory in research programmes. In particular, work on feedback theory, adaptive control and dynamic instability appear to offer great potential. At a lower level, several other general concepts have found wide acceptance; entropy modelling has been employed in slope and glacial studies and in work on sediment transfer and channel morphology, and the ergodic theorem has been several times misapplied and misunderstood. Perhaps one of the most important of these general concepts is the principle of steady-state behaviour in dynamic systems, again a subset of systems theory. Normally, the principle is used essentially to obtain solutions to already formalised theory. In this way it performs like the other principles of invariance. In geomorphology it has, for some reason, assumed a status well beyond its potential. To some extent this has been at the expense of the consideration of transient, unsteady behaviour. Since most of our perceived experiences are of transient systems, this has led to some difficulties in providing an adequate framework for testing. The widespread use of the equations of state, the principles of conservation and the equations of motion in related sciences has not been paralleled in geomorphology, and geomorphologists by-and-large seem relatively unaware of them, so that 'in the world of the blind, the one-eyed man is King'. Since these basic (if somewhat difficult) ideas will form a basis of *all* future geomorphological work, they must rank high in our list of priorities.

At a lower level, other theory has found a more restricted application. The theory of kinematic waves, for example, is now well established in glacial and hydrological studies and in channel and slope morphology.

PROBLEMS

From the reductionist viewpoint, all geomorphological theory ultimately derives from, and can be reduced to, the laws of physics and chemistry. In a certain measure the success of the theoretical work based on mechanics encourages this opinion. By way of an example, consider the success of the shallow planar failure model in unconsolidated materials. Initiated to explain the slope forms in clays in temperate environments, it has since been applied, apparently successfully, to environments as far apart as Iceland (R. J. Chandler, 1973) and Wyoming (Carson, 1971). It forms the cornerstone of the concepts of characteristic and limiting angles in weathered debris, offers a mechanism for polymodal threshold straight segments, and may even participate in the developing theory of an exponentially decaying rate of slope change.

Despite this, and many successes in fluvial dynamics, mechanical principles have still played a fairly restricted role in geomorphological explanations. There appear to be several reasons for this. Among the foremost is that the mathematics involved is extremely difficult. M. I. Taft and A. Reisman (1965) point out that, even in engineering courses at doctoral level, the techniques of solution of the three-dimensional unsteady flow equations are relatively poorly known, and if we add to them the difficulties of a two-phase medium, the models are virtually intractable. For example, solutions to the one-dimensional differential equations of gradually varying unsteady flow in natural alluvial channels can only be obtained when the channel is sufficiently straight and uniform in the reach that the flow characteristics may be physically represented by a one-dimensional model; the velocity is uniformly distributed over the cross-section; hydrostatic pressure obtains at every point in the channel; the water surface is small; the density of the sediment-laden water is constant over the cross-section and the resistance coefficient is assumed to be the same as that for steady flow in alluvial channels and can be approximated from resistance equations applicable to alluvial channels (Y. H. Chen *et al.* 1975). Even when it is possible to obtain numerical solutions to more complex equations, it has been usual to treat the sediment load and scour independently through one of the bed-load equations (J. P. Bennett, 1974).

Another source of difficulty arises from the problem of specifying all or even most of the forces in operation. The complex nature of the forces acting on a soil particle led Culling (1963) to abandon a fully deterministic model in favour of a stochastic model. The complexities involved in the hysteresis effects in the drying and wetting curves for describing the relationship between soil moisture and tension involve similar difficulties, and A. Klute (1973) has called for a revision of the fully distributed deterministic approach.

A related problem arises in the need to specify a larger and more diverse set of model parameters and to optimise these complex models for applied purposes. Of course the current inability to parameterise the models is not an argument against their development, though it has to be admitted that the models sometimes appear more empirical than those that start from an inductive procedure.

One possible way out of these difficulties is to adopt a coarser scale of resolution which avoids considering the detailed mechanics. Such a procedure has been used by hydrologists in developing the linear storage concept for catchment behaviour or the kinematic cascade, and by geomorphologists in considering the basic relationships of hydraulic geometry as 'given' and trying to work from them (Thornes, 1977). An alternative strategy has been to employ stochastic or mixed models, such as event-based models (P. Todorovic and Woolhiser, 1974) or Kalman-filtering techniques. These shift the emphasis away from 'individual' and precisely formulated behaviour towards an 'aggregate' or 'normative' formulation.

If these difficulties lead away from the adoption of microscopic mechanical modelling, why is indigenous theory so weak? Partly the answer is to be found in the weak formalisation of latent theory. L. C. King's (1950) theory of global pediplanation is very weakly formalised compared with the more recent theory of plate tectonics. Science has better theory for the collapsing galaxies millions of years old than for collapsing dolines of post-glacial times.

The problem of the content of theory, however, goes well beyond the mere act of formalisation. The shift from an historical paradigm in the 1950s to an inductive one in the mid-1960s did little to encourage the theoretical work then in progress. The 'measurement movement', encouraged at least in geology and geography by the apparent virility of statistics, also favoured a shift towards empirical explanation, fostered by the contemporary leaders in the subject. In addition, geomorphologists faced the dilemma that the 'theory' available did not appear capable of assimilating the newly developed work on processes. This situation may be compared with that in biology, where the new developments in genetics and population dynamics were easily incorporated into existing wisdom. In geomorphology, the major cyclical theory, that of W. M. Davis, was under attack and the alternative theory failed to provide adequate text against which a stimulating interaction between theory and field could take place. In the late 1970s the position is very different and it would be a great pity if the flowering of theory was nipped in the bud by a complete shift in emphasis to 'relevant' or 'applied' research.

The final group of problems relates to verification and refutation of theory. J. L. Monod (1975) describes the theory of evolution as 'secondary' theory; he says "it is a theory aimed at accounting for a phenomenon that never has been observed, namely evolution itself". In many respects the subject matter of geomorphological theory, which should ultimately explain the origin and development of surface forms, presents similar problems. This situation arises from the speed of operation of geomorphological processes, the problems of overlapping domains and the failure to provide an adequate text. The insistence by the empiricist that we must find *the* characteristic slope form and *the* graded profile in order to satisfy theoretical prediction has been a source of both friction and discouragement. It is in the nature of theory that the basic entities may sometimes not be perceived. This is not to say that the text for refutation need not be provided.

PROSPECTS

Having reviewed the state of development of theoretical geomorphology and outlined some of the problems of theory content, acceptability and formulation, we may now ask the questions of what fields seem to offer most promise for the future and how the problems of verification (or refutation) may best be overcome.

The foremost need is for the development of theoretical models which, on the macroscale (i.e. drainage-basin scale), will provide the pattern of development of the landscape, together with the transfers of water and sediment through it. Such models can have great potential for explanation but also raise great problems of verification. I have just mentioned an attack on this approach exemplified by the work of Kirkby (1971) and Luke (1974). In addition, Bennett (1974) has outlined an overall framework based on two coupled sub-systems, the slope system modelled by the L. D. Meyer and W. H. Wischmeir (1969) type of approach and a channel system based on a set of transport laws. Originally this seemed to hold out great promise. It does appear, however, that it will have to face formidable difficulties in the selection of parameters if realistic results are to be obtained. Another type of attack on the problem is to attempt to simulate the whole system digitally in order to obtain solutions over time and space. The work by F. Ahnert (1976) and A. Armstrong (1976) is of this type. The basic problems here are the formidable amount of computing required to obtain adequate scaling of the processes, the difficulties involved in decisions about the time and space scales and, particularly, the difficulties of verification. A further type of theoretical model that might be used requires the specification of distribution functions for interrelated events in a variety of catchments, in order to build up a meaningful stochastic model. Inputs to such a system will require the magnitude and frequency of generating processes, such as rainfall, to be specified, and from these and from basin functions for network distribution, slope length and so on, a series of outflows will be generated. Some progress has already been made along these lines by Todorovic and Woolhiser (1974).

Characteristically the earlier works mentioned allow a very general model, in which water transfer, sediment transfer and topographic change are combined, to be formulated. These are, in many respects, long-term theoretical structures. The later models mentioned are for the short-term, relating to individual storm events or seasons.

At a higher level of resolution, an important set of theoretical problems involve the growth of channel networks. The earlier work failed to lead towards a more general model of network development and change, or even for that matter to a useful description of networks having sedimentological and hydrological significance. In this sense, work that has stressed the integrated character of the networks in relation to water and sediment flows (such as that of A. Calver *et al.*, 1972) may prove more helpful. An alternative view has recently been expressed by J. C. Smart and C. Werner (1976). There is a real prospect here of a link with theoretical work on the hydraulic geometry of natural channels (Thornes, 1974). In this way Euclidean network properties, sediment and water transfer and hydraulic geometry will be linked.

The third major growth area in theoretical fluvial work concerns the evolution of channel geometry in space and time through the adoption of numerical solutions to the equations for unsteady flow and through the analysis of stability relations in the channels. In this respect the conditions for growth and development of perturbations in the channel system have been shown theoretically to be highly pertinent to the form of the channel and its pattern (G. Parker, 1976; V. M. Ponce and K. Mahmood, 1976). There is a direct and important link here with observations of catastrophic behaviour across thresholds (Schumm, 1973), and with the persistence through time of the effects of changes of régime (Allen, 1974) or extreme events (Thornes, 1976). 'Unsteady' geomorphology is likely to prove the keypoint of fluvial and slope studies in the next decade.

Three other matters seem destined to be the foci of a good deal of theoretical investigation. Because of the recent developments in fluvial geomorphology there should be rapid progress in our understanding of river terraces. This crucially important but highly neglected topic is the key to many of the evolutionary problems discussed in the first half of the century. Secondly, the theory of debris production, a basic constraint on longer-term evolution and the relationship between form and lithology, is likely to expand considerably. Among the British work, that of C. D. Curtis (1976) and Kirkby (1976b) seems to indicate the likely character of developments in this area. Finally, there appears to be a great need for *basic* research in the application of existing process models to everyday problems. In particular, the development of control theory and its application to soil erosion, channel sedimentation and the like seems to hold great promise.

The subject seems poised at the verge of considerable advances in the theory of processes which should lead to understanding and perhaps even some answers to the classical problems of landform evolution. These changes will also lead to an explosion in the potential application of the subject to everyday problems. They will also make heavy demands on field survey and resources of raw data. Geomorphologists adjusted quickly and surprisingly efficiently to the statistical and process paradigms. The forthcoming changes will be more difficult, but much more fundamental and rewarding.

REFERENCES

Abbott, M. B. (1966), *An introduction to the methods of characteristics* (Thames and Hudson, London), 216 pp.

Ackers, P. (1964), 'Experiments on small streams in alluvium', *Trans. Am. Soc. civ. Engrs* **90**, HY4, 1–37.

Ahnert, F. (1976), 'Brief description of a comprehensive three-dimensional process-response model of landform development', *Z. Geomorph., Suppl. Bd* **25**, 29–49.

Allen, J. R. L. (1974), 'Reaction, relaxation and lag in natural sedimentary systems: general principles, examples and lessons', *Earth Sci. Rev.* **10**, 263–342.

Armstrong, A. (1976), 'A three-dimensional simulation of slope forms', *Z. Geomorph., Suppl. Bd* **25**, 20–8.

Bakker, J. P. and Le Heux, J. W. N. (1950), 'Theory on central rectilinear recession of slopes', *Proc. K. ned. Akad. Wet.* Ser. **B**, **53**, 1073–84 and 1364–74.

Bennett, J. P. (1974), 'Concepts of mathematical modelling of sediment yield', *Wat. Resour. Res.* **10(3)**, 485–92.

Boulton, G. S. (1972), 'The role of thermal régime in glacial sedimentation', *Inst. Br. Geogr. Spec. Publ.* **4**, 1–19.

Brunsden, D. (1973), 'The application of systems theory to the study of mass movement', *Proc. I.R.P.I. Conf. on natural slopes, stability and conservation* (Naples, Cosenza).

—— and Jones, D. K. C. (1976), 'The evolution of landslide slopes in Dorset', *Phil. Trans. R. Soc.* **A**, **283**, 605–37.

Bull, W. B. (1975), 'Allometric change in landforms', *Bull. geol. Soc. Am.* **86**, 1489–98.

Callendar, R. A. (1969), 'Instability and river channels', *J. Fluid Mech.* **36(3)**, 465–80.

Calver, A., Kirkby, M. J. and Weyman, D. (1972), 'Modelling hillslope and channel flows', in R. J. Chorley (ed.), *Spatial analysis in geomorphology* (Methuen, London), 197–218.

Carslaw, H. S. and Jaeger, J. C. (1959), *The conduction of heat in solids* (Oxford Univ. Press), 2nd ed.

Carson, M. A. (1971a), 'Application of the concept of threshold slopes to the Laramie Mountains, Wyoming', *Inst. Br. Geogr. Spec. Publ.* **3**, 31–48.

—— (1971b), *The mechanics of erosion* (Pion, London), 174 pp.

Chandler, R. J. (1973), 'The inclination of talus, arctic talus terraces and other slopes composed of granular materials', *J. Geol.* **81(1)**, 1–14.

Chen, Y. H., Holly, F. M., Mahmood, K. and Simons, D. B. (1975), 'Transport of materials by unsteady flow', in K. Mahmood and V. Yevjevich (eds.), *Unsteady flow in open channels,* **1**, 313–63.

Chorley, R. J. (1962), 'Geomorphology and general systems theory', *U.S. geol. Surv. Prof. Pap.* **500–B**, 10 pp.

—— (1964), 'Geography and analogue theory', *Ann. Ass. Am. Geogr.* **54**, 127–37.

—— and Kennedy, B. A. (1971), *Physical geography: a systems approach* (Prentice-Hall, London), 370 pp.

Culling, W. E. H. (1963), 'Soil creep and the development of hillside slopes', *J. Geol.* **71**, 127–61.

Curtis, C. D. (1976), 'Stability of minerals in surface weathering reactions: A general thermochemical approach', *Earth Surface Processes* **1(1)**, 63–70.

Dacey, M. F. (1968), 'The profile of a random stream', *Wat. Resour. Res.* **4(3)**, 651–4.

Drewry, D. J. (1972), 'A quantitative assessment of dirt-cone dynamics', *J. Glaciol.* **11**, 431–46.

Fisher, O. (1866), 'On the disintegration of a chalk cliff', *Geol. Mag.* **3**, 354–6.

Harvey, D. (1969), *Explanation in geography* (Arnold, London), 521 pp.

Henderson, F. M. (1936), 'Stability of alluvial channels', *Trans. Am. Soc. civ. Engrs* **126(1)**, 657–85.

Johnson, A. (1970), *Physical processes in geology* (Freeman, Cooper and Co., San Francisco), 577 pp.

Jones, D. K. C. (1974), 'The influence of the Calabrian transgression on the drainage evolution of south-east England', *Inst. Br. Geogr. Spec. Publ.* **7**, 139–58.

Kirkby, M. J. (1971), 'Hillslope process-response models based on the continuity equation', *Inst. Br. Geogr. Spec. Publ.* **3**, 15–30.

—— (1976a), 'Tests of the random network model and its application to basin hydrology', *Earth Surface Processes* **1(3)**, 197–213.

—— (1976b), 'Deterministic continuous slope models', *Z. Geomorph., Suppl. Bd* **25**, 1–19.

Klute, A. (1973), 'Soil water theory and its application in field studies' in R. R. Bruce (ed.), *Field soil-water régime* (*Soil Science Society of America Spec. Publ.* 5), 9–36.

Li, R. M., Simons, D. B., and Stevens, M A. (1976), 'Morphology of cobble streams in small watersheds', *Proc. Am. Soc. civ. Engrs, J. hydraul. Div.* **102, HY8**, 1101–17.

Luke, J. C. (1972), 'Mathematical models for landform evolution', *J. geophys. Res.* **77(14)**, 2460–4.

—— (1974), 'Special solutions for non-linear erosion problems', *J. geophys. Res.* **79(26)**, 4035–40.
—— (1976), 'A note on the use of characteristics in slope evolution models', *Z. Geomorph., Suppl. Bd* **25**, 114–19.
Meyer, L. D. and Wischmeir, W. H. (1969), 'Mathematical simulation of the process of soil erosion by water', *Trans. Am. Soc. agric. Engrs* **12(6)**, 754–62.
Monod, J. L. (1975), 'On the molecular theory of evolution', *in* R. Harré (ed.), *Problems of scientific evolution* (Oxford University Press), 104 pp.
Nye, J. F. (1965), 'The flow of a glacier in a channel of rectangular, elliptic or parabolic cross-section', *J. Glaciol.* **5**, 661–90.
—— (1969), 'A calculation of the sliding of ice over a wavy surface using a Newtonian viscous approximation', *Proc. R. Soc.* **A, 311**, 445–67.
—— (1971), 'Glacier sliding without cavitation in a linear viscous approximation', *Proc. R. Soc.* **A, 315**, 381–403.
Parker, G. (1976), 'On the cause and characteristic scales of meandering and braiding in rivers', *J. Fluid Mech.* **76(3)**, 457–80.
Ponce, V. M. and Mahmood, K. (1976), 'Meandering thalwegs of straight alluvial rivers', *Rivers 36* (American Society of Civil Engineers), 2, 1418–42.
Scheiddegger, A. E. (1970), *Theoretical geomorphology* (Springer-Verlag, New York, 2nd ed.), 435 pp.
Schumm, S. A. (1973), 'Geomorphic thresholds and complex response of drainage systems', in M. Morisawa (ed.), *Fluvial geomorphology* (Proceedings 4th Ann. Geomorphology Symposium, Binghamton, New York), 299–309.
Shreve, R. L. (1967), 'Infinitely topologically random channel networks', *J. Geol.* **75**, 179–86.
Smart, J. C. and Werner, C. (1976), 'Applications of the random model of drainage basin composition', *Earth Surface Processes* **1(3)**, 219–35.
Smith, T. R. (1974), 'A derivation of the hydraulic geometry of steady-state channels from conservation principles and sediment transport laws', *J. Geol.* **82**, 98–104.
—— and Bretherton, F. P. (1972), 'Stability and the conservation of mass in drainage basin evolution', *Wat. Resour. Res.* **8(6)**, 1506–24.
Statham, I. (1976), 'A scree slope rockfall model', *Earth Surface Processes* **1(1)**, 43–63.
Taft, M. I. and Reisman, A. (1965), 'The conservation equations: a systematic look', *J. Hydrol.* **3**, 161–79.
Thornes, J. B. (1971), 'State, environment and attribute in scree slope studies', *Inst. Br. Geogr. Spec. Publ.* **3**, 49–64.
—— (1974), 'Speculations on the behaviour of stream channel widths', *London School of Economics, Graduate School of Geography, Discussion Paper* No. **49**, 28 pp.
—— (1976), 'Semi-arid erosional systems: case studies from Spain', *London School of Economics, Geography Department Paper* No. **7**, 88 pp.
—— (1977), 'Channel changes in ephemeral streams—observations, problems and models', in K. J. Gregory (ed.), *River channel changes* (Wiley, London), 318–35.
Todorivic, P. and Woolhiser, D. A. (1974), 'A stochastic model of sediment yield for ephemeral streams', *Proc. USDA-IASPS Symp. statist. Hydrol., USDA Misc. Publ.* 1275.
Weertman, J. (1972), 'General theory of water flow at the base of a glacier or ice sheet', *Rev. Geophys. Space Phys.* **10**, 287–333.
Woldenburg, M. J. (1972), 'The average hexagon in spatial hierarchies', in R. J.

Chorley (ed.), *Spatial analysis in geomorphology* (Methuen, London), 323–54.

Woolhiser, D. A. (1975), 'Simulation of unsteady overland flow', in K. Mahmood and V. Yevjevich (eds.), *Unsteady flow in open channels* (Water Resources Publications, Ft. Collins), 2, 485–509.

Yang, C. T. (1971), 'Potential energy and stream morphology', *Wat. Resour. Res.* **7(2)**, 311–22.

Young, A. (1963), 'Deductive models of slope evolution', *Nachr. Akad. Wiss. Göttingen* (Math. Phys. Kl.) **11**, 45–70.

3

EL ASUNTO DEL ARROYO

LUNA B. LEOPOLD
(*University of California, Berkeley*)

I sit in the shade of a piñon tree with the breeze of early morning in my face, looking down on a little valley called El Asunto del Arroyo. To the east towers the dark mass of the Sangre de Cristo Range, and to the west the low, blue hulk of the Jemez. The Rio Grande, which I cannot see, runs in between. The low hill on which I sit is mostly unconsolidated coarse sand intermixed with gravel lenses, but over all the hilltops is a discontinuous sheet of gravel.

Through the scattered trees I see a broad, nearly flat floor of alluvium. The drainage-way down the centre is a sandy wash some 10 feet (3 m) wide, dry except during storms. The general valley floor is a terrace, long ago abandoned as the flood plain of the wash. Cut in the valley floor is an inner valley, the surface of which is apparently the flood plain of the present ephemeral channel. The terrace tread is separated from the inner channel by a scarp, averaging 6 feet (2 m) high, vertical in many places, and exposing the alluvium that underlies the terrace. This exposed alluvium is tan or light brown in colour, mostly sand, with lenses of silt and of fine gravel.

El Asunto del Arroyo: the arroyo business, or the affair; the matter at hand; thus, the business of immediate consequence. I do not know the origin of the name. It may derive from some incident on the Old Santa Fe Trail which passed nearby. Perhaps there was some confrontation between the local Mexican gentry and the troops of the Army of the West, *c*. 1846, for in this and adjacent valleys, the invading American army pastured their riding horses, the grass in the vicinity of Santa Fe itself being too sparse after a couple of centuries of grazing by the Spanish.

But its name conjures up a variety of geomorphological as well as historical questions. The arroyo matter or arroyo business includes elements of many unsolved geomorphological problems, and it pinpoints several of paramount importance.

Asunto del Arroyo is a channel that has existed at the different elevations represented by its terraces but it is also discontinuous, that is, the channel is deeper upstream than down and it ends downstream in a flat fan through which no channel exists. To understand its history and the processes by which it assumed its present form, it is necessary to know why a stream aggrades, what determines the gradient during aggradation and during relative stability, how base level affects the bed elevation and the profile, and how a change in the relative amounts of sediment and discharge affects hydraulic factors including gradient.

It is true that we have 'explanations' for each of these matters but any detailed consideration leads one to the conclusion that they are neither satisfactory nor complete—in fact they are not really explanations at all. The purpose of this essay is not to offer solutions to these dilemmas but to rephrase the questions in need of answers. I do not have the answers. Perhaps the reason lies in the fact that we geomorphologists have bypassed the most difficult questions and we have spent too much effort on matters of much less importance. To explain this point of view, some field examples will be described, all of which have elements in common with Asunto del Arroyo.

Many channels are carved in and flow through alluvium deposited by the same river in a trench cut in bedrock at some earlier stage in its history. The original trench was presumably eroded in the bedrock when the ancestral river had more power to cut and less sediment to carry than it had in the later depositional stage. What does this mean quantitatively? How much less sediment was available in the eroding than in the aggrading stage? Was the difference in sediment quantity or in water discharge and how much difference in each?

Though the questions cannot be answered, an example may help to clarify the nature of the problem. Walker Creek drains a basin of 69 square miles (179 km^2) in the Coast Range of California, 30 miles (48 km) north of San Francisco. It is ephemeral, mostly dry in summer and has a small but continuous flow in the rainy season of winter. Like many, the stream flows in alluvium through most of its length but bedrock is exposed in the channel bed through a short central reach. The valley floor is generally broad and flat consisting of an alluvial terrace standing 15–30 feet (5–10 m) above the present stream bed. An inner lower terrace averaging 8 feet (2·4 m) above the bed is present through two-thirds of the stream length. This inner terrace was the floodplain in historic time, about 1900, but in the present century, the stream has degraded about 5 feet (1·5 m) in the headwater areas and aggraded in the downstream third of the basin length (W. W. Haible, 1976, in preparation). The aggradation downstream can be documented by progressively buried fence posts of known age, and local residents provide a history of degradation upstream (Figs. 3.1 and 3.2).

Haible has shown that the cutting upstream and deposition downstream have resulted in a flattening of the mean stream gradient from 0·0024 to 0·0023 over the lower 16 miles (25 km) of stream length. Though the 5 feet (1·5 m) of degradation and 4 feet (1·2 m) of deposition seem large when the results are seen in the field, the length involved is great, so the gradient has decreased only 0·0001 or 4 per cent. The recent degradation is believed by Haible to result from overgrazing of the upper basin during early decades of this century, resulting presumably in greater flood peaks and increased sediment available for transport. The variation of discharge, storm to storm and year to year, is large, more than a hundred fold, but the gradient was altered an insignificant amount. Can the hydraulic processes in the channel react to a change in slope as small as 2 per cent of the original value?

Coyote C Arroyo near Santa Fe, New Mexico, is an example on a smaller scale that can be described in quantitative terms. In an area of 14 inches (350 mm) of precipitation, this ephemeral basin of 0·064 square miles in drainage area (0·166 km^2) consists of an ungullied headwater reach and a gullied

FIG. 3.1. Channel of Walker Creek near Tomales, California, in aggrading reach. The valley flat in the right of the picture has built up 4 feet (1·2 m) in the present century but the channel maintains its geometry.

FIG. 3.2. Fence posts in aggrading valley of Walker Creek have recently been physically lifted 1 foot (0·3 m) by the rancher to keep the fence at normal height. Dr Asher Schick points to ground level on post before it was lifted.

reach downstream of a vertical headcut 12 feet (3·7 m) high. The basin area is sketched in the upper part of Figure 3.3. Some small discontinuous gullies occur in the upper half. In the lower half, six cross-sections were established and have been re-surveyed every few years. Their locations are also shown on Fig. 3.3.

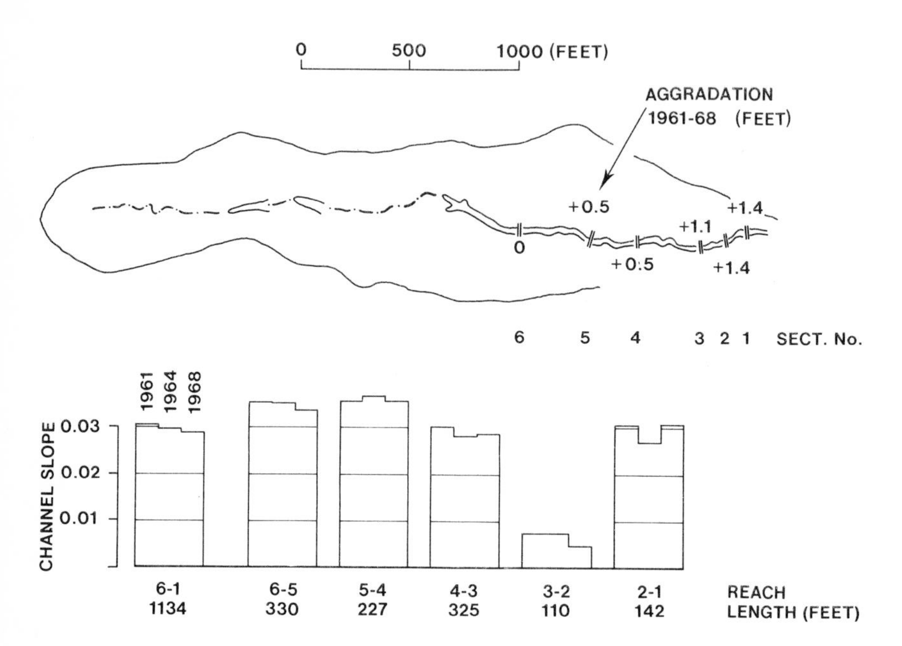

FIG. 3.3. Map of basin (above) of Coyote C Arroyo near Santa Fe, New Mexico. Channel cross-sections 1–6 are shown with the amount of aggradation in the period 1961–68. Histograms (lower) show channel bed gradient in various years between pairs of cross-sections.

Cross-section 3 is shown in Figure 3.4 as it existed in 1961, 1968 and 1974. In the 7-year record, 1961–68, the bed aggraded 1·4 feet (0·4 m) at its lower end. A more recent survey shows that the aggradation was still continuing in 1974. On the map in Figure 3.3 near each cross section is a figure representing the number of feet of aggradation in the 7-year period. At Section 6, though the bed changed in elevation from year to year, the net change after 7 years was zero, or no change. Amounts of aggradation increase irregularly downstream.

In the lower part of Figure 3.3 are bar graphs showing values of bed gradient at each of the 3 years, 1961, 1964, 1968, for each of the channel reaches between sections. The bar graph at the left is for the whole distance from Section 6 to Section 1. The distances between sections are also written along the bottom line. The gradients are all about 0·03 except for the short reach between Sections 2–3 which is less steep. The aggradation has resulted in a slight flattening of gradient, 1·4 feet (0·4 m) in 1134 feet (346 m). But as in Walker Creek, this change is but a small percentage of the original gradient, in this case about 2 per cent of the mean slope 0·03. Again the question may be posed: may one consider this change in gradient significant or merely fortuitous? If it is merely

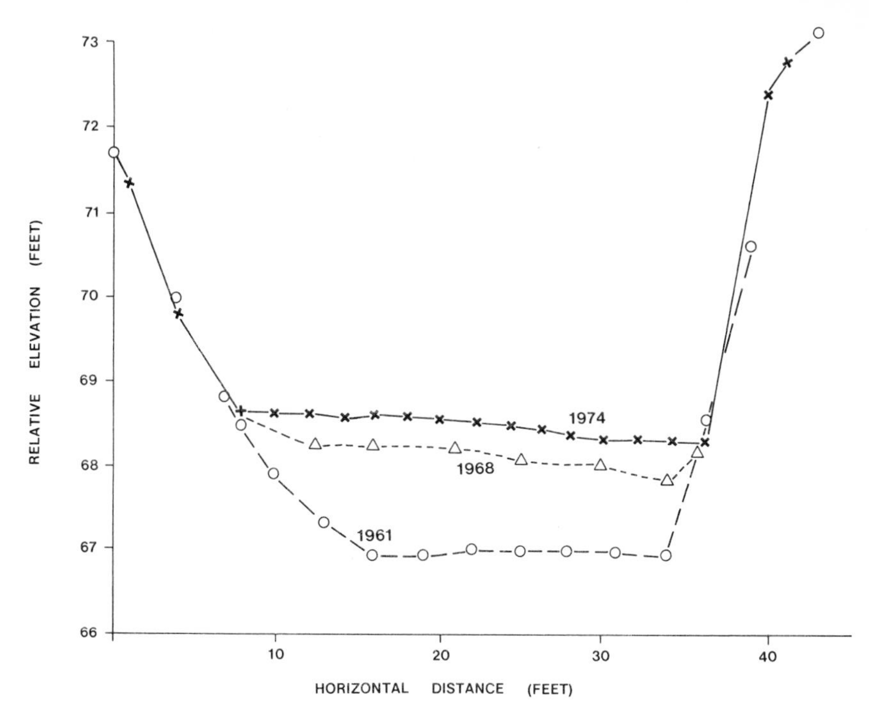

FIG. 3.4. Surveyed cross-sections at Sect. 3 on Coyote C Arroyo during the period 1961–1974.

random, then one may conclude that the gully erosion that formed the present channel below the headcut produced the same bed gradient as is characterised by a period of aggradation. What determines the value of this constant gradient? Discharges have no doubt significantly changed from the cutting to the aggrading period. One might suppose that grain size of bed material governs the bed slope for it has probably not changed significantly from the cutting to the aggrading period.

Such an explanation is shown to be too simple when one considers the discontinuous nature of the whole gully. This aspect is demonstrated better by a nearby basin, Arroyo Falta, also draining a long narrow basin in the same unconsolidated sand-gravel material of the Santa Fe Formation. A segment of the profile and one cross-section of Arroyo Falta are shown in Figure 3.5. The gully begins at a headcut shown by the steep part of the profile at Station 3 + 50 (Fig. 3.6). The valley floor, upper line of the profile, has a nearly straight profile with a gradient of 0·037. The profile has been surveyed several times. Figure 3.5 shows the surveys of 1962 and 1974.

The valley floor of Arroyo Falta was unchannelled about the turn of the century and the discontinuous gully formed probably in the first two decades of the present century. But between 1962 and 1974, headcut retreat has been negligible and the gully bed has aggraded. The gully depth changes from 7·5 feet (2·3 m) at the headcut to zero at station 8 + 00 where the sediment eroded from

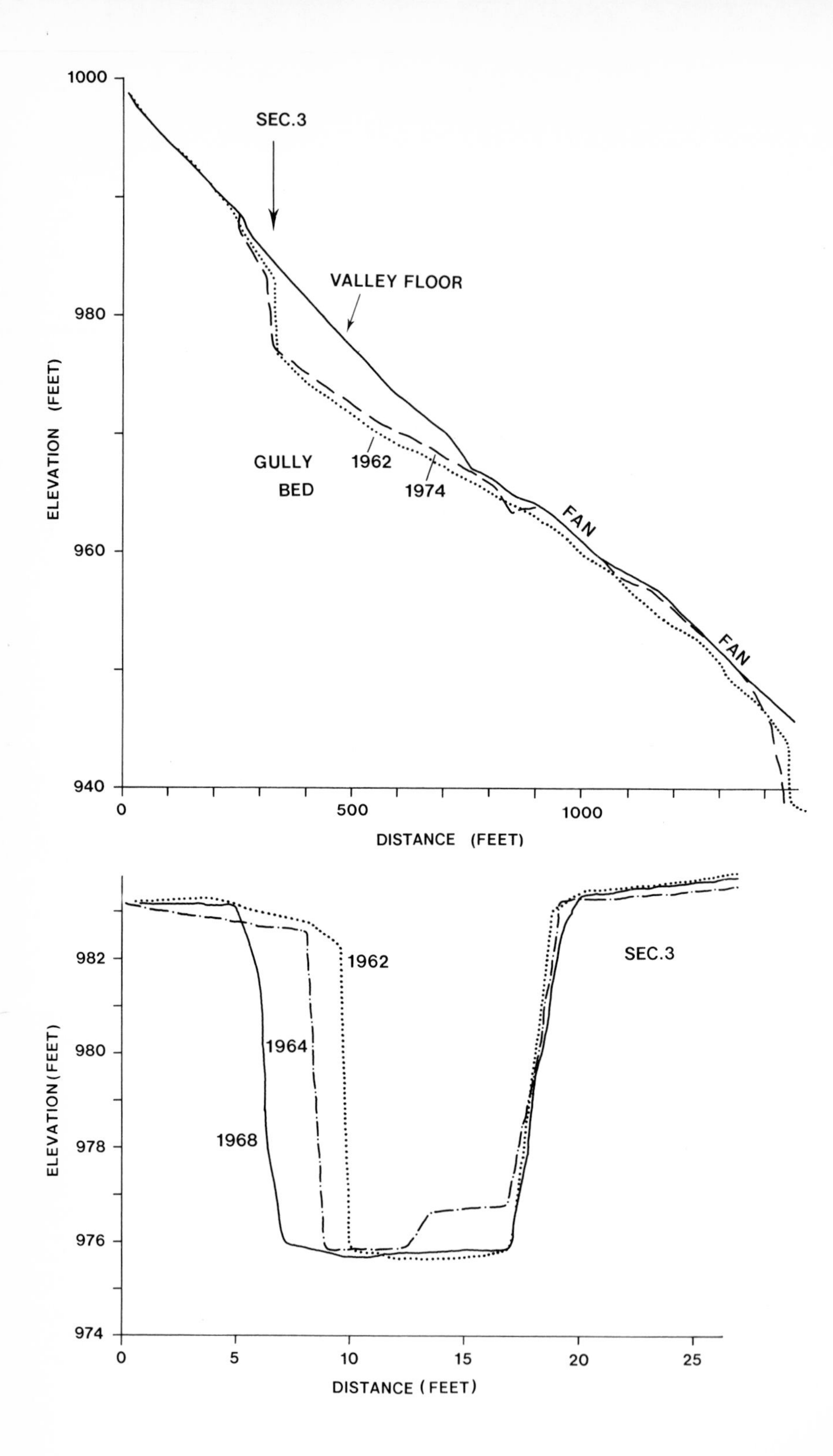
1000
SEC.3
VALLEY FLOOR
980
ELEVATION (FEET)
GULLY
BED
1962
1974
FAN
960
FAN
940
0
500
1000
DISTANCE (FEET)
982
1962
SEC.3
980
1964
ELEVATION (FEET)
1968
978
976
974
0
5
10
15
20
25
DISTANCE (FEET)

FIG. 3.6. Head-cut in discontinuous gully of Arroyo Falta.

the gully spreads as a low fan over the valley floor. In the period between these surveys the fan has increased in thickness by 1·0 feet (0·3 m) (Fig. 3.7).

The phenomenon of the discontinuous gully is inherently interesting because it poses the question, why is the gradient of the gully less steep than the valley floor? Several authors have discussed discontinuous gullies and consistently failed to deal with this key feature. If the grain size of bed sediment controls the gradient, the gully should have the same gradient as the valley floor because the same sediment is involved. A developing gully should then be continuous for long distances along the valley and, if deepening is progressive, the gully ought to deepen uniformly along the entire length. This does not occur.

In an earlier paper (Leopold and J. P. Miller, 1956) I argued that, over time, a discontinuous gully will widen with a resultant decrease in local depth of flow for a given discharge which, to carry the same sediment load, would require an increase in gradient. This would account for the fact that in the history of trenching of alluvial valleys in the past century, initial gullying was in many cases discontinuous, but as the arroyo increased in width and depth, bed gradient increased until at coalescence of several formerly discontinuous gullies a through or continuous arroyo results, but is of nearly constant depth, or the gradient is equal to that of the valley floor.

FIG. 3.5 (see left). Profile and a cross-section of a discontinuous gully, Arroyo Falta near Santa Fe, New Mexico.

FIG. 3.7. Upper end of fan in discontinuous gully of Arroyo Falta. Location is at station 800 of Figure 5. Channel which is deep just upstream has virtually disappeared at this point.

Arroyo Falta appears to be following this pattern, for between 1962 and 1974 widening has occurred as shown by the cross section, Figure 3.5, and at the same time somewhat greater aggradation has occurred upstream than down. But as in Coyote C Arroyo, this observed changed in bed gradient is small, 0·022 to 0·026, or 18 per cent. Whether this is of importance is not known.

In the same paper I pointed out that the bed gradient is a slope of deposition, not erosion, for the progressive uphill retreat of the gully results in uphill migration of the plunge pool. The pool progressively fills as it moves to more upstream portions. This feature is also exemplified by Arroyo Falta, but the significance of a depositional gradient as opposed to a gradient developed by bed erosion is not known.

An alternative explanation suggests itself. The valley floor of Arroyo Falta has a slope of 0·037 and the gully bed a mean slope of 0·026. Assume it is not sediment size that governs the slope but rather the shear stress needed for initial grain motion. When the alluvium of the valley floor was being deposited it is probable that the alluviating valley was covered with grass that progressively grew up through the accumulating layers of sediment, always providing a soil binder and thus a surface resistant to erosion. This would require a larger shear stress to move the same sediment size, and in order to keep the sediment in motion and spread out over the valley, a relatively steep slope would be required. When, however, this valley floor is gullied, the gully bed is composed of loose sand and silt lacking the binding action of grass. Thus the gradient required for initial motion of the loose grains is smaller than was the case of the aggrading valley floor. This difference is observed in Arroyo Falta. The valley floor is covered with grass, and though the density is low, less than 20 per cent crown cover, some protection against erosion is offered. In the gully there is no vegetation. It might be supposed that the unprotected gully bed requires a smaller gradient for initial sediment motion than the vegetated valley floor.

This hypothesis is denied by another line of evidence. In 1960 we built a small dam across each of two small channels in the same locality as Coyote C and Falta arroyos. The profile of sediment deposited behind these barriers has been resurveyed over the years. Sediment was deposited behind these dams immediately, and the reservoir space was filled to spillway level at the end of the first year.

Big Sweat Dam discussed here is on a small sandy wash at a drainage area of 3·6 acres (1·46 ha). The dam was built of concrete and its lip over which all water and sediment must pass is 1·3 feet (0·4 m) above the original bed. Several intermediate profiles are available but in Figure 3.8 four of them are plotted:

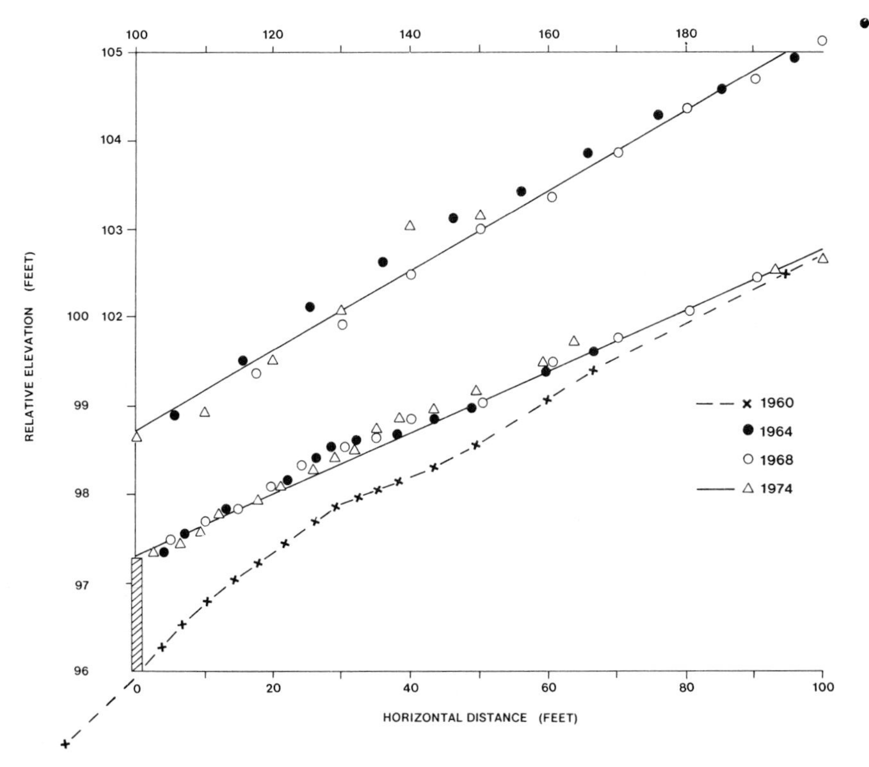

FIG. 3.8. Profiles above Big Sweat Dam near Santa Fe, New Mexico. Original channel bed before dam (1960). Profile of sediment wedge behind dam has not changed from 1961 to 1974. Lower part of diagram is profile from dam to 100 feet (30·5 m) above dam, and upper profile is a continuation from 100 to 200 feet (30·5–61 m) above dam.

the original profile in 1960 and three others run in 1964, 1968 and 1974, a span of 14 years. The reservoir created was filled to the top of the dam by October 1961, 13 months after construction. The sediment wedge consists of loose gravelly sand. It assumed a gradient less steep than that of the undammed wash. The stream profile is slightly concave. The profile of the sediment wedge intersected the original profile at station 1 + 100 or 100 feet (30·5 m) upstream of the dam, and still farther upstream the profile, though varying in detail over time, has maintained the same slope as it had originally. Average gradients in reaches at differing distance upstream from the dam are listed in Table I.

TABLE I

Gradient at three dates, sediment behind Big Sweat Dam

Reach Distance	*original 1960*	*1968*	*1974*
0+00 to 1+10	0·0425	0·0335	0·0335
1+10 to 2+20	0·0454	0·0454	
2+20 to 3+30	0·0523	0·0523	

It might be reasoned that behind the dam the water has an opportunity to spread out over a greater width than in the original channel, so for the same discharge, width is larger and depth smaller. Visually the roughness or resistance to flow seems comparable on the sediment wedge and in the original channel. The decrease in relative depth, that is the depth/grain-size ratio, tends to increase the efficiency of sediment transport, but how this change should affect gradient is not clear. Furthermore, why the gradient changed by the particular amount, from 0·0425 to 0·0335, a decrease of 21 per cent, is not known.

If channel widening results in decrease of gradient in this instance, the discontinuous gully described earlier is at variance with this conclusion, for the discontinuous gully appears to have a low slope when it is narrow and increases its slope as gully-width increases.

Construction of a dam is a case of a rising base level. The installation of the dam causes local base level to rise to the height of the spillway. But upstream of the deposited sediment wedge, it appears that the channel is quite oblivious to this change of base level. The effect of the higher base level did not propagate upstream even with the passage of time. The same conclusion was reached in the study of much larger dams on semi-arid channels (Leopold, M. G. Wolman and Miller, 1964, p. 261).

The sand and fine gravel comprising the sediment wedge behind Big Sweat Dam is identical to that on the channel floor in the undisturbed channel upstream of the reservoir deposit, and neither are protected by vegetation. Therefore the shear stress necessary for initial motion should be the same.

A second line of evidence seems not to support the concept that soil building by vegetation results in a steeper gradient than where the channel is free of vegetation but is composed of the same alluvial material. This is the nature of the valley and its channel during active aggradation. I had at one time visualised valley aggradation as taking place by sheet flow spreading over the whole valley floor in the complete absence of a definite channel. Further field experience leads to quite a different conclusion, that is, aggradation of a valley floor takes place by widespread overflow beyond the limits of a definite channel which maintains its dimensions during the process of deposition. Information on the process is scarce because it requires proof of progressive aggradation within the observational period. Two examples have been studied where the proof exists. In both instances the evidence of aggradation exists in several generations of fence posts that have been buried. The first is in Walker Creek, near Tomales, California, where Haible (1976, in preparation) showed that valley aggradation of 4 feet (1·2 m) had taken place during a period of 60 years. One fence line installed in the first decade of the century has been buried so that the posts now protrude above the valley floor less than a foot (0·3 m). New fence posts were installed and the rancher keeps them at normal height by physically lifting them with a tractor every other year or so (Fig. 3.2).

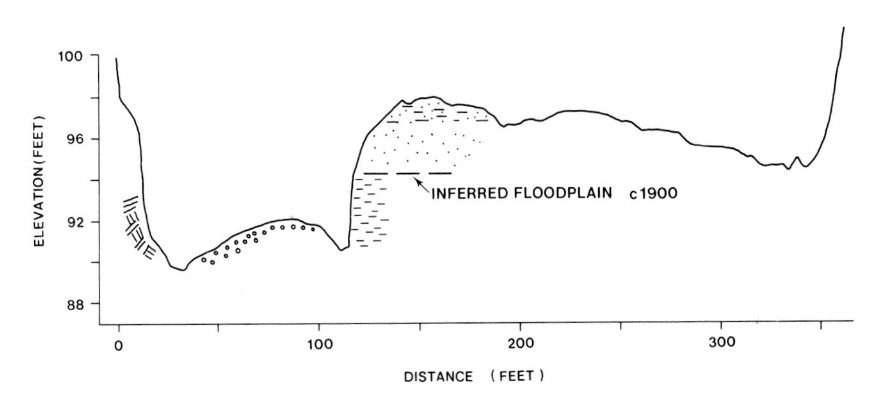

FIG. 3.9. Cross section across valley of Walker Creek near Tomales, California, where aggradation has built the valley floor 4 feet (1·2 m) since 1900.

In this area where aggradation is proven, the cross section shown on Figure 3.9 is typical. The drainage area of Walker Creek at that point is 66 square miles (171 km^2). The channel cross-section has the following dimensions and these are compared with dimensions for average or non-aggrading channels in the same region and for the same drainage area.

	Top width of channel (bankfull) (feet)	*Mean bankfull depth (feet)*	*Mean cross-sectional area (square feet)*
Walker Creek, aggrading	130	6	720
Normal channels of region	110	6·3	720

The second area is in Nebraska and includes Whitehead and Prairie Dog Creeks near Montrose, Nebraska, drainage areas respectively of 21 and 13 square miles (54 and 34 km^2). The progressive aggradation was first noted by the several generations of fence posts by H. V. Peterson and R. G. Hadley (1960) who constructed a cross-section of the pre-aggradation valley by a series of holes dug to the original valley floor. The aggradation amounts to 5 feet (1·5 m) in 30 years. The channels in the aggrading reach, maintained despite continual upbuilding of overbank areas, typically have a top width of 20 feet (6 m), bed width 2 to 5 feet (0·6–1·5 m) and mean depth of 4 feet (1·2 m). Both the Californian and Nebraskan examples show definite channels maintained during valley aggradation and channel gradient about equal to valley-floor gradient so that the channels are continuous, not extinguishing downstream by gradual shallowing (Figs. 3.10, 3.11, 3.12).

FIG. 3.10. Channel existing after half a century of continual aggradation; Whitehead Creek near Montrose, Nebraska. View is upstream. Cottonwood trees at right have been buried by several feet of silt.

We have, then, the following facts:

When a stream degrades and leaves its former flood plain as a terrace, the new gradient may be steeper or less steep, but yet sub-parallel. The change in gradient is a small percentage of the earlier one. This also holds if the stream degrades in the upper and aggrades in the lower reaches.

Aggradation in a reach of channel takes place by simultaneous deposition on the channel bed and over the adjacent valley floor. The channel in the process of aggradation is not obliterated. In fact it keeps a width and depth rather similar to others considered in quasi-equilibrium.

The relations in the hydraulic geometry, that is the width, depth, velocity, slope and roughness as function of discharge, are similar in aggrading, degrading

FIG. 3.11. Young cottonwood trees buried in several feet of silt on aggrading valley floor of Prairie Dog Creek, near Montrose, Nebraska.

and quasi-equilibrium channels. There is enough scatter of points representing rivers in a region that non-equilibrium examples cannot be separated from graded streams from the hydraulics alone.

When an ephemeral channel begins to degrade in an alluvial valley, it often begins as a discontinuous gully having a bed gradient less steep than the valley floor, and therefore it shallows downstream ending in a channelless fan. But with time such a gully tends to steepen its gradient, extending its length, and thereby tending to coalesce with others upstream and down. This increase in gradient appears to be associated with gradual channel widening.

However, when a barrier across a channel lifts the local base level, the sediment wedge deposited behind the barrier always has a flatter gradient than the reaches upstream and down. The gradient of deposition does not progressively increase with time, but becomes stable after only a few years at a value varying from 30 to 70 per cent of the original gradient of the undammed channel. The flatter gradient is usually associated with the larger width caused by the wedge of sediment filling the original trapezoidal valley to a height dependent on the height of the barrier.

These observations bear on the general problem of grade of a channel and the slope of the longitudinal profile. The present view of this matter is best expressed by J. H. Mackin (1948) who said that a graded stream delicately adjusts its slope to provide, with available discharge and channel characteristics, just the velocity required for transport of the load supplied from the drainage basin.

FIG. 3.12. Two generations of fence posts, left mid-distance, as aggradation buried the fence. Log in foreground is flood debris indicating that water inundates the whole valley. The log lodged against the upstream side of the fence.

Most of us accept the general argument for we cannot at the moment improve upon it. But it can be examined in the light of the facts enumerated above. The headwaters of the Walker Creek are degrading, indicative, one presumes, of a deficiency of available sediment. The precept given above would lead to the expectation that the gradient would tend to decrease and velocity to decrease thereby. The slope has decreased, but by only a few per cent. Any adjustment of velocity downward would be small indeed for it varies only as the square root of slope.

The lower part of Walker Creek is aggrading, presumably having more sediment than it can carry in equilibrium. But there too the slope is decreasing, also by a small amount.

Coyote C Arroyo is aggrading in the lower portion. One might suppose that the slope would tend to increase in order better to carry the supplied load, but in fact the slope is decreasing slightly.

Arroyo Falta developed, by deposition during the head-cut retreat, a bed-slope flatter than that of the valley floor. The gully erosion, a reversal from valley aggradation, implies too little sediment introduced from upstream and thus the gradient would tend to be low. This is in accord with the expectation. But as such a discontinuous gully widens, its slope tends to steepen with time. A concurrent increase in load to be transported is implied but the source of and reason for such an increase in load are not obvious.

When local base level is lifted, the gradient of deposition is less than that of the original channel. This implies a decrease in load supplied but in fact the load has not changed. To explain the small slope as a result of local widening of the flow prism is counter to the observed increase in slope as a discontinuous gully widens.

Thus these field observations cannot all be fitted into the simple explanation that slope adjusts to provide a velocity necessary to carry imposed load. How the various hydraulic and morphological factors actually change in response to changes in discharge and load, whatever their cause, constitutes an unsolved problem, even with all the new data of the past quarter-century. Continued neglect of such fundamental problems in our field while lavishing attention on details having little generality must mean that each of us should re-examine our priorities.

REFERENCES

Haible, W. W. (1976), 'Stream profile changes with time on a California coastal stream' (in preparation).

Leopold, L. B. and Miller, J. P. (1956), 'Ephemeral streams–hydraulic factors and their relation to the drainage net', *U.S. geol. Surv. Prof. Pap.* **282-A**, 37 pp.

–– Wolman, M.G. and Miller, J. P. (1964), *Fluvial processes in geomorphology* (Freeman, San Francisco), 522 pp.

Mackin, J. H. (1948), 'Concept of the graded river', *Bull. geol. Soc. Am.* **59**, 463–512.

Peterson, H. V. and Hadley, R. G. circa 1960 (undated), 'Physical features of an aggrading alluvial valley', *Abstract, U.S. geol. Surv. Denver* (Colorado).

4

FLUVIAL PROCESSES IN BRITISH BASINS

THE IMPACT OF HYDROLOGY AND THE PROSPECT FOR HYDROGEOMORPHOLOGY

K. J. GREGORY
(*University of Southampton*)

More help is wanted in various directions; for example in the measurement of velocities and other observations and in the collecting of samples of water. Everything is prepared, and it only remains to find volunteers who are able to make observations with sufficient accuracy and are willing to take the trouble.
A. STRAHAN. *Geographical Journal,* 1909, **34**, p. 650.

Although the landscape of Britain is now fashioned by rain and rivers, it is a curious paradox that this fact has not been reflected in geomorphological research until the last 20 years. This anomaly may have arisen because research tended to study unusual and distinctive elements of the British landscape and study was effected by use of, and developments from, the Davisian model. Studies of denudation chronology paralleled by investigations of glacial geomorphology were predominant in British geomorphological activity until the late 1950s and the subsequent growing interest in landscape processes was manifested first around the coast rather than at inland locations. With the realisation that the landforms of Britain were the outcome of alternating Pleistocene morphogenetic systems (C. A. Cotton, 1957), the time was opportune for more enquiries about the character of fluvial landforms and their generating processes.

Neglect of studies of fluvial processes until 1960 may be understood in relation to several interacting forces. The strands of geomorphological thought that were donated by nineteenth- to twentieth-century research (R. J. Chorley, A. J. Dunn and R. P. Beckinsale, 1964) did not obviously provide an approach that could become predominant after 1900. There was a paucity of records of river flow in Britain and geomorphological enquiries in the first half of the twentieth century were geologically-based rather than founded upon engineering principles and open to mathematical methods. The data that were available tended to remain unused in geomorphological research. Geomorphological enquiries therefore adopted the Davisian model to the extent that it came to dominate research for the first half of the twentieth century. It has been suggested that the pendulum of fashion has affected geomorphology just as it has affected other sciences

(M. A. Carson, 1971) and so it is not surprising that the fashionable period for the Davisian model in the first half of the twentieth century should be succeeded by critical review of the model and by the institution of studies that were to set new fashions in the subject.

There were, however, developments up to 1960 which provided some of the foundations for subsequent fluvial geomorphology. These foundations were of three kinds: developments of, reactions to, and parallels with studies of denudation chronology. Numerous *developments* of the Davisian model were realised as the denudation chronology of certain areas of Britain was elucidated. Terrace sequences were evaluated and this not only involved study of valley evolution but also of the extent and inter-relationships of the most recent terrace stages, so that low terraces and flood-plain terraces were often identified (e.g. R. S. Waters and R. H. Johnson, 1958; A. Straw, 1968). Analysis of the longitudinal profiles of rivers and of the reconstructed profiles of former valley floors, although requiring careful analysis (e.g. Chorley, 1958), provided useful insight, and not least among the developments of denudation chronology were the attempts to reconcile rates of erosion with the everlasting hills (D. L. Linton, 1957) and to formulate a denudation chronology for southern Britain in a framework compatible with its recent geological history (E. H. Brown, 1961). Whereas most developments of denudation chronology were empirical and field-based, the use of hardware models was pioneered by W. V. Lewis (1944, 1945) who simulated the way in which river long profiles are related to sediment discharge and to base-level changes. *Reactions* to the Davisian approach can be found in some studies that were distinctive in technique and purpose. In the 1950s, G. H. Dury (1954) began work on underfit streams, identifying them in the Cotswolds (Fig. 4.1) and employing data on contemporary fluvial processes and analyses of valley infills to proceed towards their explanation. Such studies led naturally to investigations of flood frequency (Dury, 1958). At a time when studies of slopes were receiving more interest, the headwaters of drainage networks in the southern Pennines were investigated by B. T. Bunting (1961). A few years before the hydrological significance of these investigations was fully appreciated, it was shown that percolines may be detected at the upper extremes of drainage lines (Fig. 4.1), that they had widths of 30–90 m and lengths of 600 m in the area north of Matlock, Derbyshire, and that they are lines of downslope transmission of water.

Parallel to these two groups of foundations for later work were studies by geomorphologists and other scientists. Scattered throughout the geographical literature up to 1960 were accounts of specific events demonstrating the contemporary significance of fluvial processes. The floods of the Lyn in August 1952 prompted several papers (C. Kidson, 1953; C. H. Dobbie and P. O. Wolf, 1953; G. W. Green, 1955) and the effects of the floods of south-east Scotland were detailed by A. T. A. Learmonth (1950). He showed how the rainfall of mid-August 1948 led to bridge destruction and to localised damage by flooding and mud (Fig. 4.1) and advocated more recording of the behaviour of rivers. Valuable information can be obtained from the records of past events, and the *Flood Studies Report* (*Natural Environment Research Council,* 1975) has recently summarised a large amount of data. There is scope for subsequent investigations of flood-produced landforms several decades after their formation

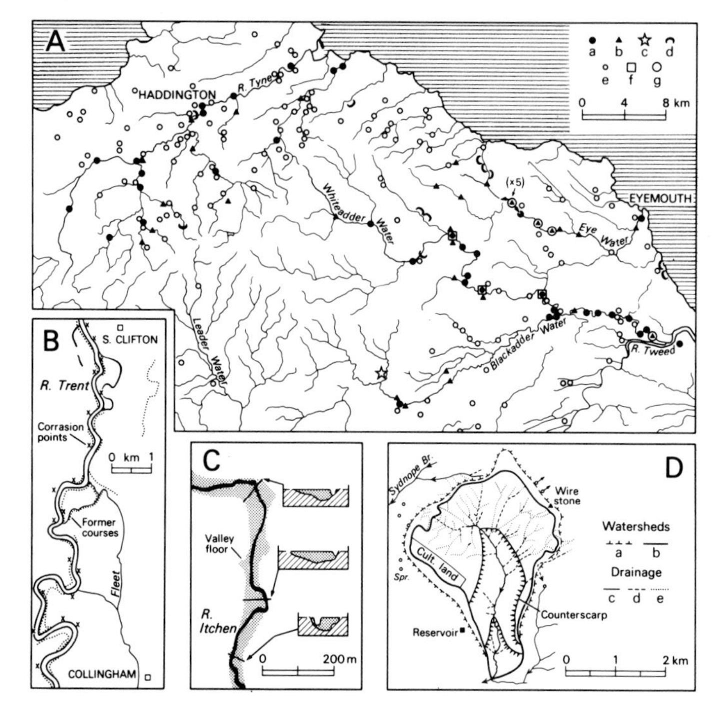

FIG. 4.1. Emerging foundations for fluvial geomorphology in Britain. Studies of the effects of large storms are illustrated by the result of heavy rainfall on south-east Scotland in A (Learmouth, 1950). Key: a – damage by flooding and mud; b – bridge destroyed; c – timber sheep bridges destroyed; d – damage to roads, railways, reservoirs by slumps, undercutting, etc.; e – minor damage to roads; f – damage by floating timber and heavy structures; g – possibly damaged by floating timber. The possibility of identification of river-channel changes from historical information is shown in B (Smith, 1910). The recognition of misfit streams is depicted in C (Dury, 1958). The components of drainage of head-water areas in the southern Pennines is represented in D (Bunting, 1961).
Key: a – morphological watersheds; b – hydrological watersheds; c – permanent stream; d – intermittent stream and invisible feature; e – percolines.

so that we can establish the permanence of fluvial features. Less dramatic changes of fluvial landforms have also been detected for many areas and in an early study of the River Trent, B. Smith (1910) showed how changes of river course could be reconstructed from historical information (Fig. 4.1). The general consequences of the taming of streams were reviewed by G. W. Lamplugh in 1914, thus anticipating the much later acknowledgement of the significance of man in affecting fluvial processes. In one study of the vegetation of the Peak district, C. E. Moss (1913) had compared the position of stream heads on maps of different dates and had concluded that a general extension of 1·2 km had taken place between 1830 and 1870. Work by hydrologists and engineers prior to the revival of interest in fluvial geomorphology in Britain included two notable contributions. M. Nixon (1959) analysed the relationship between channel size and discharge for twenty-nine gauging stations and J. E. Nash (1960) proposed general relationships between hydrograph parameters and catchment characteristics.

Notwithstanding the paucity of studies of fluvial geomorphology prior to 1960, there are several ways in which significant contributions had been made and in some cases these anticipated the prospect for more recent advances. The *Bibliography of British geomorphology* (K. M. Clayton, 1964) refers particularly to work before 1960 and less than 20 per cent of all entries could be broadly classified as fluvial including the effects of past fluvial processes; a mere 5·9 per cent were concerned with fluvial processes. The growth of interest since 1960 in fluvial processes (Table I) can be resolved into two components. The impact of hydrology was registered after 1960 and since 1970 has been complemented by an increasing geomorphological awareness which may be styled the impact of hydro-geomorphology.

TABLE 1

Increase in studies of fluvial geomorphology in Britain

Source	Date	Studies devoted to contemporary fluvial processes	Studies dominantly fluvial in character including studies of present and past processes and of fluvial landforms, but excluding studies of slopes.
Bibliography of British Geomorphology (ed. K. M. Clayton)	1964	5·9 per cent of all entries	less than 20 per cent of all entries
Register of Research, British Geomorphological Research Group	1963	3 per cent	18 per cent
Register of Research, British Geomorphological Research Group	1971	16 per cent	21 per cent
Register of Research, British Geomorphological Research Group	1975	16·3 per cent	27·7 per cent

THE IMPACT OF HYDROLOGY

In the 1960s the impact of hydrological techniques and approaches was registered on an expanding fluvial geomorphology. This impact was possible because of the atmosphere created within the subject by influences from both within and beyond Britain. Internal forces included the increasing awareness of water and its significance, emphasised by the droughts of 1959 and the floods of 1960, and also the growing interest in sedimentology (J. R. L. Allen, 1965). The number of gauging stations increased substantially in this decade (Table II) and the establishment of the Natural Environment Research Council in 1964 and the

TABLE II

Numbers of gauging stations in Britain

Data published for water year	Number of gauging stations
1935–36	28
1945–1953	81
1953–1954	102
1954–1955	116
1955–1956	128
1956–1957	147
1957–1958	174
1958–1959	188
1959–1960	205
1960–1961	238
1961–1962	271
1962–1963	311
1963–1964	364
1964–1965 Supplement 1965	415
1966–1970	782
Stations existing:	
January 1969	819
August 1970	962
January 1975	*c.* 1200

creation of the Institute of Hydrology in 1966 were possible catalysts for developments.

In view of these internal influences it is not surprising that fluvial geomorpology in Britain after 1960 developed along hydrological lines. However, this tendency was emphasised by broader and sometimes external influences. Disenchantment with the cycle of erosion had been revealed in some reviews (e.g. Chorley, 1965). By the end of the 1960s it was apparent that a closer integration of physical and social research was possible and volumes such as *Water, Earth and Man* (ed. Chorley, 1969) reflect this trend while one geographer had placed principles of hydrology (R. C. Ward, 1967a) before a geographical audience.

A further external stimulus was derived from the work of hydrologists and fluvial geomorphologists in North America. The American fluvial movement slowly produced a following in Britain, and an increasing number of citations to the work of R. E. Horton (1945) (and less frequently to his papers of 1932 and 1933) began an interest which was further stimulated by publications emanating from Columbia University (A. N. Strahler, 1964). An important milestone in fluvial geomorphology was provided by S. A. Schumm and R. W. Lichty's (1965) reconciliation of timeless and timebound approaches when they demonstrated that fluvial geomorphology may be investigated in terms of cyclic, graded, or steady time. Also in the 1960s the inauguration of the International Hydrological Decade and the proposals for Vigil Network measurements found favour in Britain (O. Slaymaker and Chorley, 1964). The importance of man's effects which could be investigated by Vigil experiments was placed in a wider context by Brown (1970) and expressed in terms of control systems by Chorley (1971).

The environment had thus been created for further studies of fluvial processes but these studies could be effected only by greater familiarity with the necessary techniques. More use was made of empirical field techniques (e.g. *British Geomorphological Research Group, Technical Bulletins*) and the tools of analysis were more readily available as familiarity with statistical and quantitative methods increased (Chorley, 1966). The impact of hydrology has really been registered as a series of individual impacts. At least four of these can be distinguished, namely, morphometry, studies of streamflow and river flow, analyses of erosion, and investigations of fluvial landforms.

Morphometry

Studies of morphometry in Britain were undertaken for nearly 20 years before a summary of objectives and methods was presented (V. Gardiner, 1974a), although a programme of research had been proposed by the British Geomorphological Research Group in the 1960s. Studies of drainage basin morphometry in Britain were initially content to illustrate the 'laws' of morphometry (e.g. Chorley, 1957a) but led towards attempts to relate morphometric parameters to a climate-vegetation index (Chorley, 1957b) and to derive a generalised drainage basin for Dartmoor. This was compared with a generalised basin for the Unaka Mountains, North Carolina and Tennessee, an area with a contrasted rainfall intensity (Chorley and M. A. Morgan, 1962). After 1965, most researchers appreciated the statistical basis of the 'laws' of basin morphometry and so more attention was directed towards other measures of drainage basin form. Therefore attempts have been made to develop measures of drainage basin character that are directly related to drainage basin process, so that more realistic measures of drainage basin shape were proposed (Chorley, D. E. J. Malm and H. A. Pogorzelski, 1957; M. G. Anderson, 1973). The desirability of improved indices that can incorporate network structure and pattern has been shown (e.g. R. I. Ferguson, 1975b).

Despite the decline of interest in the morphometric analysis of stream nets, the search for meaningful and easily used indices of the topographic character of drainage basins (e.g. M. D. Newson, 1975) was complemented by three other developments stemming from basin morphometry. These were represented by studies of network topology, of stream network extent and of network dynamics. Studies of the topology of stream networks based upon British data have led to the proposal of new methods of link analysis (e.g. R. S. Jarvis, 1972). There is still a need for methods of drainage net identification which can readily be applied to discharge estimation (e.g. Gregory and D. E. Walling, 1973) and perhaps related to hydrograph generation as attempted elsewhere by A. J. Surkan (1969). Stream network extent was initially defined using a method imported from the United States whereby a stream network was defined according to contour crenulations plus the blue lines represented on available topographic maps. In Britain the incidence of dry valleys is so spatially significant that a method of stream network definition which relies upon contour crenulations necessarily incorporates many elements of earlier networks (Gregory, 1966). It therefore appears that the stream network shown on existing topographic maps is the most appropriate basis for network definition for British basins. However, the most suitable map to use is not always easily determined and for the same area significant differences in the extent of the drainage network may be revealed by maps of different scales and editions. Differences exist between editions of the same scale and the 1:25 000 maps show different stream networks on the Provisional and the Second Series (Fig. 4.2). A. Werritty (1972)

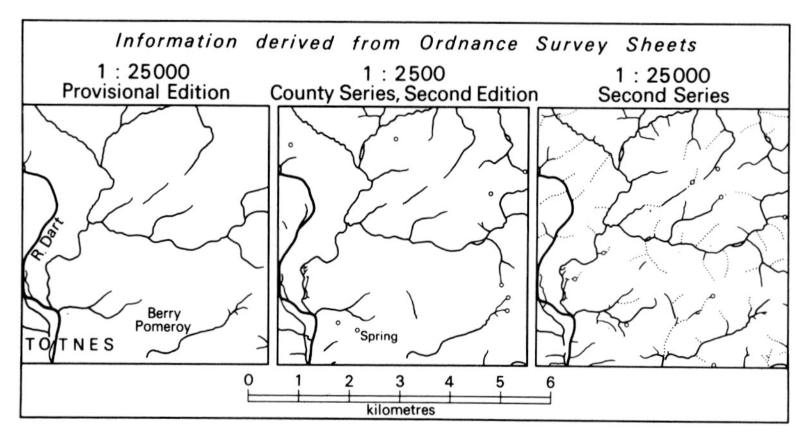

FIG. 4.2. Drainage networks according to map scale and map edition. The same area of south Devon is depicted on the three diagrams, dotted lines representing dry valleys indicated by crenulate contours.

has proposed that the 1:2500 County Series of maps may provide the best approximation to the actual stream network, but this may not always be the case and the 1:2500 maps are not easily used in studies of large areas. Comparison of three mapped networks for a small area in south-east Devon shows that the 1:2500 Second Series map provides the most extensive network (Fig. 4.2). The need for a definitive statement of the drainage network for several areas of Britain was highlighted by the U.K. Floods Study (Institute of Hydrology, 1975) although the scope for the use of a regression equation to adjust the values of stream frequency obtained from maps of Ireland at a scale of 1:63 360 to be comparable with British values was illustrated by Newson (1975). To compare densities of stream networks in several areas of Britain (Gregory, 1976a), further investigations are still required.

Whereas the drainage network was envisaged statically until the late 1960s, empirical studies of channel networks revealed the dynamics of the stream system. In an analysis of two small catchments in south-east Devon, Gregory and Walling (1968b) showed that drainage networks varied with discharge and antecedent conditions over a density ranging from 0·6 to 3·0 km/km^2 in a single basin. This analysis was relevant to interpretations of the relation between drainage density and streamflow, and indicated that the relationship between the two measures is the same for peak flows as for low flows. Subsequently, weekly measurements of stream length have been related to discharge from the River

Ray experimental catchment (K. Blyth and J. C. Rodda, 1973). Despite the earlier demonstration of the character of stream heads (Fig. 4.1) and studies of piping (A. Jones, 1971), comparatively few studies have investigated the dynamic composition of stream networks although the composition of a specific network in terms of perennial, intermittent and ephemeral streams was attempted for south-east Devon (Gregory, 1971b) and the components of the network tributary to Swildon's Hole, Somerset were analysed (J. Hanwell and Newson, 1970) following the Mendip floods of 1970.

Future progress in the morphometric analysis of stream networks in Britain has not only to proceed from a clear resolution of network definition but also to include interpretation of the structure and dynamics of networks and to acknowledge the existence of different types of stream channel. For too long, stream networks have been envisaged as purely a spatial organisation: we must now proceed to classify types according to their dynamics and composition. This need is underlined by the facts that the dynamics govern the way in which a basin can respond to extreme events, and the type of stream channel and the pattern of drainage lines directly affect hydrograph parameters.

Streamflow and river discharge

Greater appreciation of the methods of measuring discharge and drainage basin processes was evident in the 1960s and is reflected in the content of books produced (Ward, 1967a; Gregory and Walling, 1973). Analysis of existing river discharge records in relation to controlling characteristics offers considerable scope. This was indicated by a study of central Wales (G. M. Howe, Slaymaker and D. M. Harding, 1967) where the flood frequency of the Severn was shown to be greater after 1942 than previously. This was ascribed to an increase in amounts of storm rainfall and to the increase of drainage density by as much as 0·6 km/km^2 following afforestation. Subsequently it was possible for Rodda (1967) to produce a parametric model relating peak discharge of several recurrence intervals to rainfall intensity and basin indices (Table III). The increasing availability of discharge records (Table II) offers the possibility of further analyses of river discharge in relation to catchment characteristics. However, such analyses

TABLE III

Mean annual flood in British catchments (after Rodda, 1967) based on data from 26 catchments

$\log Q_{2\cdot33} = 1{\cdot}08 + 0{\cdot}77 \log A + 2{\cdot}92\, R_{2\cdot33} + 0{\cdot}81 \log D$

where $Q_{2\cdot33}$ = peak discharge in the mean annual flood (cubic feet per second)

A = basin area (square miles)

$R_{2\cdot33}$ = mean annual daily maximum rainfall (inches)

D = drainage density (miles per square mile)

Note: 1 cubic foot per second = 0·028 m^3 s^{-1}
1 square mile = 2·590 km^2
1 inch = 25·400 mm
(correct to 3 places of decimals)

do not have to rely upon the hydrograph parameters as previously employed; demodulation techniques can be applied to the total discharge record, as demonstrated for the Wye and Hamstall Ridware by Anderson (1975). Because the basins with long-established records were often the largest and because the fluvial geomorphologist sought to explain the character of the runoff-generating process, field investigations have been focused on limestone areas, on small catchments, on components of the water balance, and on recent changes in the production of runoff.

It was in limestone areas that fluvial geomorphology, disguised as karst geomorphology, first began to involve significant measurements of processes. Experiments were initially confined to tracing the origin of water discharged at resurgences but later the possibilities of assessing gross erosion were appreciated. Because dramatic fluctuations of discharge were not likely, it was often possible to estimate mean discharge values from infrequent measurements, and in addition to using the velocity-area method, the salt-dilution technique was employed successfully in the southern Pennines (A. F. Pitty, 1966). In addition to the groundwater and water actually in stream channels, it has become customary to distinguish swallet or conduit water from percolation water. The residence time, or time that groundwater is in contact with rock, has been analysed (Pitty, 1968), the catchment area of Mendip springs has been identified using spores and dyes (T. C. Atkinson and D. P. Drew, 1974) and the groundwater flow in limestone areas has been resolved into conduit flow and percolation water. In the Mendips it has been shown (Atkinson and Drew, 1974) that groundwater flow, as turbulent flow in conduits, comprises less than 30 per cent of the total, whereas more than 70 per cent is percolation water. Although more recent work has required continuous discharge measurements, and more measurements of streamflow in limestone areas are required (Newson, 1971), the approach is capable of being applied to other areas where groundwater plays a dominant part.

Small catchments became the vehicle for the expanding interest in fluvial processes in the late 1960s. Their attractions included the advantages of manageable size and the possibility of providing explanations more readily than larger gauged catchments. The considerations involved in the design of small catchment experiments (Ward, 1967b) include cost, time available for experiment, and research and teaching requirements. Although it was desirable to avoid unnecessary duplication of experiments there were no means of looking at catchment aims and objectives overall. The objective of one group of experiments was to provide information on the water balance of a single catchment so that the Catchwater Drain in Holderness (Ward, 1967b) was equipped with recurring rain gauges, a wooden flume, a climatological station including evapotranspirometer, and sites for soil-moisture measurements by neutron probe for a period of 15 years. An alternative approach was to investigate several small basins in less detail and often in relation to a specific problem. Thus, in south-east Devon, five small basins were instrumented (Gregory and Walling, 1968a) as a basis for an investigation of the effect of catchment characteristics upon discharge and sediment production. This approach has the merit that it is strictly purpose-orientated, analysis is easier because results are obtained using comparable methods, but the level of instrumentation is often not as great as in single watershed experiments. Many experimental catchments were inaugurated in the 1960s and the distribution of some of these is illustrated in Figure 4.3. In addition a considerable investment has been made in catchments instrumented by the Institute of Hydrology. At Plynlimon (Fig.4.3, No. 36) the headwaters of the Wye used for sheep-grazing are being compared with the adjacent headwaters of the Severn, two-thirds of which are under coniferous forest. The Coal Burn catchment (Fig. 4.3, No. 12) will reveal the effects of the Forestry Commission's deep ploughing and planting, and two small catchments at Milton Keynes (Fig.4.3, No. 31) are designed to elucidate the hydrological changes brought about by building the new city. There is obviously scope for the investigation of specific problems within the context of small catchments and in future work it is desirable that

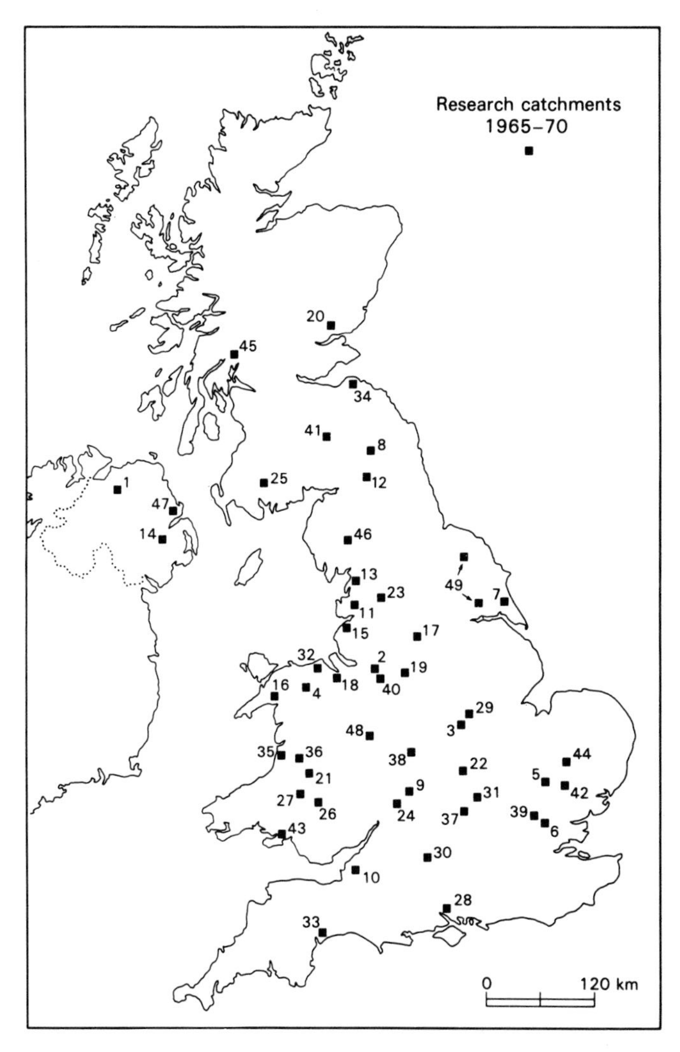

FIG. 4.3. Research Catchments, 1965–70 (based upon *Natural Environment Research Council,* 1971).

existing catchments are fully utilised as locations for other detailed experiments.

Achievements from studies of small catchments include advances in knowledge of runoff production. In the 0·021 km^2 East Twin Brook basin, Somerset, the generation of throughflow was examined (D. R. Weyman, 1970). It was shown that stormflow was derived from the 10–45 cm horizon and was controlled by the upslope extent of the saturated area, whereas base flow was derived from the 45–75 cm horizon supplied by slow unsaturated flow. Later studies showed (Weyman, 1974) that the only 'storm runoff' generated was from direct channel precipitation, and that saturated and unsaturated throughflow produce the main response of the basin to rainfall. The upper part of the basin studied produces a storm hydrograph by surface and near-surface runoff operating over up to 32 per cent of the basin area. Small catchments have thus been able to show not only the several types of runoff that may contribute to the streamflow hydrograph but also the sources of these types. It has thus gradually been appreciated that only a small proportion of the basin generates the storm hydrograph. In a catchment in south-east Devon, Walling (1971a) was able to suggest the extent of the areas that contributed to storm hydrographs under different antecedent conditions and detailed mapping of surface water frequency has been achieved in the East Twin catchment (M. J. Kirkby and Weyman, 1972). Further progress has come from experimental catchments as the water balance has been estimated more precisely for several small areas (Table IV). A comparison of methods of modelling the water balance from the Catchwater Drain catchment indicated that the Thornthwaite climatic water balance is potentially useful and that it compares favourably with more sophisticated methods (Ward, 1972).

Studies of changing basins have sometimes been possible, employing existing discharge records, and G. E. Hollis (1974) showed for the 21·4 km^2 clay catchment of the Canon's Brook, Harlow, Essex how discharge changed from 1950 to 1968 as the New Town grew to cover 16·6 per cent of the basin. He showed that the frequency of summer peaks increased although the winter peaks did not, that summer flood peaks have increased in magnitude up to 11·5 times, and that urbanisation does not appear to have affected the hydrograph of floods with a return period of at least 20 years. Analysis of discharge from a small catchment on the margin of Exeter, Devon, before and during building activity and during the early stages of urbanisation (Gregory, 1974) showed how peak flows had increased by up to three times their former values, how lag times had decreased to less than half their original amounts, and how runoff had increased in accordance with the proportion of the basin rendered impervious owing to the progress of urbanisation.

Considerable scope remains for analysing other changes as revealed by the discharge record from small catchments, and in particular for analysing the hydrological effects of total urbanisation. The effects of different land-use practices, particularly of land drainage and of afforestation, need to be identified. More generally there is still a need to determine runoff-producing mechanisms in small instrumented watersheds, but much can still be achieved by continuing to employ the existing catchments (Fig. 4.3) rather than proceeding to instrument new ones. The most recent review of hydrological research in the United Kingdom (Natural Environment Research Council, 1976) demonstrates a considerable

TABLE IV

Water balance of instrumented catchments

(a) *Catchwater Drain water-balance components (after Imeson and Ward, 1973)*

Three-year balance, October 1966 – September 1969

Precipitation	*Discharge*	*Evapotranspiration*	*Change in groundwater storage*	*Residual*
2073 mm	785 mm	1346 mm	– 70 mm	+ 12 mm

(b) *Evaporation 1966 in Catchwater Drain (after Pegg and Ward, 1972)*

Method of measurement			*Method of estimation*		
U.S. Weather Bureau Class A Pan	British Meteorological Office Sunken Pan	Evapotranspirometer	Thornthwaite estimate	Penman *Et* albedo 0·25	Penman *Eo* albedo 0·05
418·2 mm	574·0 mm	473·7 mm	620·5 mm	413·8 mm	513·6 mm

(c) *Catchments in south-east Devon, Figure 2, Number 33* (after Walling, 1971)

7 July 1967 – 6 July 1968

Catchment Number	*Rainfall*	*Runoff*	*Loss*	*Δ Storage*
1	1033 mm	467 mm	551 mm	+ 15 mm
3	1033 mm	443 mm	590 mm	0
4	1033 mm	649 mm	384 mm	0
5	1033 mm	525 mm	488 mm	+ 19 mm

range of hydrological activity and new developments could economically occur within the existing framework wherever possible.

Analyses of erosion

Contributions to the study of erosion in British drainage basins have included, first, attempts to estimate gross yield from the basin and secondly, studies that have looked at the components of the stream load in terms of solutes, suspended load and bedload. Although the errors inherent in calculating and interpreting average rates have often been emphasised, it is necessary to have some idea of erosion rates as a basis for spatial interpretations of the present, for estimates of the consequences of the past, and for predictions of the future. Until the 1960s there were few attempts to assess erosion rates for British basins partly because virtually no long-term measurements of river load had been made in Britain at river gauging stations, despite early geographical demonstrations of the need for, and relevance of, such measurements (A. Strahan, 1908, 1909, 1911). Opportunity to estimate erosion in British basins has sometimes been provided when a reservoir has been at a very low level, or was drained, so that a survey of the included sediment could be made. More recently, other estimates of erosion have

TABLE V

Some published rates and characteristics of denudation (mm/100 years) in river basins in the British Isles

(a) *Reservoir Studies*

Site	Rate	Source
Tributary to Strines Reservoir, Yorkshire	300 mm 1016 mm from steep slopes	Young (1958)
Tributary to Cropston Reservoir, Leicestershire	12·9 mm	Cummins & Potter (1967)
Tributary to Catclough, Northumberland	114 mm	Hall (1967)
Tributary to North Esk Reservoir, Midlothian	10 mm	Lovell, Ledger, Davies & Tipper (1973)

(b) *Sediment and Solute Discharge*

	In solution	*Solid load*	*Total*	
Tyne, Bywell	23·5 mm	44·1 mm	67·6 mm	Hall (1967)
Derwent, Eddybridge	34·8 mm	82·2 mm	117·0 mm	
Clyde, Daldowie		21·2 mm		Fleming (1969)
Avon, Fairholm		64·1 mm		
Whitecart, Hawkhead		43·5 mm		
Estimates for several rivers	4·2–167·7 mm			Douglas (1964)
		Suspended load		
Central Wales		106–1050 mm (Average rates: 366mm, 452mm)		Slaymaker (1972)
Montgomeryshire,				
Ebyr North	2·05 mm	0·30 mm		Oxley (1974)
Ebyr South	1·55 mm	1·55 mm		
Slapton Stream			23·0 mm	Troake & Walling (1973)
South-east Devon			26·2–98·3 mm	Walling (1971)
Catchwater Drain	51·8 mm	3·6 mm	55·4 mm	Imeson & Ward (1972)

(c) *Limestone Basins*

Site	Rate	Source
Fergus basin, Ireland	51·0 mm	Williams (1963)
Southern Pennines	75–83 mm	Pitty (1968)
Cheddar	50–102 mm	Smith & Newson (1974)

(d) *Basin Components*

Site	Rate	Source
Gully sites, Grains Hill	16 000 mm	Harvey (1974)
Rill and sheet wash, Peak District	1376 mm	Evans (1971)

Stream erosion, Co. Clare

Aggressivity of water	*Distance from beginning of stream*	*Rate*	
very aggressive	365 m	5000 mm	High
aggressive	820 m	4000 mm	and
nearly saturated	2560 m	500 mm	Hanna (1965)

Aggressivity of water	*Distance from beginning of stream*	*Rate*	
Stream bank erosion			
Bradgate Brook and tributaries to Cropstone Reservoir, Leicestershire	30 mm per year (average rate of all eroding stream banks		Cummins and Potter (1972)
Northern Ireland, two small rivers	30–60 mm per year most October–January		Hill (1973)
Bollin-Dean, Cheshire	up to 900 mm per year		Knighton (1974)
(e) *Sediment Sources*			
Hodge Beck, North York Moors	Suspended sediment discharge	192·1 mm	Imeson (1974)
	sediment production from unvegetated moorland	3810 mm	
	Channel erosion in moorland region	52·9 mm	
	Channel erosion in Lias region	21·2 mm	
Cheddar solute load from	Catchment of sinking stream	1·26 %	Smith and Newson (1974) from T.C. Atkinson
	Cave stream passages	0·05 %	
	Soil profile above limestone surface	3·62 %	
	Soil rock interface	60·08 %	
	Bedrock interstices	34·99 %	
	Total:	100·00 %	

South-east Devon: Percentage of annual load from five catchments (Walling, 1971)

Suspended Load	*Dissolved Load*	*Bedload*
20·7–53·6 %	45·4%–76·8 %	0·5–2·5 %

been derived (Table V) and these have been provided as studies of erosion have proceeded in limestone areas, in small catchments, in larger basins and in areas subject to change. Most of the rates of erosion in Table V are expressed for convenience as millimetres of lowering per 1000 years but they are usually based upon small areas and upon short periods of record. Other problems underlining the need for caution when interpreting such rates include the techniques of field measurement and subsequent calculation, the variability of rates within areas studied and the extent to which the study period is representative of longer periods.

Studies of limestone areas prompted some of the earliest attempts to measure erosion and although early attempts were based upon average figures (e.g. P. Williams, 1963), more frequent sampling was then adopted (Pitty, 1966), and solute rating curves were employed (Newson, 1917). Some of the most recent work in carbonate terrains has revealed the sources of the dissolved load (Table V, Cheddar) and experiments using tablets of limestone suspended in streams have indicated that mechanical erosion occurs, thus prompting the conclusion that the significance of solution may have been over-estimated (Newson, 1971).

Whereas studies of karst areas initially concentrated upon solutes and later embraced the solid load, studies elsewhere only later incorporated measurements of the dissolved load. Recommendations of the Vigil Programme to use monument cross-sections, scour chains, marked pebbles and infrequent sampling provided some results (e.g. Slaymaker, 1972; see Table V) but it soon became apparent that more detail was required. This detail was provided by taking water samples over the entire range of discharge and by using traps in small streams to capture bedload (Gregory and Walling, 1971). Analysis of the empirical data involved constructing rating curves that related, for example, suspended sediment concentration to discharge. Although such rating curves could afford a basis for comparison of several basins (e.g. Walling and Gregory, 1970), more frequent sampling showed that distinct curves could be produced for the rising and falling limbs of the hydrograph (N. C. Oxley, 1974) and that when a pumping sampler was used to provide yet more detail, a cloud of points is revealed which exposes the limitations of a single rating curve for a particular site (Walling and A. Teed, 1971). The impact of hydrology upon studies of sediment and solute movement in streams was largely realised when investigations attempted to characterise the sediment and solute load with detail comparable to that expected of the stream discharge record. Weekly water samples from the Yare, Tud, Wensum and Tas in East Anglia (A. M. C. Edwards, 1974) showed marked depletions in the spring and summer concentrations of dissolved silicon. These depletions were found to be unconnected with hydrological events and were assumed to be due to the assimilation of silicon by diatoms. Once more detail of sediment and solute hydrographs is available, it will be possible to apply time-series methods in data analysis. For the River Stour in eastern England, Edwards and J. B. Thornes (1973) used regression, spectral and cross-spectral techniques to examine the trends and periodicity in a 20-year record of weekly observations of eight water quality variables.

The greatest advance in the study of sediment and solute loads at a specific gauging station has undoubtedly been realised with the advent of automatic samplers and continuous monitoring apparatus utilised, for example, by Walling (1975). This development has exposed the inadequacy of infrequent or even regular interval samples because the suspended sediment hydrographs, and the solute hydrographs or 'chemographs', may not imitate the streamflow hydrograph in any general manner. In particular, the hysteretic effect can lead to substantial differences between sediment and solute hydrographs, in terms of timing and magnitude, when compared with the streamflow trace. Thus two events in the Rosebarn catchment (Fig. 4.4) show how a rating loop expresses the concentration-discharge relation for suspended sediment and, furthermore, how the single rating curve is a poor representation of the relationship even when drawn for a single event. Differences in the timing of solute chemographs are also likely along many streams, and Walling and I. D. L. Foster (1975) have drawn attention to the variability that can occur as illustrated by the Back Brook (Fig. 4.4).

Small areas have been employed for studies of sediment and solute production because the details of dynamics can be established and the effects of basin changes can be documented. Results from small basins have illuminated the relative importance of the different components of the river's load. In the

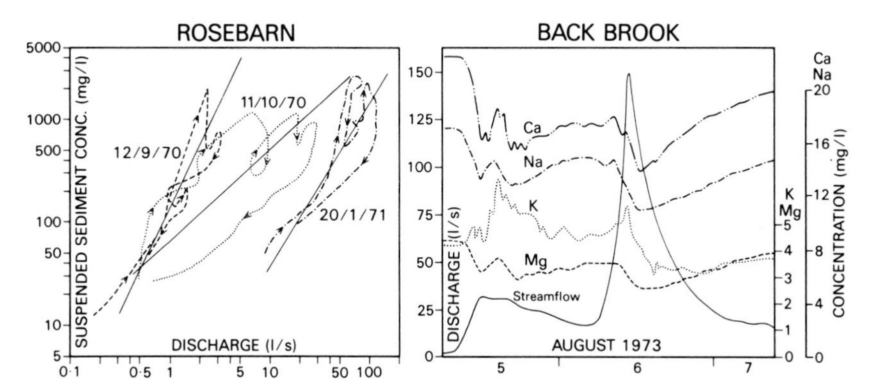

FIG. 4.4. Temporal patterns of sediment and solute production. For the Rosebarn catchment, three rating loops are shown (Walling, 1974); the solid lines depict linear rating relationships fitted to the values for each storm. Ionic concentrations in relation to streamflow for Back Brook are based upon Walling (1974) and Walling and Foster (1975).

Catchwater Drain basin, A. C. Imeson and Ward (1972) showed that solutes represented 93·7 per cent of the total load in one year, and in five basins in south-east Devon the suspended load represented between 20·7 and 53·6 per cent of the total, the dissolved load comprised 45·4 to 76·8 per cent of the total, leaving bed load between 0·5 and 2·5 per cent (Walling, 1971b). The major proportion of the stream's load is removed as solutes but, whereas this removal occurs throughout much of the year, suspended sediment and bedload have been shown to be removed in small fractions of the year. Thus in south-east Devon catchments as much as 46 per cent of the suspended load could be moved in 0·1 per cent of the time (Walling, 1971b) and from the Catchwater Drain 98 per cent of the bedload was moved by the five largest discharges of the year (Imeson and Ward, 1972). Analysis of the results from small areas can enable the sources of sediment and solutes to be identified. Although solute sources have been identified in studies of limestone areas (Table V), it has more recently been possible to suggest the contribution of unvegetated moorland and channel erosion to the production of suspended sediment from Hodge Beck (Table V; Imeson, 1974). In mid-Wales, analysis of a 0·54 km^2 basin showed that sediments were derived almost entirely from streambank bluffs in solifluction deposits and that solutes are closely related to antecedent rainfall and to flow routes through soils, peat deposits and pipes through the catchment (J. Lewin, R. Cryer and D. I. Harrison, 1974). Knowledge of sediment and solute sources must be the basis for interpreting suspended sediment hydrographs and chemographs. Results from Devon catchments have shown how certain solute species increase rather than decrease in concentration during storm events and such dilution effects, together with lag effects, need to be analysed against the background of the spatial pattern of sources.

Temporal variations in catchment condition have been analysed from small catchments. Considerable scope for this type of analysis remains but progress has been made in examining the effects of afforestation on sediment supply (Institute of Hydrology, 1974) and the effects of building activity. In the small basin on the margin of Exeter, techniques of analysing sediment and solute discharge have been refined (Walling, 1974a). Using these techniques, it has been shown (Walling, 1974b) that sediment concentrations increased five-fold and sediment loads between five and ten times, when only 25 per cent of the area was affected by building activity.

The adoption of a hydrological approach to study of water quality was seen as a future research need during 'a view from the watershed' in 1973 (Institute of Hydrology, 1973). By 1976 the impact of hydrology was evident in studies of sediment and solute production, and catchment response can now be expressed by sediment hydrographs and by chemographs for individual ions. Future studies will need to look at sediment changes during transport. Empirical studies have been reported (A. Whittel, 1973; Gregory and R. A. Cullingford, 1975) and the broad theoretical implications have been outlined (Allen, 1974). As the task of modelling sediment and solute production is further refined, the spatial pattern of flows will require further theoretical elucidation. Hitherto, we have representations of the patterns of processes but documentation of the processes themselves is not so easily realised.

Fluvial landforms

Study of fluvial landforms tended to develop after the other three lines of enquiry noted above. The impact of hydrology was registered by studies of the relationships between process and landform. Although related studies had emerged from the parametric models devised for discharge (e.g. Table II), subsequent studies were made of channel geometry, of channel pattern and of networks.

Channel geometry and discharge were related in an important paper by Nixon (1959). Data from twenty-nine gauging stations in England and Wales allowed régime equations, which related bankfull discharge to width, depth, velocity and cross-sectional area of water, to be formulated. Few studies of hydraulic geometry have been undertaken in Britain but in an analysis of gauging station data, Richards (1973) showed that the commonly-used power relations may not be the most appropriate and that higher-order polynomials are sometimes preferable. In an analysis of variations in hydraulic geometry as discharge changes, A. D. Knighton (1975) showed how, in the case of a channel with cohesive banks and no marked downstream variation in bank erodibility, the downstream rate of width increase is a function of discharge while the rate at a station is controlled largely by the composition of the bank material and especially the silt-clay content. He went on to suggest that the deposition of non-cohesive sediment in the form of point bars and central islands provides a means whereby the stream can increase the rate of change of width at a cross-section. This suggests that meandering and braided reaches could be distinguished from the straight ones by the *b* exponent. To provide further understanding of the relation between channel capacity and stream discharge, A. M. Harvey (1969) surveyed seventy-one cross-sections from three streams in southern England. He concluded that variations in bankfull frequency occurred both along and between streams and this was interpreted in the light of the occurrence of flood régime stream segments and baseflow streams that are adjusted to rarer floods (see page 50). Further work is required on channel capacity, and the suitability of the annual series for discharge analysis is questionable. Bed and bank sediment must be considered because coarse bedload is an important influence on the shape of channels in the Hodder basin of the western Pennines (D. N. Wilcock, 1971).

Work on channel patterns was spearheaded in Britain by G. H. Dury. In the course of his work on underfit streams (1958, 1964), it was necessary to establish the present relationship between meander wavelength and discharge to demonstrate the underfitness of present streams and to indicate the discharges likely in these meandering valleys. It has recently been suggested that the drainage of the upper Severn has been so strongly affected by climatic change that some of its landforms are disharmonious with the present climate (Dury, C. A. Sinker and D. J. Pannett, 1972). The existence of the buried channels in the Cotswolds has been debated (Beckinsale, 1969-70) and a spate of discussion was precipitated by discovery elsewhere of 'active' valley meanders and dispute over the superimposition hypothesis for incised bedrock meanders. This discussion has certainly exposed our inadequate knowledge of the significance of sediment (Ferguson, 1973), the control of bedrock rather than alluvial meanders (Kirkby, 1972), the variability of bankfull capacity along reaches and between streams, and the fact that a single discharge value (B. A. Kennedy, 1972) cannot express the range of discharges that control meander dimensions.

Further work on the relative significance of these different factors is necessary as we proceed towards the eventual model required to explain channel patterns, and some of this work may be effected by empirical studies of processes (e.g. R. D. Hey and C. R. Thorne, 1975).

Studies of networks have not fully registered the impact of hydrology. Some studies of gully development have been undertaken and these derived from studies of peat erosion (e.g. J. Radley, 1962) and from work by ecologists (J. H. Tallis, 1973). Although morphometric analysis has been employed unsuccessfully to ascertain whether two types of gully could be differentiated in the Peak district (Mosley, 1972) and the nature and rate of erosion have been established for New Forest gullies (C. G. Tuckfield, 1964), few studies have looked at gully growth in relation to discharge and sediment production. Gullies cut in till at Grains Gill in the Howgill Fells allowed the seasonal variation of processes of surface erosion by runoff, mudflow activity and the free fall of particles to be identified (Harvey, 1974). Studies of the modification of stream networks by gully development are only recently being set in the context of a hydrological approach. This is necessary in the future and the study of network contraction, leading to the production of dry valleys, should also be visualised in the context of our hydrological understanding of drainage basin dynamics (Gregory, 1971a).

The period 1960-70 may thus be seen as a time when the impact of hydrology on studies of the four themes identified above was felt. This impact necessarily involved gaining familiarity with existing hydrological techniques and developing new ones, especially for studies of sediment and solute production.

THE PROSPECT FOR HYDROGEOMORPHOLOGY

New terms should be introduced with caution but A. E. Scheidegger (1973) has defined hydrogeomorphology as the study of landforms caused by the action of water. The term includes the study of other water-produced landforms such as coastal ones but it stresses the need to concentrate upon the interface of hydrology and geomorphology. Whereas forms and processes were measured during the 1960s in a variety of ways it is increasingly acknowledged that processes must be analysed in relation to controlling and dominant basin characteristics, and that studies of form must concentrate upon fundamental landform indices. Developments that reveal some of the prospects for hydrogeomorphology include analysis of spatial patterns, identification of significant basin characteristics, and improved representation of fluvial landforms in relation to analysis of landform changes.

Spatial patterns

Identification and description of spatial patterns is one of the fundamental features of a geomorphological approach and so it is desirable that spatial analysis of fluvial landforms and processes, and eventually of fluvial form-process assemblages, is attempted. There have been surprisingly few attempts to map spatial variations of river flow in Britain. An early attempt (Linton, 1959) outlined the possibilities and, as more data have become available, it has been possible to produce a range of maps (Ward, 1968). Numerous indices of streamflow can be employed for mapping and spatial patterns may be established using mean runoff figures, discharge ratios, time of occurrence of extreme flows and the correlation between precipitation and runoff. Each of these maps reflects some of the controlling variables in a particular way, and the runoff ratio (Fig. 4.5A) shows how a higher percentage of precipitation reaches the streams in the higher, wetter and more impermeable north-west of Britain than in the south-east. There is further scope for characterising the spatial pattern of runoff in Britain, possibly including its temporal variation during the period of records, and there is also scope for the development of hydrological maps. Although the published geomorphological map (Gregory and Brown, 1975) contains some hydrological information there is a need to develop a scheme appropriate for Britain and to implement it in the form of a hydrological map.

Three other types of spatial pattern have been illustrated. Components of the water balance have been mapped, and as a result of applying Penman's equation over grid squares, a mapping technique was developed from grid-square information and employed to produce maps of potential evapotranspiration, mean actual evapotranspiration and mean annual runoff in part of south-west England, 1960–69 (A. M. Foyster, 1973). A second type of mapping has emerged from studies of drainage-basin morphometric variables. From an analysis of twenty-one variables of 379 second-order basins from 1:10 560 Regular Edition maps for north-west Devon, Gardiner (1974b) produced maps of individual variables such as relief ratio, of hybrid variables such as ruggedness, and also of combinations of factors to give a hydrological assessment and the pattern of regionalisation (Fig. 4.5C). The hydrological assessment was based upon altitude because it is related to precipitation which is in turn related to discharge, and upon drainage density because it affects the timing and magnitude of runoff response. Although such maps are in their early stages of development and the hydrological limitations of the grid square have to be overcome, the technique offers the basis for differentiating morphometric patterns that are of intrinsic interest and also of relevance to the pattern of fluvial processes. Further

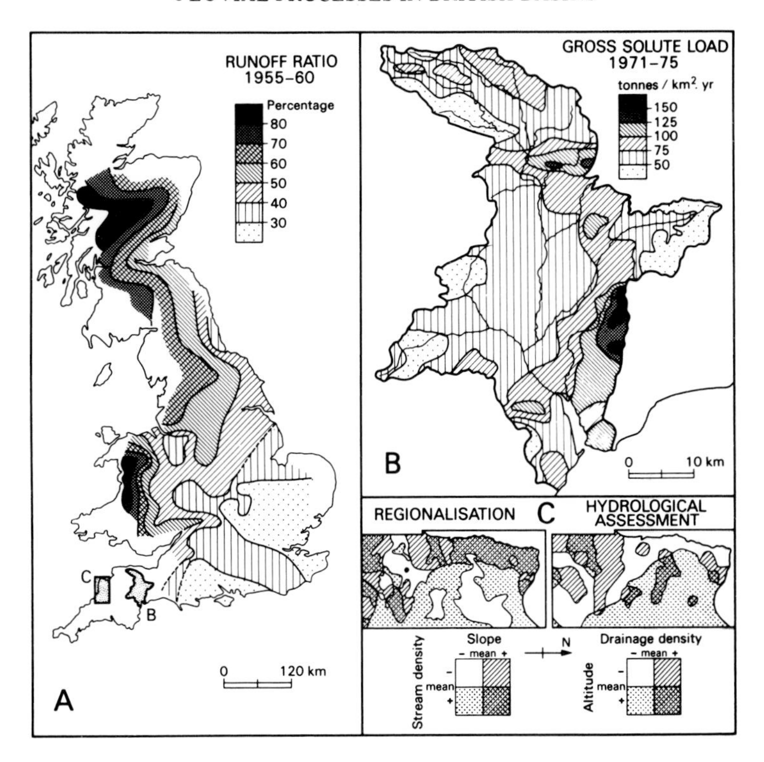

FIG. 4.5. Spatial patterns of hydrogeomorphology. The Runoff Ratio in A is based upon Ward (1968). Gross Solute Load of the Exe basin in B is derived from work by D. E. Walling and B. Webb, and two classifications of terrain based upon drainage basin morphometry in C are after Gardiner (1974).

progress may be made in mapping dominant parameters: maps of estimated drainage density have been produced for Dartmoor (Gardiner, 1971) and for south-west England (Gregory and Gardiner, 1975), and a map of drainage density for south-east England has been produced to illustrate a photo-mechanical technique based upon line-weight and a whirling illuminator (Gardiner, 1974c).

A third approach to the identification of spatial patterns has been achieved by mapping water quality. Within the Exe basin, water samples were collected from more than 500 sites during stable low flows in July 1971 (Walling and B. W. Webb, 1975). The spatial pattern of specific conductance was found to be closely related to lithology but land use was also found to exert a significant influence. A provisional estimate of gross solute load (*Lg*) was calculated for each sampling point (Fig. 4.5B) from

$$Lg \text{ (tonnes/km}^2\text{/year)} = \frac{R \text{ (mm)} \times Cm \text{ (mg/1)}}{1000}$$

where: $Cm = SC \times C \times K$
C = constant to convert total conductivity to total solute concentration (value of 0·65 was used)

K = constant to correct observed value of total solute concentration to the approximate discharge-weighted mean concentration ($K < 1.0$). Values were derived for gauging stations where solute concentrations had been measured frequently.
SC = observed values of specific conductance
R = mean annual runoff

The gross solute loads were then converted to estimated denudation rates (mm/1000 years) by making allowance for atmospheric solute inputs. The spatial pattern revealed gives rates varying from 2 to 100 mm/1000 years (*cf.* Table V) and the pattern conforms broadly to the contemporary distribution of relief (Walling and Webb, 1975). Such patterns of estimated chemical denudation are revealing not only because they suggest higher rates than those estimated for many limestone areas, but also because they provide a way whereby the results of analyses of contemporary processes can be interpreted directly in relation to landform development.

Significant basin characteristics

Parametric models relating indices of fluvial processes to basin characteristics were frequently used in the decade beginning in 1960. Very often large numbers of basin parameters were measured and it has subsequently been appreciated that several significant parameters, each of which expresses the way in which the basin functions, must be used. Indeed, C. Kirby and Rodda (1974) have argued that parameters with real physical meaning must be introduced. Such parameters should not only be meaningful but ideally they should be easily expressed and determined, and this was a consideration in the selection of indices incorporated in the Flood Studies Report (NERC, 1975). Analyses of data should in future be concerned with large catchments as well as the small ones favoured (p. 00 above) and so parameters should be pertinent to flood routing (R. K. Price, 1974) and to sediment- and solute-routing methods. In a study of water movement through or across limestone areas of the Gower peninsula, it has been shown qualitatively how the pattern of water movement is controlled by the state of development of the karstic drainage systems (D. P. Ede, 1975).

Further empirical studies of the relation between stream hydrographs and the network parameters should be possible but in addition it is necessary to develop theoretical models for channel flow, and eventually for sediment production as well. Such models must acknowledge the way in which the area contributing the runoff changes with antecedent conditions and with input, and also reflect the way in which the flow network is composed and structured. To proceed towards modelling flows based upon hillslope and channel flows, A. Calver, Kirkby and Weyman (1972) modelled the hillslope hydrograph and then used this as the input for channel flows which were modelled by the continuity equation as:

$$\frac{dQ}{dx} + \frac{dA}{dt} = q\,(t)$$

where Q = stream discharge and A = channel cross section. A basin hydrograph may then be computed using the Manning equation. Drainage density may be useful in this context although Calver, Kirkby and Weyman (1972) preferred channel frequency because it specifies the number of channels at set distances from the basin outflow and indicates where most of the source areas may be found. Further use of an index of stream power as employed by Richards (1972) may be a potentially useful development.

Fluvial landforms and landform adjustments

The dimensions, significance and stability of fluvial landforms have attracted more attention since 1970. Although palaeohydrology and river metamorphosis (Schumm, 1969) have not yet been fully explored in Britain, Hey (1974) has proposed that a river has five degrees of freedom, namely: width, depth, slope, velocity and plan shape, and that these may be regarded as dependent variables whereas input of water, input of sediment, bed and bank material, and the valley slope are independent variables. These two groups of variables are related by five equations including continuity, flow resistance, bed-load and bank-competence equations, and these account for the way in which a river adjusts to changes of the independent variables. Some changes in water, sediment, solute distribution and availability have been induced by man so that it is necessary to establish the natural changes and to differentiate them from anthropogenic changes. To proceed towards this objective it is necessary to have improved methods for characterising fluvial landforms, to establish the nature and magnitude of changes taking place, and to determine the character and extent of induced changes.

Although the editor of *Area* concluded in 1972 that the debate on meanders was closed, this debate, which had embraced several aspects of channel landforms, will certainly continue. Continuation is necessary to proceed towards a more complete understanding of the morphometry of fluvial landforms. Despite the studies of hydraulic geometry we have insufficient understanding of the spatial pattern of stream-channel geometry. One way in which this understanding may be approached is by examining the relationship between channel cross-sectional area and drainage area (*cf.* Fig. 4.8), a relation that has been investigated for the river Dart and its tributaries in Devon (C. C. Park, 1975). Further progress may be made by incorporating values of channel slope in the relationship, and data from tributaries of the Ure and Nidd have been used to show that drainage area (Ad), channel cross-sectional area (X) and slope (S) are related by an expression of the form $Ad = X/\sqrt{S}$ (Gregory and Park, 1976) and that map-slope values appear to be as effective as field measurements (Park, 1976).

Channel patterns have been further investigated by field measurements of change. Empirical measurements (Table V) have been undertaken by Knighton (1974) and by A. R. Hill (1973) in northern Ireland who showed that erosion processes vary considerably on a single bank as well as between streams. Lateral corrasion and hydraulic action were found to be the main processes affecting the lower parts of stream banks, and frost action was important on some exposures of till. Although empirical measurements provide much needed data on rates of river-channel change in Britain, the short period over which they can be made may conceal long-term change. Some insight is therefore provided by the detailed documentation of specific changes such as meander cut-offs inspired by high discharges (e.g. R. H. Johnson and J. Painter, 1967; Mosley, 1975a) but the potential of maps, remote sensing and documentary evidence to indicate channel change is now beginning to be realised (Lewin, 1972). Thus in a study of the river Endrick, channel change was isolated by comparing 1:10 560 maps for 1896 and 1957 (B. J. Bluck, 1971) and the direction of point-bar growth deduced (Fig. 4.6A). Considerable scope for the analysis of channel change remains but this must be based upon improved representation of the river trace. Whereas earlier studies relied upon simple parameters such as meander wavelength, it has been shown that non-uniformity of environmental conditions leads to differences in meander irregularity as well as in sinuosity and dominant wavelength (Ferguson, 1975a). Direction-change spectra have been shown to provide better wavelength estimates for nineteen British rivers and wavelength is nearly linearly related to the square root of the 1 per cent duration discharge.

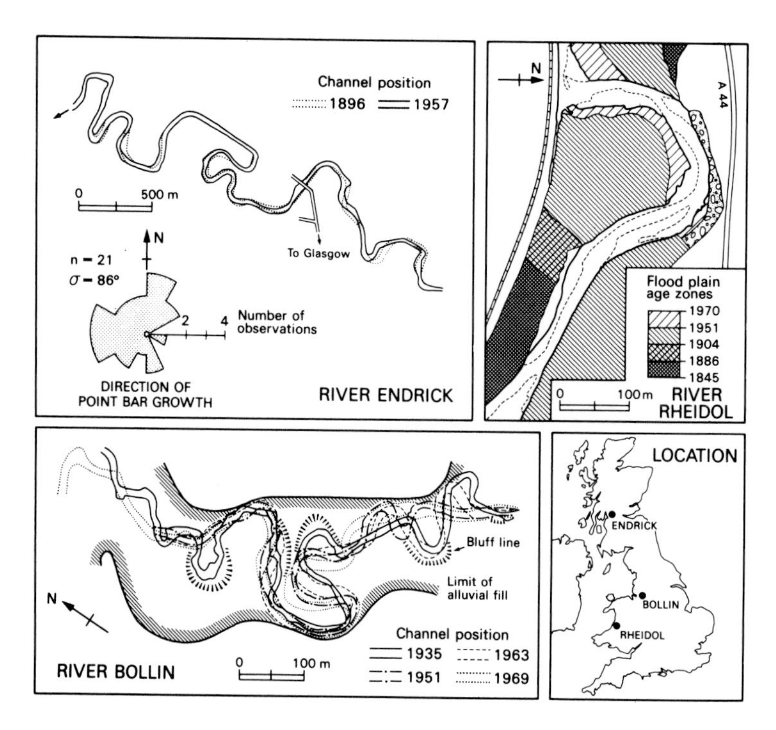

FIG. 4.6. River channel and flood-plain changes. Work on the River Endrick was by Bluck (1971); on the Rheidol by Lewin and Davies (1974); and on the Bollin by Mosley (1975).

Similar progress has not yet been made with regard to flood-plain morphology. In the Rheidol valley, flood-plain renewal may reach some 560 m^2/km of valley per year and it has been possible, using maps and aerial photographs, to reconstruct the location of the active river channel on up to eight occasions in 130 years (B. E. Davies and Lewin, 1974). For a loop of the Rheidol (Fig. 4.6B) it was possible to reconstruct the flood-plain zones and also to show that contamination of soils was greatest during the periods when lead mining was most active. Flood-plain morphology has received little attention but the inherent diversity of detail that may occur is illustrated by surveys of flood-plain geometry for four years in Wales by Lewin and M. M. M. Manton (1975). They showed that migration plains may have a relief of as much as 2·5 m and studies of flood-plain geometry are necessary as a basis for understanding patterns of flooding. Detailed mapping of the valley-floor morphology for the upper Exe and Barle valleys has shown the existence of at least three distinct levels within a range of 1–2m (B. Cant, 1974). When the significance of such features is understood this may be useful in relation to estimating the flood hazard from morphological evidence.

Some idea of the rates of change of fluvial landforms in Britain is slowly being acquired, and at the same time there is a greater appreciation of flood-plain composition, channel patterns and channel geometry. Further progress towards the clear understanding of the relation between fluvial processes and landforms is desirable and a model of fluvial landform adjustment throughout the basin may assist in the prediction of change. Studies of channel geometry, of channel pattern and of networks have hitherto been conducted rather separately but they must be integrated more closely for the progress of research (Gregory, 1976c).

Until there is a better understanding of the overall fluvial model, it may be somewhat premature to investigate the effects of changes in fluvial processes. However, these changes are substantial and have been ignored for too long. Changes induced by man over the last 2000 years at least must have had effects upon fluvial landforms but a study of south-east Wales (C. B. Crampton, 1969) is one of the few that has identified terrace deposits and other consequences of deforestation. More recently, the effects of afforestation, land-use pressure through grazing, river regulation and land drainage should have produced hydrogeomorphological as well as hydrological effects. The most significant effects for river channel forms may have been caused by flow regulation following reservoir construction and by increased runoff from urbanisation. Anderson (1975) showed that a significant drop in the annual amplitude of the demodulated discharge series for the river Blythe occurred in 1952–54; it subsequently remained consistently below the pre-1950 level, and the period 1952–54 coincided with the impounding of the Blythe. Whereas the damming of reservoirs can lead to a reduction in the frequency of peak flows, urbanisation has been shown to lead to larger peak discharges (Gregory, 1976b) and also to more frequent ones. In Figure 4.7, the monthly pattern of rainfall is contrasted with the monthly frequency of hydrographs from the Rosebarn catchment on the margin of Exeter. This shows how the number of events recorded by the streamflow trace is now 2·7 times greater than the number before building activity and urbanisation. Some progress has been made in identifying the landform adjustments consequent upon such hydrological changes and the modified ergodic hypothesis has been employed to indicate the likely magnitude of channel capacity downstream from reservoirs and downstream of urban areas. Clatworthy reservoir has a substantial effect on discharges of the Tone below the dam (Gregory and Park, 1974) and comparison of channel capacity above and below the dam (Fig. 4.8A) in relation to drainage area indicates that capacity immediately below the dam is less than half of that expected. This effect persists downstream until the catchment area is over twice that draining to the reservoir. Streams flowing through the urban

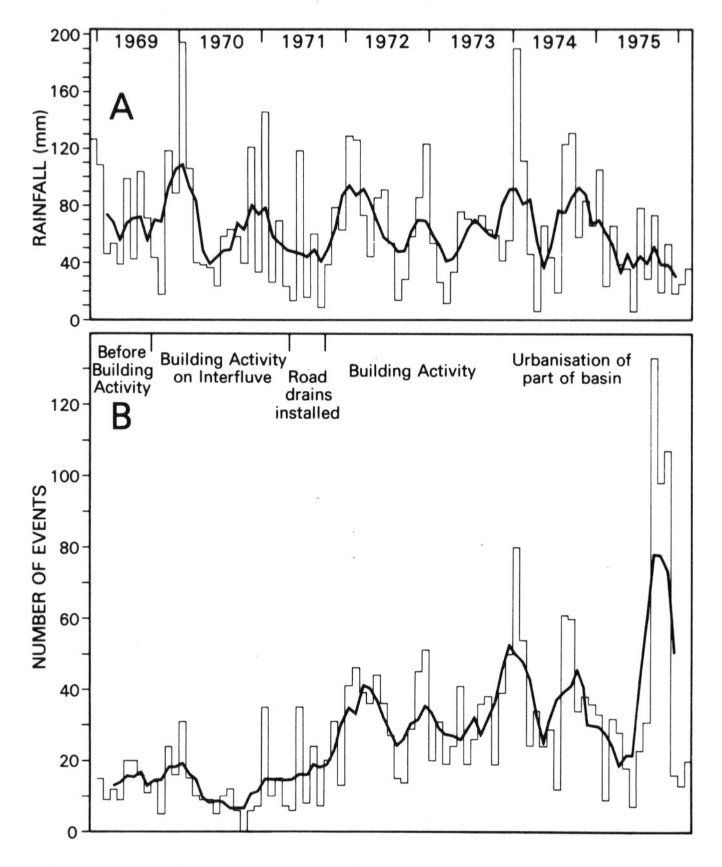

FIG. 4.7. Building activity, urbanisation and catchment response. For a small catchment on the margin of Exeter, Devon, monthly rainfall and 5-month running means in A are contrasted with the number of detectable hydrographs per month in B where the heavy line also represents the 5-month running means.

area of Catterick camp may be compared above, within and downstream from the urban area (Fig. 4.8B). The capacity of the urban channels averages 1·75 times the expected values, compared with 2·25 times the expected values below the urban area.

Such examples illustrate the way in which recent fluvial landform adjustments have occurred in Britain. The spatial significance of such adjustments may not have been fully realised. A small gully in Devon has been shown to have developed in 25–30 years as a result of installing the road-drainage system (Gregory and Park, 1976b) and Dury (1973) has detected a significant decrease in channel widths below the points of abstraction of river water through mill leats. Changes of channel pattern have also been ascribed to the increasing effects of human activity. A reach of the river Bollin, Cheshire, appeared to be stable between 1872 and 1935 (Mosley, 1975b). After 1935, channel shifting increased considerably, there were seven meander cut-offs, and channel sinuosity declined significantly from 2·34 to 1·37 (Fig. 4.8C). It was proposed that one or

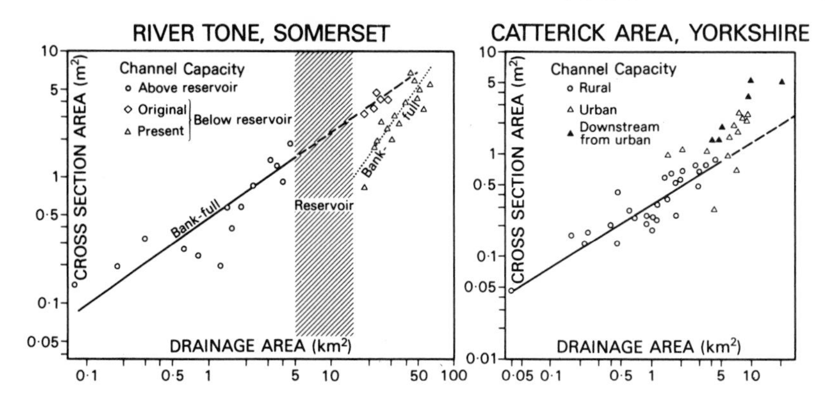

FIG. 4.8. Adjustments of river-channel capacity. The relation between channel cross-sectional area and drainage area above Clatworthy reservoir on the river Tone is found to continue below the dam, although the contemporary bankfull capacities are now significantly less since reservoir construction in 1959. The reduced flood frequency downstream from the dam is analysed by Gregory & Park (1974).

A similar relation for rural channels upstream from Catterick Garrison in Yorkshire indicates that many of the channels within, and downstream from, the urban area have channel capacities greater than expected. The enlarged channels result from the urban runoff which has increased as the stormwater drainage system of Catterick Garrison has been developed since 1900.

two flood events enabled the reach to cross an 'equilibrium threshold' and this new equilibrium has been maintained as a result of human activity expressed through urbanisation and agricultural drainage.

Such studies of recent adjustments of fluvial landforms open up extended prospects for hydrogeomorphology. Further studies are required on such topics as the closer integration of geometry, pattern and network; the expression of meaningful catchment parameters; the more complete representation and analysis of water and sediment discharge; and the use of dating techniques such as dendrochronology. In British fluvial geomorphology we have perhaps concentrated too much on the dramatic and the particular. In the future we should look for the complete range of effects and analyse why complex responses (Schumm, 1973) occur in kind and in time.

CONCLUSION

Over the past two decades, studies of fluvial processes in British basins may be resolved into those stimulated by the impact of hydrology and those motivated by the prospect for hydrogeomorphology. Although both themes will continue, we should now be able to proceed further to identify changes past, present and future. We still do not have sufficient knowledge of how and why fluvial processes operate in British basins. Further spatial analysis is desirable and the production of a hydrological map could be a worthwhile objective. The modelling of runoff generation and of sediment and solute production must be a long-term objective and although theoretical developments are necessary for this purpose they will depend upon empirical advances to indicate the components for the models. For this reason we need a better understanding of the ways in which significant basin forms can be expressed so that we can model their function. With an improved knowledge of processes, realisation of the magnitude of process change would be feasible and this could be the basis for the modelling of landform changes. It has often been contended that Britain does not have a soil-erosion problem (although this may now be questioned) but we must have had such a problem in the past. Where are the results? On permeable outcrops, dry valleys have been analysed in relation to climatic change but where are the equivalent effects on impermeable rocks?

Answers to such intriguing questions must come from closer liaison between empirical and theoretical approaches, so that it is undesirable for theoretical geomorphology to become divorced from fluvial geomorphology. It is equally desirable that applied aspects stem from the frontiers of fluvial geomorphology. Developments in the analysis and prediction of spatial patterns of fluvial processes, and of channel and flood plain morphology, should be adequately represented in future research and are pertinent to the environmental impact (Lewin, 1975). Many of the aims of the Royal Geographical Society project (Strahan, 1909) find a continuing relevance nearly 70 years later, and objectives rather than techniques can now take pride of place because the environment of Michael Frayn's *A very private life* may still lie well in the future.

REFERENCES

Allen, J. R. L. (1965), 'A review of the origin and characteristics of recent alluvial sediments', *Sedimentology* **5**, 85–191.

—— (1974), 'Reaction, relaxation and lag in natural sedimentary systems: general principles, examples and lessons', *Earth Sci. Rev.* **10**, 263–342.

Anderson, M. G. (1973), 'Measure of three-dimensional drainage basin form', *Wat. Resour. Res.* **9**, 378–83.

—— (1975), 'Demodulation of stream-flow series', *J. Hydrol.* **26**, 115–21.

Atkinson, T. C., and Drew, D. P. (1974), 'Underground drainage of limestone catchments in the Mendip hills', in 'Fluvial processes in instrumented watersheds' (ed. K. J. Gregory and D. E. Walling), *Inst. Br. Geogr. Spec. Publ.* **6**, 87–108.

Beckinsale, R. P. (1969–70), 'Physical problems of Cotswold rivers and valleys', *Proc. Cotteswold Nat. Fld Club* **35**, 194–205.

Berry, K. (1967), 'Gypsey Race, a very intermittent stream of the Yorkshire Wolds', *Weather* **22**, 434–9.

Bluck, B. J. (1971), 'Sedimentation in the meandering River Endrick', *Scott. J. Geol.* **7**, 93–138.

Blyth, K. and Rodda, J. C. (1973), 'A stream length study', *Wat. Resour. Res.* **9**, 1454–61.

Brown, E. H. (1961), 'Britain and Appalachia: a study of the correlation and dating of planation surfaces', *Trans. Inst. Br. Geogr.* **29**, 91–100.

—— (1970), 'Man shapes the earth', *Geogr. J.* **136**, 74–84.

Bunting, B. T. (1961), 'The role of seepage moisture in soil formation, slope development and stream initiation', *Am. J. Sci.* **259**, 503–18.

Calver, A., Kirkby, M. J. and Weyman, D. R. (1972), 'Modelling hillslope and channel flows', in R. J. Chorley (ed.), *Spatial analysis in geomorphology*, (Methuen, London), 197–220.

Cant, B. (1974), 'Aspects of valley-floor development in the Upper Exe basin', *Rep. Trans. Devon. Ass. Advmt Sci.* **106**, 77–94.

Carson, M. A. (1971), *The mechanics of erosion* (Pion, London), 174 pp.
Chorley, R. J. (1957a), 'Illustrating the laws of morphometry', *Geol. Mag.* **94**, 140–50.
—— (1957b), 'Climate and morphometry', *J. Geol.* **65**, 638–68.
—— (1958), 'Aspects of the morphometry of a "poly-cyclic" drainage basin', *Geogr. J.* **124**, 370–4.
—— (1965), 'A re-evaluation of the geomorphic system of W. M. Davis', in R. J. Chorley and P. Haggett (eds.), *Frontiers in geographical teaching* (Methuen, London), 21–38.
—— (1966), 'The application of statistical methods to geomorphology' in G. H. Dury (ed.), *Essays in geomorphology* (Heinemann, London), 275–387.
—— (ed.) (1969), *Water, Earth and Man* (Methuen, London), 588 pp.
—— (1971), 'The role and relations of physical geography', *Progr. Geogr.* **2**, 87–109.
——, Dunn, A. J. and Beckinsale, R. P. (1964), *The history of the study of landforms* (Methuen, London), Volume **1**, 678 pp.
——, Malm, D. E. J. and Pogorzelski, H. A. (1957), 'A new standard for estimating drainage basin shape', *Am. J. Sci.* **255**, 138–41.
—— and Morgan, M. A. (1962), 'Comparison of some morphometric features Unaka Mountains, Tennessee and North Carolina and Dartmoor, England', *Bull. geol. Soc. Am.* **73**, 17–34.
Clayton, K. M. (ed.) (1964), *A bibliography of British Geomorphology* (Philip, London), 211 pp.
Cotton, C. A. (1957), 'Alternating Pleistocene morphogenetic systems', *Geol. Mag.* **95**, 125–36.
Crampton, C. B. (1969), 'The chronology of certain terraced river deposits in the south east Wales area', *Z. Geomorph.* **13**, 245–59.
Cummins, W. A. and Potter, H. R. (1972), 'Rates of erosion in the catchment area of Cropston Reservoir, Charnwood Forest, Leicestershire', *Mercian Geol.* **4 (2)**, 149–57.
Davies, B. E. and Lewin, J. (1974), 'Chronosequences in alluvial soils with special reference to historic lead pollution in Cardiganshire, Wales', *Envir. Pollution* **6**, 49–57.
Dobbie, C. H. and Wolf, P. O. (1953), 'The Lynmouth flood of August 1952', *Proc. Inst. civ. Engrs. Hydraulics Paper no. 1*, **2 (3)**, 522–8.
Douglas, I. (1964), 'Intensity and periodicity in denudation processes with special reference to the removal of material in solution by rivers', *Z. Geomorph.* **8**, 453–73.
Dury, G. H. (1954), 'Contribution to a general theory of meandering valleys', *Am. J. Sci.* **252**, 193–224.
—— (1958), 'Tests of a general theory of misfit streams', *Trans. Inst. Br. Geogr.* **25**, 105–18.
—— (1959), 'Analysis of regional flood frequency on the Nene and the Great Ouse', *Geogr. J.* **125**, 223–9.
—— (1964), 'Principles of underfit streams', *U.S. geol. Surv. Prof. Pap.* **452–A**, 67 pp.
—— (1973), 'Magnitude frequency analysis and channel morphology', in M. E. Morisawa (ed.) *Fluvial geomorphology* (Publs. in Geomorphology, State University, New York, Binghampton), 91–121.
——, Sinker, C. A. and Pannett, D. J. (1972), 'Climatic change and arrested meander development on the river Severn', *Area,* **4**, 81–5.
Ede, D. P. (1975), 'Limestone drainage systems', *J. Hydrol.* **27**, 297–318.

Edwards, A. M. C. (1974), 'Silicon depletions in some Norfolk rivers', *Freshwat. Biol.* **4**, 267–74.

—— and Thornes, J. B. (1973), 'Annual cycle in river water quality: a time series approach', *Wat. Resour. Res.* **9**, 1286–95.

—— and J. C. Rodda, (1972), 'A preliminary study of the water balance of a small clay catchment. Results of research on representative and experimental basins', *Int. Ass. scient. Hydrol., Proc. Wellington Symposium 1970* (Paris, 1972), 2, 187–99.

Evans, R. (1971), 'The need for soil conservation', *Area,* **3**, 20–3.

Ferguson, R. I. (1973), 'Channel pattern and sediment type', *Area,* **5**, 38–41.

—— (1975a), 'Meander irregularity and wavelength estimation', *J. Hydrol.* **26**, 315–33.

—— (1975b), 'Network elongation and the bifurcation ratio', *Area,* **7**, 121–4.

Fleming, G. (1969), 'The Clyde basin: hydrology and sediment' (Unpubl. Ph.D. thesis, Dept. of Civil Engineering, Univ. of Strathclyde).

Foyster, A. M. (1973), 'Application of the grid square technique to mapping of evapotranspiration', *J. Hydrol.* **19**, 205–6.

Gardiner, V. (1971), 'A drainage density map of Dartmoor', *Rep. Trans. Devon. Ass. Advmt Sci.* **103**, 167–80.

—— (1974a), 'Drainage basin morphometry', *Br. geomorph. Res. Grp Tech. Bull.* **14**, 48 pp.

—— (1974b), 'Land form and land classification in North-west Devon', *Rep. Trans. Devon. Ass. Advmt Sci.* **106**, 141–53.

—— (1974c), 'A photo-mechanical technique for the production of drainage density maps', *Cartogr. J.* 42–4.

Green, G. W. (1955), 'North Exmoor floods, August 1952', *Bull. geol. Surv. Gt Br.* **7**, 68–84.

Gregory, K. J. (1966), 'Dry valleys and the composition of the drainage net', *J. Hydrol.* **4**, 327–40.

—— (1971a), 'Drainage density changes in South West England', in K. J. Gregory and W. L. D. Ravenhill (eds.) *Exeter essays in geography* (Univ. of Exeter), 33–53.

—— (1971b), 'River networks in Devon', *Trans. Proc. Torquay nat. Hist. Soc.* **16**, 4–11.

—— (1974), 'Streamflow and building activity', in 'Fluvial processes in instrumented watersheds (ed. K. J. Gregory and D. E. Walling), *Inst. Br. Geogr. Spec. Publ.* **6**, 107–22.

—— (1976a), 'Drainage networks and climate', in E. Derbyshire (ed.), *Climate and landforms* (Wiley, London), 291–315.

—— (1976b), 'Drainage basin adjustments and man', *Geogr. polonica* **34**, 155–73.

—— (1976c), 'Changing drainage basins', *Geogr. J.* **142**, 237–47.

—— and Brown, E. H. (1975), 'Landscape in the eye of the beholder', *Geogr. Mag.* (December 1975), **48**, 145–50.

—— and Cullingford, R. A. (1974), 'Lateral variations in pebble shape in northwest Yorkshire', *Sedim. Geol.* **12**, 237–48.

—— and Gardiner, V. (1975), 'Drainage density and climate', *Z. Geomorph.* **19**, 287–98.

—— and Park, C. C. (1974), 'Adjustment of river channel capacity downstream from a reservoir', *Wat. Resour. Res.* **10**, 870–3.

—— and Park, C. C. (1976a), 'Stream channel morphology in northwest Yorkshire', *Rev. Géomorph. dyn.* **25**, 63–72.

—— and Park, C. C. (1976b), 'The development of a Devon gully and man', *Geography,* **61**, 77–82.

—— and Walling, D. E. (1968a), 'Instrumented catchments in south east Devon', *Rep. Trans. Devon. Ass. Advmt Sci.* **100**, 247–62.
—— and Walling, D. E. (1968b), 'The variation of drainage density within a catchment', *Bull. int. Ass. scient. Hydrol.* **13**, 61–8.
—— and Walling, D. E. (1971), 'Field measurements in the drainage basin', *Geography,* **56**, 277–92.
—— and Walling, D. E. (1973), *Drainage basin: form and process* (Arnold, London), 456 pp.
Hall, D. G. (1967), 'The pattern of sediment movement in the River Tyne', *Publs int. Ass. scient. Hydrol.* **75**, 117–42.
Hanwell, J. and Newson, M. D. (1970), 'The great storms and floods of July 1968 on Mendip', *Wessex Cave Club Occas. Publ.* **1**, 72 pp.
Harvey, A. M. (1969), 'Channel capacity and the adjustment of streams to hydrologic regime', *J. Hydrol.* **8**, 82–98.
—— (1974), 'Gully erosion and sediment yield in the Howgill Fells, Westmorland', in 'Fluvial processes in instrumented watersheds' (ed. K. J. Gregory and D. E. Walling), *Inst. Br. Geogr. Spec. Publ.* **6**, 45–58.
Hey, R. D. (1974), 'Prediction and effects of flooding in alluvial systems', in 'Prediction of geological hazards' (ed. B. M. Funnell), *Geol. Soc. Lond. Misc. Pap.* **3**, 42–56.
—— and Thorne, C. R. (1975), 'Secondary flows in river channels', *Area,* **7**, 191–5.
Hill, A. R. (1973), 'Erosion of river banks composed of till near Belfast, Northern Ireland', *Z. Geomorph.* **17**, 428–42.
Hollis, G. E. (1974), 'The effect of urbanisation on floods in the Canon's Brook, Harlow, Essex', in 'Fluvial processes in instrumented watersheds' (ed. K. J. Gregory and D. E. Walling), *Inst. Br. Geogr. Spec. Publ.* **6**, 123–40.
Howe, G. M., Slaymaker, H. O. and Harding, D. M. (1967), 'Some aspects of the flood hydrology of the upper catchments of the Severn and Wye', *Trans. Inst. Br. Geogr.* **41**, 33–58.
Horton, R. E. (1945), 'Erosional development of streams and their drainage basins: hydrophysical approach to quantitative morphology', *Bull. geol. Soc. Am.* **56**, 275–370.
Imeson, A. C. (1974), 'The origin of sediment in a moorland catchment with particular reference to the role of vegetation', in 'Fluvial processes in instrumented watersheds' (ed. K. J. Gregory and D. E. Walling), *Inst. Br. Geogr. Spec. Publ.* **6**, 59–72.
—— and Ward, R. C. (1972), 'The output of a lowland catchment', *J. Hydrol.* **17**, 145–59.
Institute of Hydrology (1973), *A view from the watershed* (Wallingford Report No. 20, June 1973), **54** pp.
Institute of Hydrology (1975) *Flood studies report* (N.E.R.C., London), 5 vols.
Jarvis, R. S. (1972), 'New measure of the topologic structure of dendritic drainage networks', *Wat. Resour. Res.* **8**, 1265–71.
Johnson, R. H. and Painter, J. (1967), 'The development of a cutoff on the River Irk at Chadderton, Lancashire', *Geography,* **52**, 41–9.
Jones, A. (1971), 'Soil piping and stream channel initiation', *Wat. Resour. Res.* **7**, 602–10.
Kennedy, B. A. (1972), '"Bankfull" discharge and meander forms', *Area,* **4**, 209–12.
Kidson, C. (1953), 'The Exmoor storm and the Lynmouth floods', *Geography,* **38**, 1–9.
Kirby, C. and Rodda, J. C. (1974), 'Managing the hydrological cycle', in A.

Warren and F. B. Goldsmith (eds.) *Conservation in practice* (Wiley, London), 73–86.

Kirkby, M. J. (1972), 'Alluvial and non-alluvial meanders', *Area,* **4**, 284–8.

—— and Weyman, D. R. (1972), 'Measurements of contributing area in very small drainage basins', *Dept. of Geography, Seminar Pap. Ser. B, Univ. of Bristol,* **3**, 15 pp.

Knighton, A. D. (1973), 'Riverbank erosion in relation to streamflow conditions, River Bollin Dean, Cheshire', *E. Midld Geogr.* **5**, 416–26.

—— (1974), 'Variation in width-discharge relation and some implications for hydraulic geometry', *Bull. geol. Soc. Am.* **85**, 1069–76.

—— (1975), 'Variations in at-a-station hydraulic geometry', *Am. J. Sci.* **275(2)**, 186–218.

Lamplugh, G. W. (1914), 'Taming of streams', *Geogr. J.* **43**, 651–6.

Learmonth, A. T. A. (1950), 'The floods of 12th August 1948 in south-east Scotland', *Scott. geogr. Mag.* **66**, 147–53.

Lewin, J. (1972), 'Late-stage meander growth', *Nature, Lond.* **240**, 116.

—— (1975), 'Geomorphology and environmental impact statements', *Area,* **7**, 127–9.

——, Cryer, R. and Harrison, D. I. (1974), 'Sources for sediments and solutes in mid-Wales', in 'Fluvial processes in instrumented watersheds' (ed. K. J. Gregory and D. E. Walling), *Inst. Br. Geogr. Spec. Publ.* **6**, 73–86.

—— and Manton, M. M. M. (1975), 'Welsh floodplain studies: the nature of floodplain geometry', *J. Hydrol.* **25**, 37–50.

Lewis, W. V. (1944), 'Stream trough experiments and terrace formation', *Geol. Mag.* **81**, 241–53.

—— (1945), 'Nick points and the curve of water erosion', *Geol. Mag.* **82**, 256–66.

Linton, D. L. (1957), 'The everlasting hills', *Advmt Sci.* **14**, 58–67.

—— (1959), 'Mapping river water in England and Wales 1955–6', *Nature, Lond.* **183**, 714–16.

Lovell, J. P. B., Ledger, D. C., Davies, I. M. and Tipper, J. C. (1973), 'Rate of sedimentation in the north Esk reservoir, Midlothian', *Scott. J. Geol.* **9**, 57–61.

Mosley, M. P. (1972), 'Gully systems in blanket peat, Bleaklow, north Derbyshire', *E. Midld Geogr.* **37**, 235–44.

—— (1975a), 'Meander cutoffs on the River Bollin, Cheshire in July 1973', *Rev. Géomorph. dyn.* **24**, 21–31.

—— (1975b), 'Channel changes on the River Bollin, Cheshire, 1872–1973', *E. Midld Geogr.* **6**, 185–99.

Moss, C. E. (1913), *Vegetation of the Peak District* (Cambridge), 236 pp.

Nash, J. E. (1960), 'A unit hydrograph study, with particular reference to British catchments', *Proc. Instn civ. Engrs* **17**, 249–82.

Natural Environment Research Council (1975), *Flood study report,* 180 pp.

Natural Environment Research Council (1976), *Hydrological research in the United Kingdom, 1970–75* (London), 180 pp.

Newson, M. D. (1971), 'A model of subterranean limestone erosion in the British Isles based on hydrology', *Trans. Inst. Br. Geogr.* **54**, 55–70.

—— (1975), 'Mapwork for flood studies. Part I: selection and derivation of indices', *Inst. Hydrol. Rep.* **25**, 52 pp.

Nixon, M. (1959), 'A study of the bankfull discharges of rivers in England and Wales', *Proc. Instn civ. Engrs* **12**, 157–74.

Oxley, N. C. (1974), 'Suspended sediment delivery rates and the solute concentration of stream discharge in two Welsh catchments', in 'Fluvial processes in

instrumented watersheds' (ed. K. J. Gregory and D. E. Walling), *Inst. Br. Geogr. Spec. Publ.* **6**, 141–54.
Park, C. C. (1975), 'Stream channel morphology in mid-Devon', *Rep. Trans. Devon. Ass. Advmt Sci.* **107**, 25–41.
—— (1976), 'The relation of slope and stream channel form in the River Dart, Devon', *J. Hydrol.* **29**, 139–47.
Pegg, R. K. and Ward, R. C. (1972), 'Evapotranspiration from a small clay catchment', *J. Hydrol.* **15**, 149–65.
Pitty, A. F. (1966), 'The estimation of discharge from a karst rising by natural salt dilution', *J. Hydrol.* **4**, 63–9.
—— (1968), 'The scale and significance of solutional loss from the limestone tract of the southern Pennines', *Proc. Geol. Ass.* **79**, 153–77.
Price, R. K. (1974), 'Flood routing methods for British Rivers. Discussion', *Proc. Instn civ. Engrs* **57**, Part 2, 391–7.
Radley, J. (1962), 'Peat erosion on the high moors of Derbyshire and west Yorkshire', *E. Midld Geogr.* **15**, 40–50.
Richards, K. S. (1972), 'Meanders and valley slope', *Area,* **4**, 288–90.
—— (1973), 'Hydraulic geometry and channel roughness—a non-linear system', *Am. J. Sci.* **273**, 877–96.
Rodda, J. C. (1967), 'The significance of characteristics of basin rainfall and morphometry in a study of floods in the United Kingdom', *Publs int. Ass. scient. Hydrol.* **85**, 834–45.
—— (1974), 'Water resources in the United Kingdom: a hydrological appraisal', in J. A. Taylor (ed.), *Climate, resources and economic activity* (David & Charles, Newton Abbot), 135–57.
Scheidegger, A. E. (1973), 'Hydrogeomorphology', *J. Hydrol.* **20**, 193–215.
Schumm, S. A. (1969), 'River metamorphosis', *J. Hydraul. Div. Am. Soc. civ. Engrs* **6352**, 255–73.
—— (1973), 'Geomorphic thresholds and complex response of drainage systems' in M. E. Morisawa (ed.) *Fluvial geomorphology* (Publs. in Geomorphology, State University, New York, Binghampton), 299–310.
—— and Lichty, R. W. (1965), 'Time, space and causality in geomorphology', *Am. J. Sci.* **262**, 110–19.
Slaymaker, H. O. (1972), 'Patterns of present sub-aerial erosion and landforms in mid-Wales', *Trans. Inst. Br. Geogr.* **55**, 47–68.
—— and Chorley, R. J. (1964), 'The Vigil network system', *J. Hydrol.* **2**, 19–24.
Smith, B. (1910), 'Some recent changes in the course of the Trent', *Geogr. J.* **35**, 568–79.
Smith, D. I. and Newson, M. D. (1974), 'The dynamics of solutional and mechanical erosion in limestone catchments on the Mendip Hills, Somerset', in 'Fluvial processes in instrumented watersheds' (ed. K. J. Gregory and D. E. Walling), *Inst. Br. Geogr. Spec. Publ.* **6**, 155–68.
Strahan, A. (1908, 1909, 1911), 'Reports of progress in the investigation of rivers' *Geogr. J.* **31**, 310–13; **34**, 622–50; **38**, 297–305.
Strahler, A. N. (1964), 'Quantitative geomorphology of drainage basins and channel networks', in V. T. Chow (ed.) *Handbook of applied hydrology* (McGraw Hill, New York), 4–39 to 4–76.
Straw, A. (1968), 'A Pleistocene diversion of drainage in north Derbyshire', *E. Midld Geogr.* **4**, 275–80.
Surkan, A. J. (1969), 'Synthetic hydrographs: effects of network geometry', *Wat. Resour. Res.* **5**, 112–28.
Tallis, J. H. (1973), 'Studies on southern Pennine peats. V. Direct observations

on peat erosion and peat hydrology at Featherbed Moss, Derbyshire', *J. Ecol.* **61**, 1-22.

Troake, R. P. and Walling, D. E. (1973), 'The hydrology of Slapton Wood stream. A preliminary report', *Fld Stud.* **3**, 719–40.

Tuckfield, C. G. (1964), 'Gully erosion in the New Forest, Hampshire', *Am. J. Sci.* **262**, 795–807.

Walling, D. E. (1971a), 'Streamflow from instrumented catchments in south-east Devon', in K. J. Gregory and W. L. D. Ravenhill (eds.) *Exeter essays in geography* (Univ. of Exeter), 55–81.

—— (1971b), 'Sediment dynamics of small instrumented catchments in south-east Devon', *Rep. Trans. Devon. Ass. Advmt Sci.* **103**, 147–65.

—— (1974a), 'Suspended sediment and solute yields from a small catchment prior to urbanisation', in 'Fluvial processes in instrumented watersheds' (ed. K. J. Gregory and D. E. Walling), *Inst. Br. Geogr. Spec. Publ.* **6**, 169–192.

—— (1974b), 'Suspended sediment production and building activity in a small British basin', *Publs int. Ass. scient. Hydrol.* **113**, 137–44.

—— (1975), 'Solute variations in small catchment streams: some comments', *Trans. Inst. Br. Geogr.* **64**, 141–7.

—— and Foster, I. D. L. (1975), 'Variations in the natural chemical concentration of river water during flood flows and the lag effect: some further comments', *J. Hydrol.* **26**, 237–44.

—— and Gregory, K. J. (1970), 'The measurement of the effects of building construction on drainage basin dynamics', *J. Hydrol.* **11**, 129–44.

—— and Teed, A. (1971), 'A simple pumping sampler for research into suspended sediment transport in small catchments', *J. Hydrol.* **13**, 325–37.

—— and Webb, B. W. (1975), 'Spatial variation of river water quality: a survey of the River Exe', *Trans. Inst. Br. Geogr.* **65**, 155–71.

Ward, R. C. (1967a), *Principles of hydrology* (McGraw Hill, London), 403 pp.

—— (1967b), 'Design of catchment experiments for hydrological studies', *Geogr. J.* **133**, 495–502.

—— (1968), 'Some hydrological characteristics of British rivers', *J. Hydrol.* **6**, 358–72.

—— (1972), 'Estimating streamflow using Thornthwaite's Climatic Water Balance', *Weather* **27**, 73–84.

Waters, R. S. and Johnson, R. H. (1958), 'The terraces of the Derbyshire Derwent', *E. Midld Geogr.* **2**, 3–15.

Werrity, A. (1972), 'Accuracy of stream link lengths derived from maps', *Wat. Resour. Res.* **8**, 1255–64.

Weyman, D. R. (1970), 'Throughflow on hillslopes and its relation to the stream hydrograph', *Bull. int. Ass. scient. Hydrol.* **15**, 25–33.

—— (1974), 'Runoff process, contributing area and streamflow in a small upland catchment', in 'Fluvial processes in instrumented watersheds' (ed. K. J. Gregory and D. E. Walling), *Inst. Br. Geogr. Spec. Publ.* **6**, 33–43.

Whittel, A. (1973), 'Pebble bedload studies on the River Wharfe, Yorkshire', *Rev. Géomorph. dyn.* **22**, 71–81.

Wilcock, D. N. (1971), 'Investigation into the relations between bedload transport and channel shape', *Bull. geol. Soc. Am.* **82**, 2159–76.

Williams, P. (1963), 'An initial estimate of the speed of limestone solution in County Clare', *Ir. Geogr.* **4**, 432–41.

Young, A. (1958), 'A record of the rate of erosion on Millstone Grit', *Proc. Yorks. geol. Soc.* **31**, 149–56.

5

SLOPES: 1970-1975

ANTHONY YOUNG

(*University of East Anglia, Norwich*)

THE POSITION OF SLOPE STUDIES IN GEOMORPHOLOGY

The past 5 years has not been a growth period for studies of slopes. Such growth took place in the 1960s, within the context of expansion of university research in general and geography and geomorphology in particular. Rather, 1970-75 has been a period in which slope studies held a place as one of about ten to twelve major branches of geomorphology. This had not been the case before 1960, as inspection of textbooks of that period will show; the discussions they contained, usually in a chapter on 'Weathering and mass movements' or the like, rested heavily on a few sources, notably K. Bryan (1940) and A. Wood (1942). At the end of this period of growth, and forming a convenient starting point for the present review, four books appeared devoted wholly to slopes: a monograph (M. J. Selby, 1970), two textbooks (M. A. Carson and M. J. Kirkby, 1972; Young, 1972) and a collection of readings (S. A. Schumm and M. P. Mosley, 1973).

Evidence of the current position of slope studies as an established major branch of geomorphology includes the following:

(a) In a recent review of geomorphology courses in British universities, among the ten most frequently-mentioned topics, 'slopes' was third and 'weathering and mass movement' fourth, yielding place only to 'fluvial' and to 'glacial geomorphology' (*Geophemera* 8, March 1975).
(b) A 9000-word entry by Carson, under the American title 'Hillslopes', in the 1974 edition of *Encyclopaedia Britannica*, (8, 873–80).
(c) Of the first sixteen Technical Bulletins of the British Geomorphological Research Group, five are on slopes or slope processes; while among the Special Publications series of the Institute of British Geographers, that on slopes (D. Brunsden, 1971) is one of the two best-sellers.
(d) 'Slopes' is now a section in the journal *Geo-Abstracts*, having been separated from 'Weathering and slopes' in 1972.
(e) On the international front, the Slope Study Commission of the International Geographical Union achieved a lengthy existence (1952–68) and published six Reports (listed in Young, 1972, p. 259). By a process of transmutation it became in 1972 the Commission on Present-Day Geomorphological Processes, and in 1976 the Commission on Field Experiments in Geomorphology, on each occasion with some gain in membership and no perceptible loss.
(f) Since 1970, eight collections of papers wholly or in substantial part on slopes have appeared: Brunsden (1971), A. Jahn and L. Starkel (1972), O. Slaymaker and H. J. McPherson (1972), P. Macar (1973), A. P. Schick *et al.* (1974), H.

Poser (1974), E. Yatsu *et al.* (1975) and F. Ahnert (1976a). These contain in all some 100 papers on slopes.

VOLUME OF PUBLICATION AND RESEARCH

Some indicators of the volume of publication and research on slopes in recent years are shown in Fig. 5.1. The first indicator of publications, the number of articles in the 'Slopes' section of *Geo-Abstracts A*, appears to show a decline from 1972 to 1975, falling from over 6 to under 4 per cent of all entries in section A. This decline may in part result from changes in allocation practice, with some carry-over from before 1972 when 'Weathering and slopes' contained a much wider range of material; one suspects in particular that in later years articles that would earlier have come under 'Slopes' are being allotted to 'Soil mechanics', 'Applied geomorphology' and 'Soil erosion'. The most recent years suggest a stabilisation at about eighty publications per year.

The second indicator, Selected Publications, refers to the author's own card

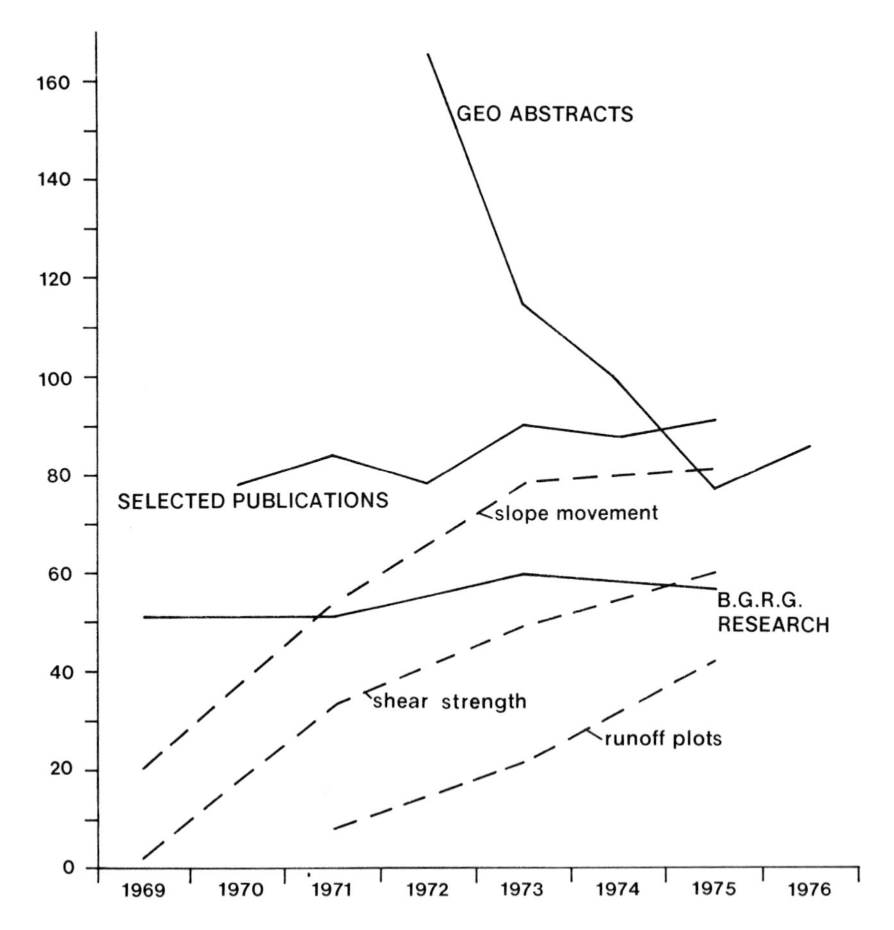

FIG. 5.1. Indicators of the volume of publication and research on slopes.

index. Since 1970 this has become selective, ignoring papers which appear to be of little moment. Given this reservation, there appears to be neither growth nor decline but a steady 80–90 publications per year (these are far from being the same as those contained in the previous indicator).

With printing costs at their present high level, it seems unlikely that the rate of publication in the second half of the 1970s will exceed that of the first. I had previously surmised a doubling rule whereby the number of papers on slopes doubles in every decade from a base of 10 in the 1910s, which would have led to 640 during the 1970s (Young, 1972, p. 15). From the two indicators above, it seems likely that the number of significant papers will prove to lie between 800 and 1000.

As regards the volume of research, information for the United Kingdom is given in the biennial editions of the British Geomorphological Research Group register *Current Research in Geomorphology*. The number of research workers supposedly active in the subject, labelled 'B.G.R.G. Research' in Fig. 5.1, has remained almost constant at fifty to sixty despite the fact that the 1969–71 figures refer to 'Weathering and slopes'. The numbers listed as employing relevant techniques, however, present a striking picture. Three techniques, all concerned with processes, show a continuous growth, until in 1975 there were forty-two people using runoff plots, sixty measuring shear strength, and eighty-one, or no less than one in five of all British geomorphologists, making field measurements of slope processes. The number using slope profiling is unfortunately subsumed in the 250 or so who employ Levelling or Surveying (indeed, the mind boggles at the 187 British geomorphologists who apparently do not use this technique).

Turning briefly to other countries a noteworthy contribution to slope studies has been made by Poland; this is in part through the work of the Centre for Physical Geography of the Polish Academy of Sciences at Krakow, whence has come a succession of publications particularly on soil erosion and landslides. Proportionally to their size, Belgium and New Zealand have each contributed more than average to the subject. The amount and quality of work from the United States has been considerable in absolute terms, although it appears that the relative importance of slope studies within geomorphology as a whole is somewhat less than in Britain. This is certainly the case in France and Germany. Hampered in world recognition by the language barrier, it is clear that research is active in Japan, not inappropriately in a country with landforms so dominated by slopes.

Taking publications and current research together, the overall position of slope studies in the period 1970–75 is clear: a constant total volume, coupled with a rapid growth of interest in processes and in techniques for their measurement. We have come far since a plaintive cry of the pre-technical period, 'In existing contributions to the study of slope development there is a preponderance of theoretical discussion, based upon general observation but not upon the detailed study of specific slopes' (Young, 1963, p. 1).

BRANCHES OF STUDY

Turning to the branches of slope studies on which published work has appeared, Table I is based on the selective card index to which I previously referred. The

TABLE 1

Classification by subject of 550 publications on slopes, 1970–75

		Number of papers
GENERAL TOPICS, TEXTBOOKS, REVIEWS		16
METHODS AND TECHNIQUES		81
of which: Instrumentation of processes	18	
Profile analysis	13	
Slope mapping	21	
PROCESSES (EXCLUDING LANDSLIDES)		180
of which: Rates of processes	132	
LANDSLIDES		66
THEORY		29
of which: Models	22	
CLIFFS AND SCREES		42
of which: Screes	28	
ANGLE AND ANGLE FREQUENCY		19
SLOPE FORM, INCLUDING PROFILE FORM		108
ENVIRONMENTS		95
of which: Specific climates	34	
Specific rock types	20	
Polar and montane zone	15	
Rain-forest zone	11	
Pediments	16	
Microrelief	16	
Regolith	15	
SLOPE EVOLUTION		4
APPLIED STUDIES		19
of which: Engineering applications	12	

Table is constructed non-exclusively, that is to say, an article about surface wash on pediments would be listed under both 'Processes' and 'Pediments'. There is, in particular, considerable overlap between works listed under 'Slope form' and those related to specific 'Environments'. The degree of selectivity is greater with respect to papers on landslides and related aspects of soil mechanics than for other topics. Under regolith, studies of soil catenas which are orientated pedologically rather than geomorphologically are excluded, although the distinction is a blurred one.

The major feature of Table I is that the number of studies of processes, inclusive of landslides, is now equal to or greater than the number of studies of slope form. Coupled with this feature is the substantial amount of attention given to instrumentation and methodology, in part, although by no means exclusively, to techniques for studying processes. Both these features confirm trends noted above.

Comparing Table I with the state of the subject prior to 1970, I have selected those aspects that have approximately held their position, those in which interest has apparently declined, and those to which more attention than formerly is being devoted.

Maintained interest

Slope mapping. This is hard to separate from geomorphological mapping in general, itself a growth area. Morphological mapping *sensu stricto* has always been based on slopes, and there has been increasing recognition within genetically-based systems that slopes are the areally preponderant landform. Slope-angle mapping, formerly the most time-consuming of all tasks in cartographic analysis, has now been automated (D. B. Waldrip and M. C. Robert, 1971; V. Gardiner and D. W. Rhind, 1974).

Screes. This special type of slope has always been popular for study, but any thoughts that it might have been worked out are dispelled by twenty-eight new studies, notable in that all but three are based on field measurements. It has been established that the characteristic scree angle of *c*. 33°–35° is not the angle of repose, but a gentler angle produced either by impact velocity, mass movements or both (J. Gardner, 1971; I. Statham, 1973; but *cf.* H. Jeffreys, 1932!) Somewhat unexpected is the reappearance of Richter slopes, rectilinear segments in bedrock standing at or slightly below the angle of repose of screes (C. A. Cotton and A. T. Wilson, 1971; Selby, 1971).

Slope profiles. Together with the associated topic of angle frequency, this technique has neither attracted the growth that might have been expected nor has it declined. The Abney level has remained the most widely-used instrument for profile survey, despite the development of the slope pantometer and the profile recorder (Young *et al.*, 1974). It has been shown that the measured lengths used in profile survey substantially affect the results obtained (A. J. W. Gerrard and D. A. Robinson, 1971). There is still interest in the methodology of profile analysis including one fundamentally new approach, that of treating a slope as a series, contrasting with the earlier approach of seeking slope units or other sections possessing some measure of uniformity (J. B. Thornes, 1972, 1973); it is not yet clear whether this new approach will produce results of comparable geomorphological significance to those yielded by slope-unit analysis. A result of much interest but as yet unknown significance is that slope-angle frequency shows an approximately log-normal distribution, irrespective of whether a landscape is formed predominantly of steep or very gentle slopes (J. G. Speight, 1971).

Environments. There have continued to be studies of slopes within particular climates and rock types, although these are not yet numerous enough to build up a general typology of slope-form characteristics. Two environments that have long been popular have remained so: pediments and cold climates (polar and montane). Results from the rain-forest zone have been of special interest in that there is reasonable confidence that present-day processes are responsible for observed form (*cf.* J. I. S. Zonneveld, 1975).

Declined

Theoretical studies. Gone is the former type of reasoning in which inferences about slope form and evolution are based upon the presumed manner of action of processes, but the same general approach lives on in the form of models (see below). Some of the earlier speculation has proved to be correct: thus many pediments probably are slopes of transport, whilst the ability of soil creep to cause rounding of convexities has never been contradicted by field evidence. But papers based on verbal reasoning alone have probably gone for good, displaced by models and by the availability of field measurements of process and form.

Slope evolution. The apparent lack of interest in the manner of slope evolution is less satisfactory, for this remains the central problem in the geomorphology of slopes. Mentions of Davis, Penck and King are infrequent. It is to be hoped, however, that this is a case of *reculer pour mieux sauter*, a realisation that no further progress was likely from a frontal attack on the problem–hence recourse to an outflanking movement through the study of processes. D. Brunsden and R. H. Kesel (1973) have attempted to measure slope evolution under different conditions of debris removal at the slope base.

Growth subjects

Process studies. Even if landslides are excluded, process studies now comprise a third of all slopes publications. Four of the six growth subjects listed here are wholly or partly concerned with processes. The existence of throughflow has been established, together with some of its relations with the two types of surface runoff, Horton and saturation (T. Dunne and R. D. Black, 1970). Casting further doubt on the reality of Horton's 'belt of no erosion' has been the measurement of surface runoff down the crest line of nose-shaped spurs (A. Yair, 1973). Classical problems have been attacked by modern methods, as in the demonstration of a continuous transition in processes at the piedmont zone separating mountain front from pediment (Kirkby and Kirkby, 1974); and surely the spirit of G. K. Gilbert must have listened with interest to the paper entitled 'Rainsplash and the convexity of badland divides' (M. P. Mosley, 1973). Among minor processes the faunal world is well represented, for besides the well-known activities of worms and termites, we now have quantitative records of the devastating geomorphological effects of moles, porcupines, isopods and the banner-tailed kangaroo rat (T. L. Best, 1972; E. Jonca, 1972; Yair, 1974).

Instrumentation. Techniques have been developed for measuring virtually all important processes: rock weathering, soil creep, solifluction, surface wash, throughflow, solution and the continuous recording of landslide movement. Soil creep has attracted the greatest attention and ingenuity, but the equally difficult task of recording rates of solution is still in its early stages.

Rates of operation of slope processes. In 1970 it was difficult to find a satisfactory collection of process measurements (Young, 1972, Tables 1–3). A review written 3 years later contained fifty-seven additional records (Young, 1974), since when a further seventy-five have appeared. There is by now the making of a climate-based typology of absolute and relative rates of processes. Records of solution obtained from the soil solution on slopes, as distinct from river water, remain sparse. Some 'world records' may be claimed. The fastest retreat recorded for any slope in consolidated rocks is 2·6 m in 4 years for a sandstone slope in the Crimean Mountains (N. S. Blagovolin and D. G. Tsvetkov, 1972) while for unconsolidated materials it is 23 m/yr for an actively undercut Mississippi river bluff (Brunsden and Kesel, 1973). The longest periods of observation of soil creep, or indeed of any process, are a 10-year record by Jahn and Cielinski (1974) and one of 12 years by Young (unpublished).

Landslides. The study of landslides, together with the associated technique of soil mechanics, has grown into a major branch of slope geomorphology, and as such is reviewed separately in this volume (D. B. Prior, 1977). It is becoming increasingly clear that landsliding is closely linked to a limited range of structural conditions, and that for any given lithology it is largely restricted to slopes above a particular angle of long-term stability. A possible exception is the rain-forest zone, where slides occur on the deep regolith developed over a variety of rock types. Removal of natural vegetation can greatly increase susceptibility to landsliding (D. H. Gray, 1970; C. L. So, 1971; A. Rapp *et al.*, 1972; F. J. Swanson and C. T. Dyrness, 1975).

Slope models. This has become the main, indeed almost the only surviving, aspect of theoretical studies of slopes. There are two fundamentally different approaches. The first, and more widely developed, is the iterative approach, based on calculating successively the effects of the assumed manner of action of processes upon an initial slope. The COSLOP model of Ahnert (1971, 1973, 1976b), first devised for profile development and later extended to three dimensions, has overtaken all else in comprehensiveness, although not before some useful climatically-based models have been obtained by H. Gossmann (1970) and a spectacular computer-produced block-diagram ciné film obtained by Armstrong (unpublished). The second approach is to compute ideal end-forms from a continuity equation, based again on the assumed manner of action of processes (Kirkby, 1971; Carson and Kirkby, 1972, pp. 107ff.). The two approaches are complementary and may prove to be convergent.

Applied studies. Work based on slopes and slope processes has played a substantial part in the general growth of applied geomorphology. The main applications are in engineering, including landslide prevention or prediction and road siting. The field of accelerated soil erosion has seen a group of studies in Tanzania destined to rank among the classical works on slopes (Rapp *et al.*, 1972). A further aspect in which slopes play a role is in studies, for planning purposes, into the aesthetics of landforms.

THE FUTURE

In conclusion, here is a list of types of research which appear likely to repay study over the next 10 years:

(a) The continued study of the mechanisims of processes, and the recording of their rates. Work is particularly needed on solution, and the necessary techniques, throughflow troughs coupled with analysis of dissolved matter in the soil solution, are now available. Process studies intended for purposes of pure geomorphology, i.e. to learn which processes have produced existing landforms, should be sited under natural vegetation only, and a clear distinction made from results obtained on cleared or farmed land, the latter being more important for practical purposes.

(b) Further application of the techniques of soil mechanics to the relatively thin regolith that covers most slopes in consolidated rocks. The hypothesis that the steepest segments not subject to superficial soil movements can be related to the angle of long-term stability in medium to coarse-textured regoliths is of the highest interest, but evidence in support of it is still sparse (Carson and D. J. Petley, 1970; Carson, 1971).

(c) Further results from the survey and analysis of slope profiles. There has been a danger that this field will be neglected amid the current vogue for process studies. In slopes, as in many other branches of geomorphology and natural science in general, the study of form is as fundamental as that of process. Techniques for profile survey and analysis are well developed and what are now needed are data from a wide variety of structural and climatic conditions. With a few exceptions (e.g. the English Chalk) we do not yet have a climate- and structure-based typology of slope-form characteristics. A specific need is for a good common-sense method of separating micro-relief from larger-scale features of slope form.

(d) The soil catena. It has for some time been apparent that the distinction between geomorphological and pedogenic processes is slight, and the growing recognition of the geomorphological importance of solution, which is but the long-recognized pedogenic process of leaching, still further decreases it. The catena is one of the major elements in the distributional pattern of soil types, and there are practical applications to agricultural management. A start has already been made in applying models to pedogenesis on slopes.

(e) Finally, there is scope for a type of study of which there is as yet no example: the employment of slope form as evidence contributing to a regional geomorphology, descriptive and genetic. The possibility of using slope form as a basic type of evidence for denudation chronology, complementing that of planation surfaces, has never been realised, although it was envisaged 20 years ago by D. L. Linton (verbally) and briefly employed by R. A. G. Savigear (1952) to suggest phases of sea-level change from the form of marine cliffs. There may be good reason for not attempting so ambitious a project until more has been established about the manner of slope retreat.

Indications of a broad research strategy emerge from the nature of recent studies and the needs of the future. The joint foundations are studies of surface processes and slope form. Given knowledge of the manner of action of processes, and their relative and absolute rates of operation under different climatic and structural conditions, then the way is clear for deduction of slope evolution and hence form by means of models, making use of paleo-climatic evidence derived from other branches of geomorphology. Comparisons between slope form as deduced from models and that surveyed in the field, already attempted (Ahnert, 1970; B. P. Moon, 1975), could then become increasingly common. By these means, knowledge of the systematics of slope evolution would progressively become more secure (bringing with it answers to that ever-present question, 'Was Davis/Penck/King right?'). The way would then be open to apply this knowledge regionally, thus contributing to the ultimate task of geomorphology, an understanding of the landforms that we see before us.

REFERENCES

Ahnert, F. (1970), 'A comparison of theoretical slope models with slopes in the field', *Z. Geomorph., Suppl. Bd* **9**, 88–101.

— — (1971), 'A general and comprehensive theoretical model of slope profile development', *Univ. Maryland Occas. Pap. Geogr.* **1**, 95 pp.

— — (1973), 'COSLOP 2–a comprehensive model program for simulating slope profile development', *Geocom Programs,* **8**, 24 pp.

— — (1976a), 'Quantitative slope models', *Z. Geomorph., Suppl. Bd* **25**, 168 pp.

— — (1976b), 'Brief description of a comprehensive three-dimensional process-response model of landform development', *Z. Geomorph., Suppl. Bd* **25**, 29–49.

Armstrong, A. (1976), 'A three-dimensional simulation of slope forms', in Ahnert, F. (ed.), op. cit. (1976a), 20–8.

Best, T. L. (1972), 'Mound development by a pioneer population of the banner-tailed kangaroo rat, *Dipodomys spectabilis baileyi* Goldman, in eastern New Mexico', *Am. Midld Nat.* **87**, 201–6.

Blagovolin, N. S. and Tsvetkov, D. G. (1972), 'The use of repeated ground photogrammetric survey for studying the dynamics of slopes', in W. P. Adams and F. M. Helleiner (eds.), *International Geography 1972* (Univ. of Toronto Press), 9–10.

Brunsden, D. (ed.) (1971), 'Slopes: form and process', *Inst. Br. Geogr. Spec. Publ.* **3**, 178 pp.

— — and Kesel, R. H. (1973), 'Slope development on a Mississippi river bluff in historic time', *J. Geol.* **91**, 576–97.

Bryan, K. (1940), 'The retreat of slopes', *Ann. Ass. Am. Geogr.* **30**, 254–67.

Carson, M. A. (1971), 'Application of the concept of threshold slopes to the Laramie Mountains, Wyoming', *Inst. Br. Geogr. Spec. Publ.* **3**, 31–48.

— — and Kirkby, M. J. (1972), *Hillslope form and process* (Cambridge Univ. Press), 475 pp.

— — and Petley, D. (1970), 'The existence of threshold hillslopes in the denudation of the landscape', *Trans. Inst. Br. Geogr.* **49**, 71–96.

Cotton, C. A. and Wilson, A. T. (1971), 'Ramp forms that result from weathering and retreat of precipitous slopes', *Z. Geomorph.* **15**, 199–211.

Dunne, T. and Black, R. D. (1970), 'An experimental investigation of runoff production in permeable soils', *Wat. Resour. Res.* **6**, 478–90.

Gardiner, V. and Rhind, D. W. (1974), 'The creation of slope maps by a photo-mechanical technique', *Area,* **6**, 14–21.

Gardner, J. (1971), 'Morphology and sediment characteristic of mountain debris slopes in Lake Louise District (Canadian Rockies)', *Z. Geomorph.* **15**, 390–403.

Gerrard, A. J. W. and Robinson, D. A. (1971), 'Variability in slope measurements. A discussion of the effects of different recording intervals and micro-relief in slope studies', *Trans. Inst. Br. Geogr.* **54**, 45–54.

Gossman, H. (1970), 'Theorien zur Hangentwicklung in verschieden Klimazonen. Mathematische Hangmodelle und ihre Beziehung zu den Abtragungsvorgangen', *Mitt. geogr. Ges. Wurzburg* **31**, 146 pp.

Gray, D. H. (1970), 'Effects of forest clear-cutting on the stability of natural slopes', *Bull. Ass. Engng Geol.* **7**, 45–66.

Jahn, A. and Cielinski (1974), 'The rate of soil movement in the Sudety Mountains', *Abh. Akad. Wiss. Göttingen* (Math. -Phys. Klasse) **29**, 86–101.

Jahn, A. and Starkel, L. (eds.) (1972), 'Geomorphic processes in cold, temperate and tropical regions', *Geogr. Polonica* **23**, 180 pp.

Jeffreys, H. (1932), 'Scree slopes', *Geol. Mag.* **69**, 383–4.

Jonca, E. (1972), 'Water denudation of molehills in mountainous areas', *Acta theriol.* **17**, 407–12.

Kirkby, A. and Kirkby, M. J. (1974), 'Surface wash at the semi-arid break of slope', *Z. Geomorph., Suppl. Bd* **21**, 151–76.

Kirkby, M. J. (1971), 'Hillslope process-response models based on the continuity equation', *Inst. Br. Geogr. Spec. Publ.* **3**, 15–30.

Macar, P. (ed.) (1973), 'Slope evolution and continental morphology', *Z. Geomorph., Suppl. Bd* **18**, 1–77.

Moon, B. P. (1975), 'The application of a mathematical model to slopes on granite terrain' (Unpubl. M.A. thesis, Univ. of Witwatersrand).

Mosley, M. P. (1973), 'Rainsplash and the convexity of badland divides', *Z. Geomorph., Suppl. Bd* **18**, 10–25.

Poser, H. (ed.) (1974), 'Recent geomorphic processes and process combinations under different climatic conditions', *Abh. Akad. Wiss. Göttingen (Math. Phys. Klasse)* **29**, 440 pp.

Prior, D. B. (1977), 'Some recent progress and problems in the study of mass-movement', in this volume, pp. 84–107.

Rapp, A., Berry, L. and Temple, P. H. (eds.) (1972), 'Studies of soil erosion and sedimentation in Tanzania', *Geogr. Annlr* **54A**, 105–379.

Savigear, R. A. G. (1952), 'Some observations on slope development in South Wales', *Trans. Inst. Br. Geogr.* **18**, 31–51.

Schick, A. P., Yaalon, D. H. and Yair, A. (eds.) (1974), 'Geomorphic processes in arid environments', *Z. Geomorph., Suppl. Bd* **21**, 106–215.

Schumm, S. A. and Mosley, M. P. (eds.) (1973), *Slope morphology* (*Benchmark papers in geology*, Stroudsberg, Pa.), 454 pp.

Selby, M. J. (1970), 'Slopes and slope processes', *N. Z. geogr. Soc., Waikato Branch Publ.* **1**, 59 pp.

—— (1971), 'Slopes and their development in an ice-free arid area of Antarctica', *Geogr. Annlr* **53A**, 235–45.

Slaymaker, O. and McPherson, H. J. (eds.) (1972), *Mountain geomorphology* (Vancouver), 274 pp.

So, C. L. (1971), 'Mass movements associated with the rainstorm of June 1966 in Hong Kong', *Trans. Inst. Br. Geogr.* **53**, 55–65.

Speight, J. G. (1971), 'Log-normality of slope distributions', *Z. Geomorph.* **15**, 290–311.

Statham, I. (1973), 'Scree slope development under conditions of surface particle movement', *Trans. Inst. Br. Geogr.* **59**, 41–53.

Swanson, F. J. and Dyrness, C. T. (1975), 'Impact of clear-cutting and road construction on soil erosion by landslides in the western Cascade Range, Oregon', *Geology* **3**, 393–6.

Temple, P. H. and Rapp, A. (1972), 'Landslides in the Mgeta area, western Uluguru Mountains, Tanzania', *Geogr. Annlr* **54A**, 157–93.

Thornes, J. B. (1972), 'Debris slopes as series', *Arct. alp. Res.* **4**, 337–42.

—— (1973), 'Markov chains and slope series: the scale problem', *Geogr. Anal.* **5**, 322–8.

Waldrip, D. B. and Robert, M. C. (1972), 'The distribution of slopes in Indiana', *Proc. Indiana Acad. Sci.* **81**, 251–7.

Wood, A. (1942), 'The development of hillside slopes', *Proc. Geol. Ass.* **53**, 128–40.

Yair, A. (1973), 'Theoretical considerations on the evolution of convex hill-slopes', *Z. Geomorph., Suppl. Bd* **18**, 1–9.

—— (1974), 'Sources of runoff and sediment supplied by the slopes of a first order drainage basin in an arid environment (northern Negev-Israel)', *Abh. Akad. Wiss. Göttingen* (*Math. -Phys. Klasse*) **29**, 403–17.

Yatsu, E., Ward, A. J. and Adams, F. (eds.) (1975), *Mass wasting. 4th Guelph symposium on geomorphology, 1975* (Geo Abstracts, Norwich), 202 pp.

Young, A. (1963), 'Some field observations of slope form and regolith and their relation to slope development', *Trans. Inst. Br. Geogr.* **32**, 1–29.

—— (1972), *Slopes* (Oliver & Boyd, Edinburgh), 288 pp.

—— (1974), 'The rate of slope retreat', *Inst. Br. Geogr. Spec. Publ.* **7**, 65–78.

—— with Brunsden, D. and Thornes, J. B. (1974), 'Slope profile survey', *Br. Geomorph. Res. Grp Tech. Bull.* **11**, 52 pp.

Zonneveld, J. I. S. (1975), 'Some problems of tropical geomorphology', *Z. Geomorph.* **19**, 377–92.

6

SOME RECENT PROGRESS AND PROBLEMS IN THE STUDY OF MASS MOVEMENT IN BRITAIN

D. B. PRIOR
(*The Queen's University of Belfast*)

Until relatively recently geomorphologists have paid little attention to mass-movement processes and landforms, except perhaps to acknowledge their existence. This is illustrated by the restricted treatment of these phenomena in most standard geomorphological textbooks.

However, in the last decade, and particularly in the last 5 years, there has been considerably more research on various aspects of mass-movement from many points of view. Undoubtedly, the main progress has been promoted by engineers, particularly those concerned with the application of soil and rock mechanics to the explanation of slope instability and the solution of practical problems. Geologists, engineering geologists and geomorphologists have also contributed to the increased volume of literature reflecting several overlapping and complementary approaches to the subject. One simple indicator of increased research activity has been the change in the way in which publications are classified in *Geo-Abstracts*. Up to 1971 all mass-movement studies were grouped under the general category 'Weathering and slopes' whereas since 1972, 'Soil Mechanics', 'Slopes' and 'Weathering' have each received separate treatment. Similarly, the most recent textbooks in geomorphology contain substantially more detailed sections on slope instability. For example, M. A. Carson (1971) and Carson and M. J. Kirkby (1972) place considerable emphasis on the stability of slopes against landsliding. R. U. Cooke and J. C. Doornkamp (1974) have also discussed mass-movement phenomena in the context of environmental management.

In Britain, increased interest and activity has also been demonstrated by several symposia, for example the Institute of British Geographers' *Slopes Symposium* (Brighton, 1971) and two meetings under the auspices of the Engineering Group of the Geological Society of London (*Engineering of slopes* (Bristol, 1972); *Mudslides and mudflows* (London, 1973)). The volume of published work is quite considerable. A bibliography on mass-movement studies in the British Isles for 1970–75 to be presented to the I.G.U. at Moscow (1976) includes over 100 entries and is not yet complete (P. H. Temple, personal communication).

The volume of work and the breadth of interest in the subject make any review necessarily very selective. It is proposed to identify some main themes within which progress has been made, but this is, of course, a subjective selection. It is readily conceded that many important recent contributions have been omitted, especially in relation to rock-slope instability. Similarly, the review and assessment of problematical aspects is limited to Britain but, where relevant, some other international work will be briefly mentioned. In Britain, we are fortunate (in one sense) in having a wide range of natural mass-movement phenomena, many of which are presently active, as well as the results of former periods of slope instability.

IDENTIFICATION OF UNSTABLE SLOPES AND CLASSIFICATION

Considerable attention has been focused on the identification, mapping and classification of mass-movement features on unstable slopes. In particular, aerial photographs are increasingly important sources of evidence for the documentation of potentially unstable areas and signs of previous instability. They may also be used to identify and list the general environmental factors that contribute to instability. For example, the recent work of J. W. Norman *et al.* (1975) is an important study which usefully illustrates the suitability of various kinds of air-photo imagery and scale of photography for the identification of landslides along the Hythe Beds escarpment in Kent. Seven basic degrees of clarity of feature were established together with a list of criteria for recognition of individual features. Evidence was presented to show a marked decrease in the successful recognition of unstable ground at photo scales smaller than 1:10 000. Further, the best results were obtained from infra-red colour film, using infra-red colour prints, together with monochrome prints from the red and infra-red parts of the spectrum (Fig. 6.1).

The field identification and mapping of unstable slopes has also been considered by a working party of the Engineering Geology Group (of the Geolo-

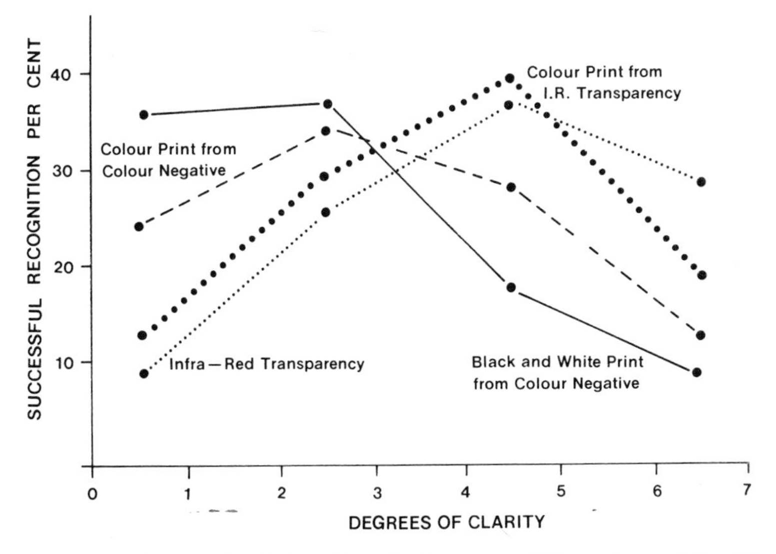

FIG. 6.1. Comparison of the clarity of landslip features on different types of film (Norman *et al.*, 1975).

gical Society of London), 1972. In their recommendations for the preparation of maps and plans, from the point of view of engineering geology, they included various geomorphological criteria. In addition to suggesting appropriate mapping symbols they also provided examples of geomorphological mapping, used primarily as a supplement to other techniques.

Indeed, the broad application of geomorphological mapping to engineering practice, using techniques developed by R. A. G. Savigear (1965) has been

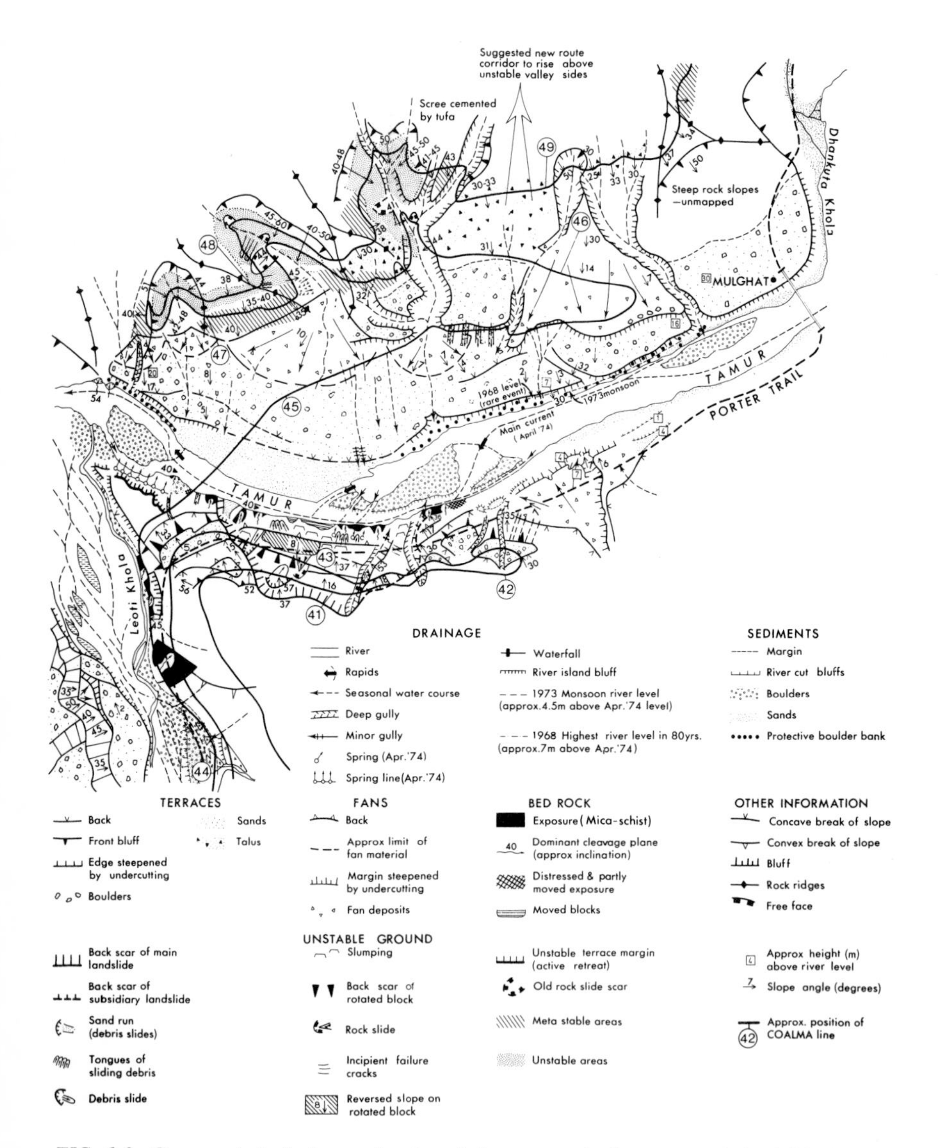

FIG. 6.2. Geomorphological mapping in relation to road alignment and landslide hazard in Nepal (Brunsden *et al.*, 1975).

strongly advocated by D. Brunsden *et al.* (Brunsden and D. K. C. Jones, 1972; Brunsden *et al.*, 1975) (Fig. 6.2). It is held that geomorphological investigations may complement established geotechnical procedures and are of special value in selecting areas for more detailed investigation; for example, areas containing evidence of former mass-movement or active slope instability. W. R. Dearman and P. G. Fookes (1974), two engineering geologists, and Brunsden *et al.* (1975) optimistically forecast that geomorphological mapping will be increasingly used by engineers in a wide variety of planning situations, not only those involving mass-movement processes. Without detracting from the ability of geomorphologists to map and identify mass-movement features, we can hardly claim this skill as our exclusive prerogative. To promote wider use of such maps and to utilise the full potential of such techniques, Brunsden *et al.* (1975) insist that the information must be presented in the most useful way for engineering interpretation. This, in turn, requires the geomorphologist to be aware of the needs of the engineer, but more particularly to be familiar with the geotechnical concepts that are fundamental to engineering practice.

A strictly morphological approach to landslide description has sometimes been adopted using length, depth and width parameters. A. W. Skempton and J. N. Hutchinson (1969), discussing the types of instability associated with clay slopes, suggest that the depth/length ratio is appropriate to describe the shape in downslope section, but add the caution that similar ratios can be associated with widely different behaviour among landslides. This indeed is a central problem with a strictly morphological approach since a single form can result from many different processes. M. J. Crozier (1973) in New Zealand has additionally suggested indices, such as flowage, displacement and tenuity, which rely on various morphological measurements. Cooke and Doornkamp (1974) have examined some of Crozier's data to determine whether selected pairs of indices might be used to discriminate between different types of landslide. The results are only partially satisfactory and of course depend on Crozier's initial distinctions between primary processes.

The morphological approach is tempting since data are fairly readily gathered and statistically analysed. It must be stressed, however, that considerable care must be exercised in inferring process from form alone. Morphological indices are perhaps only applicable at a rather general level of classification. An open question might be, to what degree has morphometry fundamentally contributed to our understanding of mass-movement processes? Certainly, the problem of a satisfactory method of classification of mass-movement phenomena still exists. Perhaps the most often-quoted scheme is that by D. J. Varnes (1958), in which three basic mechanisms—fall, slide and flow—are identified. Distinction is also made between bedrock and unconsolidated materials, and the latter are subdivided in terms of physical characteristics and water content. Approximate rates of movement are ascribed to different types of mass-movement.

Hutchinson (1968) maintains that a truly rigorous classification is difficult and prefers to sub-divide on the basis of mechanism, morphology, materials and rate of movement (Table I). In this scheme it is notable that many flow phenomena are considered as translational slides, the inference being that their further analysis can be best facilitated in this way. The point is amplified by Skempton and Hutchinson (1969) on clay-slope landslides. Certain mudflows, for example,

TABLE I

Classification of mass-movement by Hutchinson (1968)

	(1) Shallow, predominantly seasonal creep:
	(a) Soil creep
	(b) Talus creep
CREEP	(2) Deep-seated continuous creep: mass creep
	(3) Progressive creep
	(4) Freeze-thaw movements
FROZEN	(a) Solifluction
GROUND	(b) Cambering and valley bulging
PHENOMENA	(c) Stone streams
	(d) Rock glaciers
	(5) Translational slides
	(a) Rock slides: block glides
	(b) Slab, or flake slides
	(c) Detritus, or debris slides
	(d) Mudflows
	(i) Climatic mudflows
	(ii) Volcanic mudflows
	(e) Bog flows: bog bursts
	(f) Flow failures
	(i) Loess flows
LANDSLIDES	(ii) Flow slides
	(6) Rotational slips
	(a) Single rotational slips
	(b) Multiple rotational slips
	(i) In stiff, fissured clays
	(ii) In soft, extra-sensitive clays: clay flows
	(c) Successive, or stepped rotational slips
	(7) Falls
	(a) Stone and boulder falls
	(b) Rock and soil falls
	(8) Sub-aqueous slides
	(a) Flow slides
	(b) Under-consolidated clay slides

which adopt 'plug-flow' movement characteristics, are considered by Skempton and Hutchinson to be slides and they conclude that the distinction between slides and flows made by C. F. S. Sharpe (1938) is not generally valid.

A somewhat different approach has been adopted by Carson and Kirkby (1972) in which they sensibly suggest that mass-movement processes are rarely distinctly pure flow or pure slide. They also introduce a third category—'heave'—in which the slope material expands and contracts perpendicular to the surface. A triangular diagram is used, which also includes rates of movement and moisture content.

A. E. Scheidegger (1975) emphasises that it is very difficult to set up a general mechanical scheme since there may well be combinations of different mechanical processes; he rejects the subdivision in terms of fall, slide and flow, and prefers to group phenomena into characteristic classes, including surficial movements and deep mass-movements.

It is debatable whether a single scheme that includes all mass-movements is either possible or indeed desirable. Nevertheless there is no doubt that mass-movement studies are hindered by imprecise and confused terminology, mainly stemming from the difficulty of distinguishing slides from flows. In particular, the terms 'earthflow' and 'earthslide' appear to be used interchangeably; similar difficulties arise with 'mudflows' and 'mudslides', 'debris flows' and 'debris slides'. Moreover, in spite of A. Rapp's (1960) well-founded objections, the term 'debris avalanche' persists in the literature and adds to the confusion. Some rationalisation is clearly overdue. In this respect it is appropriate that work is proceeding towards a geotechnical classification of landslides (Hutchinson and Brunsden, 1974).

THE EXPLANATION OF INSTABILITY MECHANISMS

The instability of slopes, the interaction of factors involved and the actual behaviour of unstable areas can only be properly assessed by detailed evaluation of the properties of slope materials. Many geomorphologists seem to have been rather dilatory in appreciating this point and E. Yatsu (1966), for example, justifiably felt it necessary to voice criticisms in this respect. Once the appropriate properties have been determined, stability calculations can be made and compared with the actual field conditions, and the interaction of different factors studied. This approach clearly uses concepts and methods already formulated and practised by engineers.

It is not appropriate to restate at length these concepts and methods. A useful review has been provided by Skempton and Hutchinson (1969), who explain the significance of the concept of effective stress (which in turn emphasises the importance of pore-water pressure), the nature of peak and residual strength, and the distinction between drained and undrained strength. Similarly the various methods of analysing stability are described and the appropriate equations presented for the analysis of different types of landslide.

A major advance in geomorphology has been that these concepts are being increasingly recognised as relevant to the geomorphological consideration of landslide processes. The recent textbooks by Carson (1971) and Carson and Kirkby (1972) contain detailed treatments of the relationships between forces promoting slope failure and those resisting it. The factors controlling the strength of rocks and soils are discussed together with the techniques for measuring the strength of various parameters. Similarly, Carson and Kirkby (1972) examine instability in rocks and soils in terms of stability-analysis techniques, for example slab-failure in rocks, translational slides in soils and deep-seated slips. The success with which they demonstrate the relevance to geomorphological problems must be acknowledged. What is equally encouraging is that increasingly geomorphologists are acquiring the laboratory hardware with which to conduct tests and are demonstrating their willingness to apply the results of these tests.

Nevertheless, these are fairly recent developments and it is not surprising that most of the advances in the understanding of the factors influencing mass-movement phenomena in Britain have *not* been made by geomorphologists but by engineers. For example, Hutchinson (1967, 1969, 1970, 1971) has con-

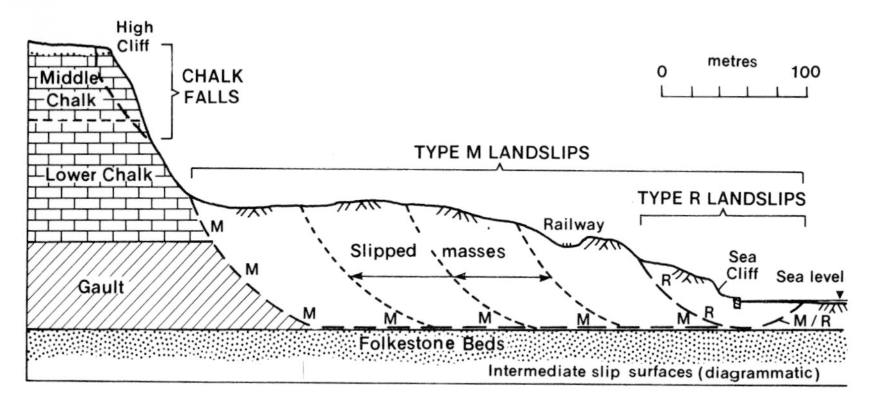

FIG. 6.3. Types of landslide at Folkestone Warren (Hutchinson, 1969).

sidered a variety of slope instability problems. The Folkestone Warren rotational slides in Gault Clay (Fig. 6.3) have been re-examined and the incidence of landslides has been related to the intensity of coastal erosion and periods of high pore-water pressure. Moreover, the results of drained shear tests and stability analyses have shown that the landslide activity is consistent with the principle of effective stress, and that shear strengths mobilised on slip surfaces within the Gault approximate to the residual (Hutchinson, 1969).

The geometry and behaviour of mudflows on London Clay cliffs has also been examined by Hutchinson (1970) in relation to the concept of the angle of ultimate stability, developed by Skempton and F. A. DeLory (1957). Those parts of the mudflows inclined at greater than the ultimate angle are seasonally active, in response to fluctuations of pore-water pressure. Those parts which are unstable at or below the ultimate angle are explained in terms of local artesian pressures (Hutchinson, 1970).

The stability of Lias slopes in Rutland and Northamptonshire has been the subject of work by R. J. Chandler (1970, 1971). Once again, stability analysis, using effective strength parameters, has proved useful together with measurements of pore-water pressure in the field. In particular, this work has shown that some of the Lias Clay slopes may be subject to active translational movements under present-day conditions, while rotational slides require higher pore-water pressures perhaps associated with cooler and wetter past climatic periods (Table II).

Skempton and Hutchinson (1969) pointed out that some types of landslide have not yet been analysed using these concepts and methods. Notwithstanding the later work already discussed, this probably remains the case. Certain types of debris slide or debris flow (Fig. 6.4) and bog-bursts (Fig. 6.5) seem to have been ignored, and undoubtedly there remains scope for future investigations of other landslide types on a wider variety of slope materials. For example, work is in progress in Ireland on mudslides on Carboniferous and Liassic shales (e.g. Prior and J. Graham, 1974).

In spite of all these studies, several related problems must be mentioned. Particularly, questions arise concerning the factors that control and influence the

TABLE II

Results of stability analyses of landslides on Lias Clay (Chandler, 1971)

Slip no.	*Surface profile*	*Piezometric surface*	*Soil density,* γ, kg/m³	*Water content, per cent*	ϕ' *when F=1, and c'=0*	*F for given value of c' and* ϕ'
						3·76(13kN/m²; 24°)
1	prior to slip	estimated	1920	32	17·9°	1·19(0; 21°)
1	present	estimated	1920	32	16·7°	1·28(0; 21°)
2	present	based on Dec. 1968	1890	34·5	21·0°	1·00(0; 21°)
3	present	based on Dec. 1968	1860	37	14·3°	1·32(0; 18·5°)
3	present	at ground level	1810	41	17·2°	1·08(0; 18·5°)

Note: The soil density was computed taking the specific gravity of the soil grains as 2·73.

strength of particular slope materials. Whether geomorphologists have the motivation or the capacity to investigate such problems is irrelevant—they cannot be ignored. For example, the exact nature of cohesion and frictional resistance, especially in terms of physico-chemical factors, is still under debate. The contribution of mineralogy, grain-size and exchangeable cations to strength of soils has continued to receive attention (e.g. G. Mesri and R. E. Olson, 1970). This leads logically to the consideration of the rate and nature of weathering. Progressive geochemical and physical changes influence strength and thereby the stability of slopes.

This aspect has also been examined by Chandler (1972) using examples from areas of Lias Clay. Progressive changes in water content and undrained strength were identified, with more weathered layers showing higher strengths at given water contents (Fig. 6.6). In terms of effective stresses some evidence was presented to show that increased weathering is associated with reduction of effective cohesion and effective friction angle. Chandler (1972) concludes that weathering rates are so slow 'as to be quite unimportant in any engineering time scale'. The question arises, to what degree do engineering and geomorphological time-scales differ? This is a point that is discussed by Carson and D. J. Petley (1970).

There seems no doubt, however, that considerable scope exists for additional studies of both physical and chemical weathering on the strength of slope materials in relation to the inherent factors that influence strength. While there are few strictly comparable examples in Britain, the recent views by I. J. Smalley (1972) on quick clays and flow slides may serve as an illustration. The traditional concept of weathering or leaching leading to high sensitivity is supplemented by Smalley's view that the short-range bonds between non-clay mineral particles are especially important to strength (Fig. 6.7). Without engaging in the debate that has accompanied this work, it is clear that the explanation of strength in relation to mass-movement processes is not yet complete.

FIG. 6.4. Debris flows on basaltic scree-slopes in Northern Ireland.

FIG. 6.5. Source area for a bog-flow (approximate length 1·7 km).

The practical difficulties of measuring strength and effective strength parameters must not be forgotten. Skempton and Hutchinson (1969) have identified ways in which the results of laboratory testing may differ from *in situ* strength, such as sampling procedures, size of sample, rate of shearing, etc. Since many landslides, especially those in clays, may involve renewed movement on pre-existing shear surfaces, residual strength parameters apply.

The measurement of residual strength using direct shear techniques, however, has been questioned. Residual tests have usually taken one of several approaches —numerous reversals, cut-plane tests or the use of remoulded slurries. Alternatively, samples containing natural shear-planes have been used. Each test has its limitations.

Recently, A. W. Bishop *et al.* (1971) have proposed the use of Ring Shear apparatus to overcome the problem of simulation of large relative displacements, uninterrupted by changes in direction. They demonstrate that, in general, there

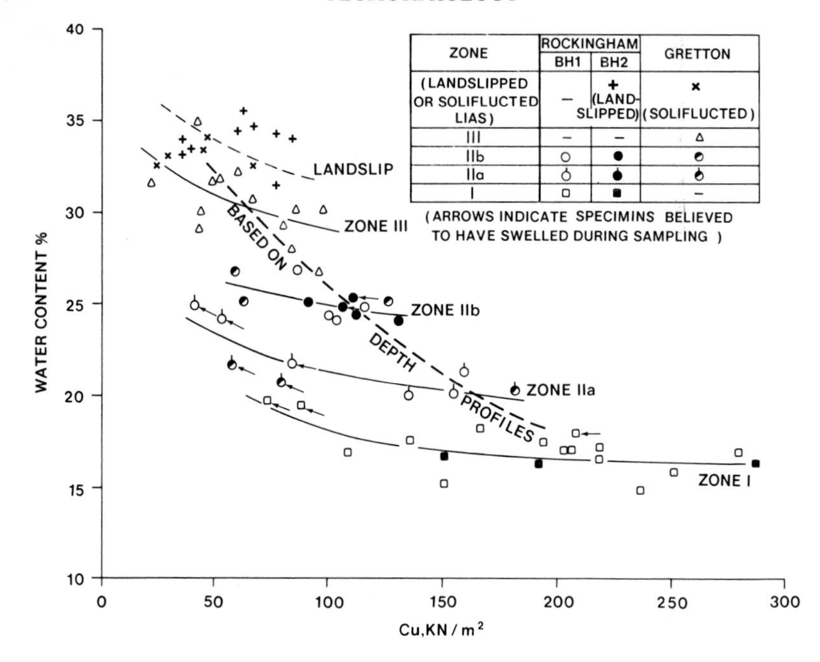

FIG. 6.6. Relationships between weathering, water content and undrained strength of Lias Clay (Chandler, 1972).

are significant differences between direct-shear and ring-shear results, although they admit, for various reasons, that the former may be of value in providing an empirical correlation with field observations. The fairly high cost of Ring Shear equipment is likely to restrict its use by geomorphologists to the more affluent laboratories. Those of us forced to rely on direct-shear techniques must be aware of the accompanying problems.

Another area of difficulty in explaining mass-movement activity is the general lack of data on pore-water pressure. A distinction should be made here between 'short-term' and 'long-term' conditions promoting instability. Considering the nature of 'cutting' slopes—either naturally (by rivers, or marine erosion, for example) or in artificial excavations—Bishop and L. Bjerrum (1960) have proposed a general model. In this, it is envisaged that there may be a reduction in pore-water pressure that initially accompanies unloading of the slope. During this phase, 'short-term' landslides may occur. There will then follow a gradual rise of pore-water pressure, leading to a reduction in strength and in the safety factor. Instability may therefore occur some time after the initial changes in slope geometry have occurred. These 'long-term' mass-movements may be viewed either on an engineer's time-scale (in tens of years) or on a much longer geological scale.

Carson and Kirkby (1972) refer to the problem of the choice of the appropriate model for pore-water pressure as the 'Achilles heel' in analysing the long-term stability of natural slopes. Skempton and Chandler (1974), considering mainly cutting slopes in London Clay, concur; they point out that 'Measurements of pore-water pressure associated with long-term slips are still rather few in number'. However, their data for some London Clay slides have shown that average values of the pore-water pressure ratio of 0·30 suggest field values of effective stress parameters that are consistent with the results of laboratory testing. Skempton and Chandler (1974) nevertheless conclude that more observations are needed with regard to pore-water pressure ratio—and, one could add, on a variety of materials and different slope geometries.

In this respect, the paper by T. C. Kenney and S. Uddin (1974) provides an interesting record from the excavation of the Kimola canal in Finland in which pore-water pressures were actually recorded, from excavation through to failure (Fig. 6.8). Additionally, K. D. Eigenbrod (1975) has reviewed pore-water pressure data from a variety of sites, and in passing mentions the relevance of the concept of equalisation of pore-water pressure to the natural long-term slope stability of river banks in the Great Plains of North America.

It is to be hoped that increased monitoring of pore-water pressure, both in areas of active landsliding and in areas of potential instability, will provide better explanations for both short- and long-term instabilities. An interesting approach might be to examine more closely the nature of climatic controls on pore-water

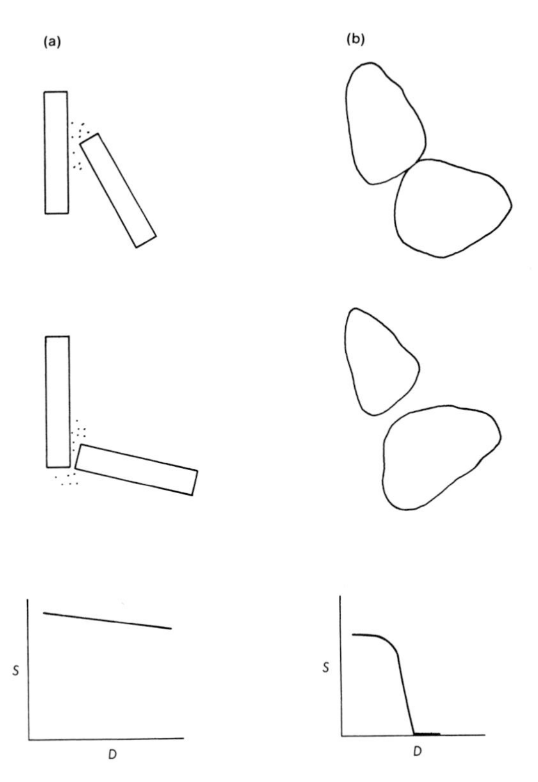

FIG. 6.7. Strength–displacement relationships for active long-range bonds and inactive short-range bonds (Smalley, 1972).

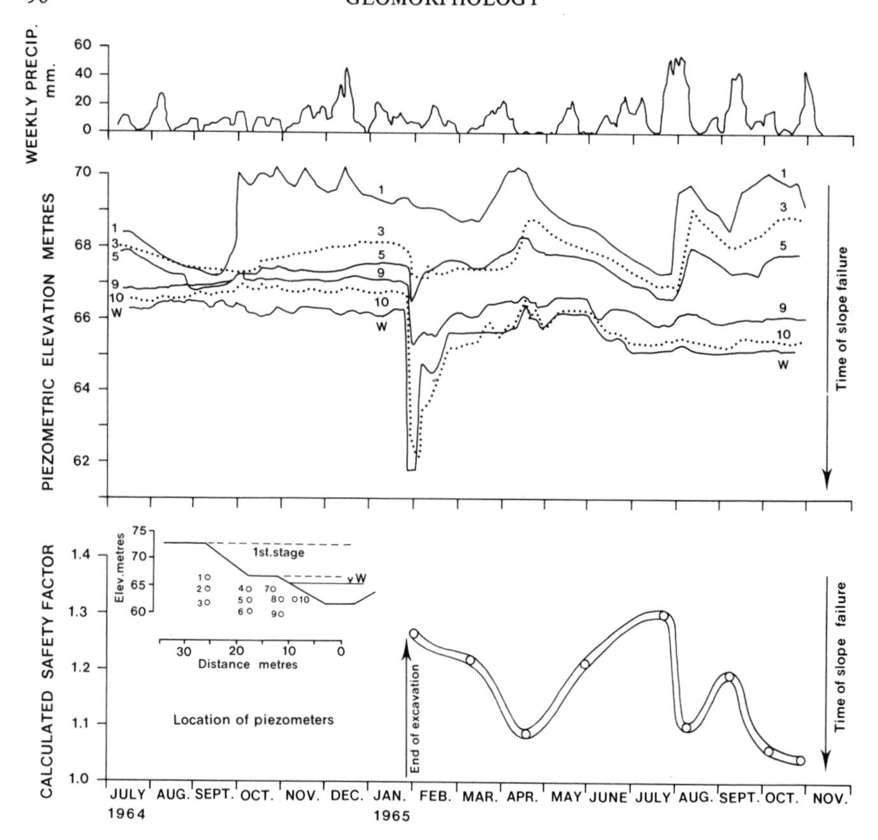

FIG. 6.8. Results of a stability study involving measurements of pore-water pressure along the Kimola canal, Finland (Kenney and Uddin, 1974).

pressure (following T. R. Harper, 1975; Fig. 6.9) to facilitate the greater use of past climatic records in considering former instability.

THE MONITORING OF UNSTABLE SLOPES

The behaviour of an unstable slope or an individual landslide can be monitored for various reasons. For example, the practising engineer may monitor slopes to check previous stability calculations. Alternatively, active mass-movement features such as mudflows, solifluction lobes and some categories of rotational slide are 'fixed site' phenomena which allow the details of changing behaviour, and the mechanics of the processes to be observed directly (Fig. 6.10). They can also be regarded as process-response systems (Brunsden, 1973) and can be analysed in terms of energy, debris and water transfers. For example, Brunsden (1973) has derived a schematic process-response model for a mudflow which emphasises the geometry, material properties, the variables that control movement, and the rate of input and throughput of mass and water (Fig. 6.11). This was designed as a basis for planning a field programme but allows attempts to be made to relate measured values of landslide behaviour to controlling factors.

Methods are needed to monitor changes of geometry and movements, and an increasing variety of methods has been applied during the past few years in Britain. J. A. Franklin and P. E. Denton (1973) provide a useful review and Soil Instruments Ltd., London, are one of several suppliers of a range of commercially available instruments. The methods may be sub-divided into those dealing exclusively with surface changes and those that provide sub-surface data; ideally, both should be measured concurrently.

Surface movements of landslides have been monitored by measuring the displacements of stakes and targets related to fixed points (e.g. Prior *et al.*, 1968; Hutchinson, 1970; Brunsden, 1973). Changes of form have been recorded by repeated ground surveying. Hutchinson (1970) has also provided interesting examples of displacement across a surface shear-plane using time-lapse ground photography. Such techniques have shown broad relationships between the patterns of movement of some types of landslide and seasonal climatic characteristics, especially availability of rainfall (Prior *et al.*, 1968; Hutchinson, 1970).

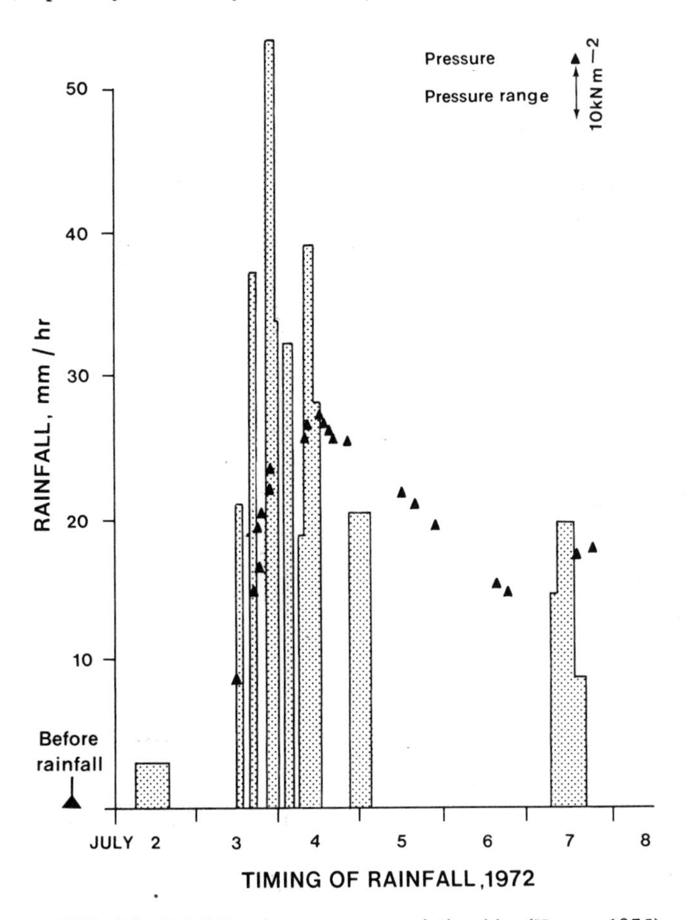

FIG. 6.9. Rainfall and pressure-wave relationships (Harper, 1975).

Attempts have also been made automatically to monitor the surface movements of mudslides (Prior and N. Stephens, 1971, 1972). While these techniques, using modified water-level recorders, are relatively crude, they have been extremely useful in detecting the wide variability of rates of movement that can occur. Complex patterns have been identified, involving pronounced accelerations and decelerations together with surges of varying magnitude (Fig. 6.12). The latter appear to be a very important and dangerous aspect of mudslide behaviour which cannot easily be detected without continuous recording techniques. Attempts have been made to link short-term rates of movement with rainfall; certainly, accelerations often accompany periods of heavy rainfall.

Sub-surface measurements of some types of mass-movement have also been obtained. Franklin and Denton (1973), describing techniques employed for rock slopes, suggest three categories of instrumentation—settlement instruments that measure vertical displacement, extensometers that measure extension or com-

FIG. 6.10. A mudslide in Northern Ireland, an example of a 'fixed' landslide site which is periodically active and allows continuous documentation of instability mechanisms.

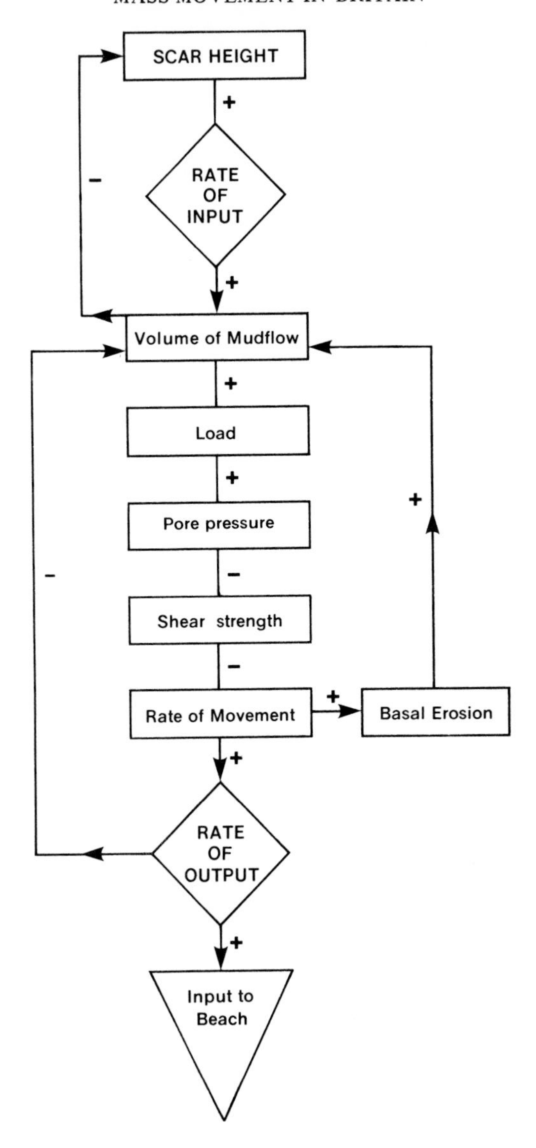

FIG. 6.11. A process-response model for a mudflow (Brunsden, 1973).

pression in boreholes, and inclinometers that measure changes in borehole alignment. An example of the latter is provided by the work of Hutchinson (1970) using a loaded-leaf inclinometer inserted into flexible tubes placed within a mudflow. This showed the distribution of movement on a vertical profile to be nearly constant with depth, interpreted as 'plug-flow' and indicating that for the period of observation this particular mudflow was behaving as a translational slide.

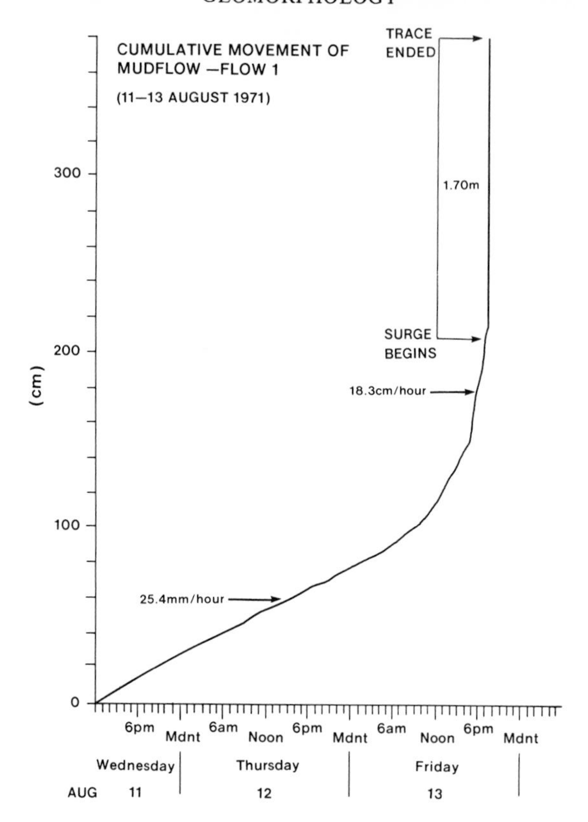

FIG. 6.12. Automatic recorder trace for mudslide movement showing surge (Prior and Stephens, 1971).

The problems associated with the field monitoring of mass-movement behaviour and associated factors such as pore-water pressure should not be underestimated. Recently, for example, attempts were made to examine in more detail the nature of surges of a mudflow in Ireland in relation to the undrained loading concept advanced by Hutchinson (1970; Hutchinson and R. K. Bhandari, 1971). The difficulties encountered provide a good example of the practical problems of field monitoring.

First, we must consider the undrained loading concept since it probably represents a most significant advance in the understanding of the mechanics of a wide variety of mass-movement phenomena. Hutchinson points out that where landslides occur on bi-linear concave slope profiles there is a tendency for rapid movements on the steeper slope to produce rapid loading on the lower-angle slope. This loading has been observed to generate high pore-water pressures and these in turn can produce failure at angles below the ultimate angle of stability for the particular material (Fig. 6.13). Moreover, in particular situations the mechanism can generate dangerously rapid movements of large amounts of material. Hutchinson has suggested that this mechanism might well account for

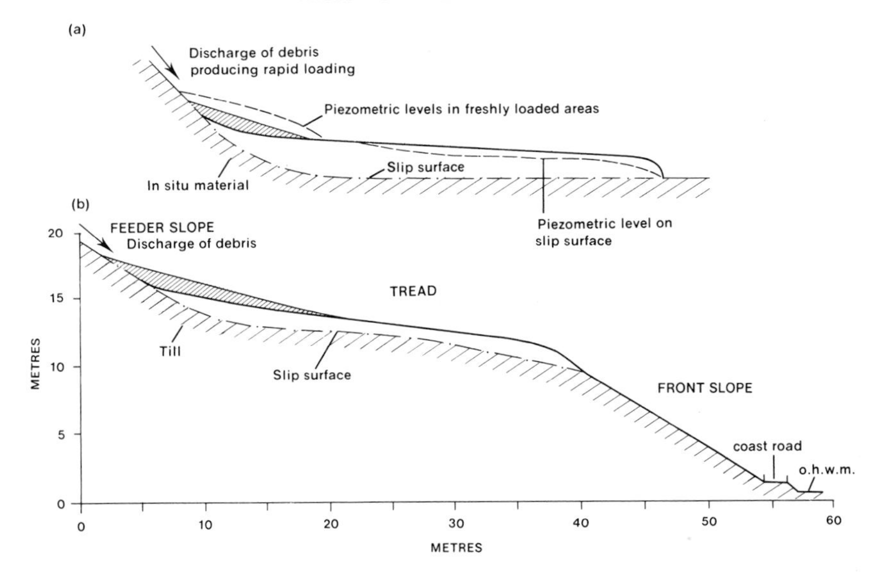

FIG. 6.13. Comparative profiles showing diagrammatic geometry of (a) mudslide type studied by Hutchinson and Bhandari (1971) and (b) mudslide at Minnis North, Northern Ireland (Hutchinson *et al.*, 1974).

the characteristics of fossil solifluction sheets, slush avalanches and submarine slides, as well as mudflows. More recently he has indicated that it may be fundamental to the explanation of one of the largest landslides recorded in historical times in Peru, 1974, where a slide of volume 10^9 m^3 moved at 120–140 km/hour (Hutchinson and E. Kojan, 1975).

One mudflow site in north-east Ireland appeared to provide an opportunity for further field investigation of the undrained loading mechanism (Hutchinson *et al.*, 1974) (Fig. 6.13). A primary objective was simultaneously to monitor surface movements and pore-water pressures as a result of loading. Movement was measured by Munro water-level recorders, and electrical diaphragm piezometers, linked to a Peekel micro-strain recorder, were installed to measure pore-water pressures. In spite of near-continuous measurements and the presence of an eyewitness, an abrupt loading surge took place of such magnitude and rapidity that pore-water pressure changes could not be manually monitored on the Peekel recorder (Fig. 6.14). A particular feature was the unpredictable nature of the surge and the apparent absence of discernible creep movements on the loading slope before failure.

It seems clear that field monitoring of landslide behaviour and the precise documentation of such mechanisms requires the development of increasingly sophisticated methods. The inherent variability of rates of movement of many types of landslide requires automatic methods which are accurate at small displacements—to detect, for example, pre-failure creep—yet are capable of recording large, rapid displacements without being destroyed. In this respect, work is proceeding in Belfast on the development of potentiometric and transducer techniques for movement recording (W. B. Whalley, personal communication).

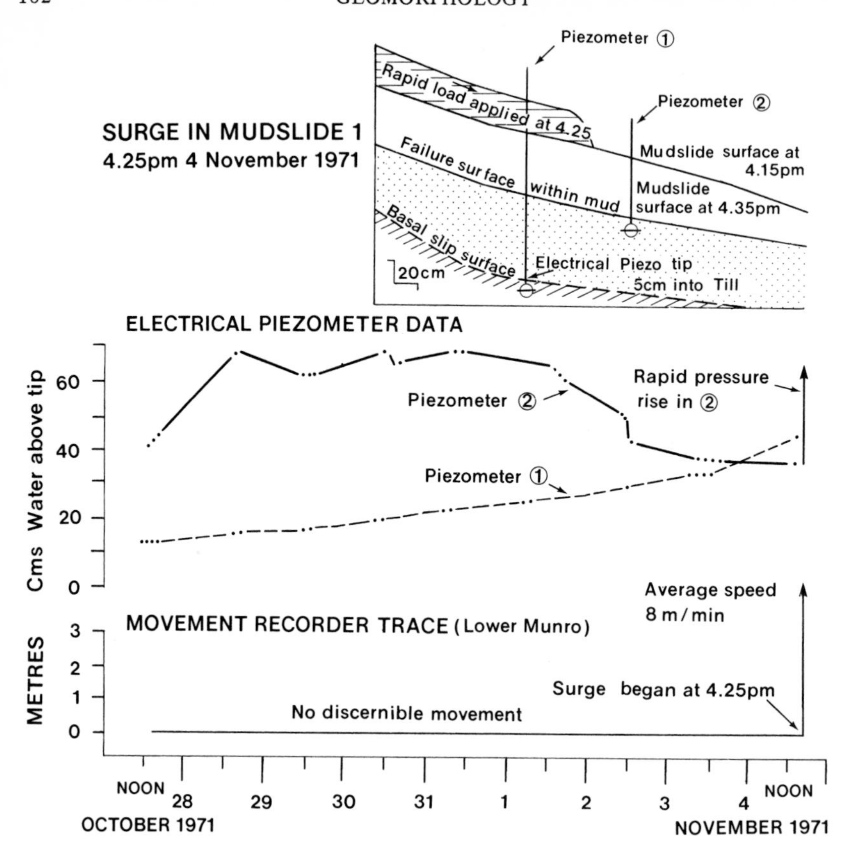

FIG. 6.14. Piezometer and movement records for a surge at Minnis North, Northern Ireland (Hutchinson *et al.*, 1974).

Similarly, the automatic monitoring of pore-water pressure may be required for the exact determination of the pressure changes that accompany slope failure. Increasingly, we must envisage the use of multi-channel data loggers linked to networks of electrical piezometers. Clearly, the availability of financial support will be a crucial factor in determining the rate of progress of such field studies. Nevertheless, there remains a need for further field studies in Britain which attempt to elucidate the basic mechanisms by which failure occurs, on a wide variety of materials and on various types of landslide. The mechanics of undrained loading was indeed detected by such detailed field observations.

The still-obscure relationships between sliding and flowing of materials on slopes have yet to be properly documented. Skempton and Hutchinson (1969) admit that 'flows' form a rather neglected and little understood group of movements. Those landslides that appear to combine the characteristics of sliding, plug-flow and viscous flow must be rigorously examined in the field to determine the factors and conditions that contribute to the appropriate changes in rheological behaviour. Recent work by A. M. Johnson (1970) and associates on alpine debris flows in America appears to have made considerable contributions towards understanding flow characteristics, if only because of their willingness to accept the applicability of the Bingham rheological model.

In Britain, the mechanics of some mudflows and debris flows on steep slopes has probably only been approximated. Likewise, the nature and magnitude of pre-failure creep awaits further field study. Where field monitoring of processes proves difficult, laboratory simulation may offer complementary information. With the exception of various experiments at Manchester using a high-speed centrifuge (e.g. T. R. Stacey, 1974) and mudflow simulation by J. B. Thornes and Brunsden (personal communication) and by I. D. L. Foster (1973) there has been little such work in Britain. This contrasts with the laboratory experiments by Johnson (1970) in the United States, G. Einsele *et al.* (1974) on sub-aqueous slides and K. J. Hsu (1975) on Sturzstroms (Fig. 6.15). It is to be hoped that increased work on laboratory simulation will accompany field instrumentation of active landslide processes.

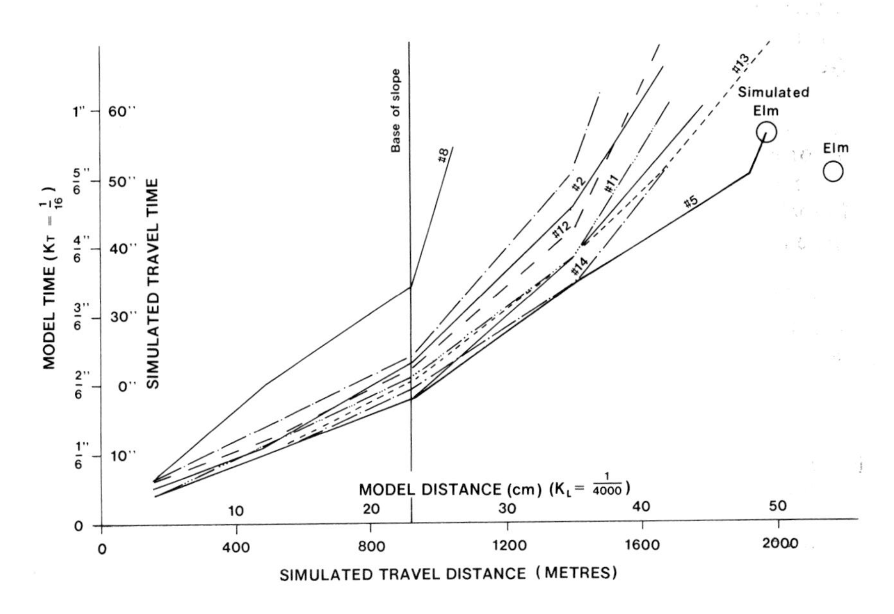

FIG. 6.15. Kinematic analysis of experimental sturzstroms (Hsu, 1975).

CONCLUSION

In this rather incomplete review it has been demonstrated that there has been considerable progress in mass-movement studies in Britain in recent years and that many problems await solution. It is also self-evident that many of the contributions are attributable to engineers. The inevitable question concerning the possible role of geomorphologists in future studies of these processes arises. Cooke and Doornkamp (1974) suggest that the particular contribution of geomorphology lies in predicting areas susceptible to landsliding and in identifying the controlling environmental characteristics of landslides. Indeed, engineers may be better qualified to do precisely these tasks, and to some extent the literature proves this!

One distinction might be that geomorphologists need to apply the results of study of contemporary slope instability to the problems of slope form and evolution over much greater time-scales than are normally of interest to the engineer. The work of Carson and Petley (1970) and W. C. Rouse (1975) on threshold slopes, and of Hutchinson (1973) on slope evolution dominated by landsliding on London Clay, illustrates this point. Alternatively, geomorphologists may find a role in considering the overlap of processes that operate on most slopes—the interaction of mass-movement with other hillslope processes—which leads ultimately to the development of particular slope forms. Or we might proceed along the lines suggested by Yatsu (1966) in which geologists, geomorphologists, engineering geologists and engineers combine to explore the overlapping boundaries of their subjects, the 'grey' areas of common interest. This might well be achieved by inter-disciplinary teams engaged in elucidating the precise mechanisms by theoretical (experimental) and field studies.

Geomorphologists cannot, however, complacently and vaguely assume that they have a contribution to make: they must attempt rigorously to define this contribution, to pursue it and keep testing it. Moreover, it is quite clear that geomorphologists interested in such problems must, albeit painfully, make themselves more familiar with the relevant fundamental concepts and methods of soil and rock mechanics and, further, must provide the appropriate course structures for the training of future geomorphologists.

Acknowledgements

The author thanks Dr D. Brunsden, King's College London and Dr W. B. Whalley, Queen's University Belfast for their critical evaluation of the original manuscript and for constructive additions.

REFERENCES

Bishop, A. W. and Bjerrum, L. (1960), 'The relevance of the triaxial test to the solution of stability problems', *Proc. Am. Soc. civ. Engrs, Res. Conf. on Shear Strength of Cohesive Soils*, 437 pp.

——, Green, G.E., Garga, U. K., Andresen, A. and Brown, J. D. (1971), 'A new ring shear apparatus and its application to the measurement of residual strength', *Géotechnique* **21**, 273–328.

Brunsden, D. (1973), 'The application of systems theory to the study of mass-movement', *Geologia appl. Idrogeol.* **8**, 185–207.

—— and Jones, D. K. C. (1972), 'The morphology of degraded landslide slopes in South West Dorset', *Q. J. Engng Geol.* **5**, 205–22.

——, Doornkamp, J. C., Fookes, P. G., Jones, D. K. C. and Kelly, J. M. H. (1975), 'Large scale geomorphological mapping and highway engineering design', *Q. J. Engng Geol.* **8**, 227–53.

Carson, M. A. (1971), *The mechanics of erosion* (Pion, London), 174 pp.

—— and Petley, D. J. (1970), 'The existence of threshold hillslopes in the denudation of the landscape', *Trans. Inst. Br. Geogr.* **49**, 71–95.

—— and Kirkby, M. J. (1972), *Hillslope form and process* (Cambridge Univ. Press), 475 pp.

Chandler, R. J. (1970), 'The degradation of Lias clay slopes in an area of the East Midlands', *Q. J. Engng Geol.* 2, 161–81.
— (1971), 'Landsliding on the Jurassic escarpment near Rockingham, Northamptonshire', in D. Brunsden (compiler) 'Slopes: form and process', *Inst. Br. Geogr. Spec. Publ.* 3, 111–28.
— (1972), 'Lias clay: weathering processes and their effect on shear strength', *Géotechnique,* 22, 403–31.
Cooke, R. U. and Doornkamp, J. C. (1974), *Geomorphology in environmental management* (Clarendon Press, Oxford), 413 pp.
Crozier, M. J. (1973), 'Techniques for the morphometric analysis of landslips', *Z. Geomorph.* **17**, 78–101.
Dearman, W. R. and Fookes, P. G. (1974), 'Engineering geological mapping for civil engineering practice in the United Kingdom', *Q. J. Engng Geol.* 7, 223–56.
Eigenbrod, K. D. (1975), 'Analysis of the pore pressure changes following the excavation of a slope', *Can. geotech. J.* **12**, 429–40.
Einsele, G., Overbeck, R., Schwarz, H. U. and Unsold, G. (1974), 'Mass physical properties, sliding and erodibility of experimentally deposited and differentially consolidated clayey muds', *Sedimentology,* **21**, 339–72.
Foster, I. D. L. (1973), 'A laboratory experiment to analyse some aspects of debris flow', (Unpubl. undergraduate dissertation, Univ. of London, King's College).
Franklin, J. A. and Denton, P. E. (1973), 'The monitoring of rock slopes', *Q. J. Engng Geol.* 6, 259–86.
Harper, T. R. (1975), 'The transient groundwater pressure response to rainfall and the prediction of rock slope instability', *Int. J. Rock Mech. Min. Sci. Geomech., Abstr.,* **12**, 175–9.
Hsu, K. J. (1975), 'Catastrophic debris streams (Sturzstroms) generated by rockfalls', *Bull. geol. Soc. Am.* **86**, 129–40.
Hutchinson, J. N. (1968), 'Mass-movement' in R. W. Fairbridge (ed.) *Encyclopaedia of earth sciences* (Reinhold, New York), 688–95.
— (1969), 'A reconsideration of the coastal landslides at Folkestone Warren, Kent', *Géotechnique* **19**, 6–38.
— (1970), 'A coastal mudflow on the London clay cliffs at Beltinge, North Kent', *Géotechnique,* **20**, 412–38.
— (1973), 'The response of London clay cliffs to differing rates of toe erosion', *Geologia appl. Idrogeol.* 8, 221–37.
— Bhandari, R. K. (1971), 'Undrained loading, a fundamental mechanism of mudflows and other mass-movements', *Géotechnique* **21**, 353–8.
— Brunsden, D. (1974), 'Mudflows: a review and classification' (abstract), *Q. J. Engng Geol.* 7, 328.
— Kojan, E. (1975), 'The Mayunmarca landslide of 25 April, 1974', *Unesco Rep.* No. 3124/RMO.RD/SCE (Paris).
—, Prior, D. B. and Stephens, N. (1974), 'Potentially dangerous surges in an Antrim mudslide', *Q. J. Engng Geol.* 7, 363–76.
Johnson, A. M. (1970), *Physical processes in geology* (Freeman, Cooper and Co., San Francisco), 571 pp.
Kenney, T. C. and Uddin, S. (1974), 'Critical period for stability of an excavated slope in clay soil', *Can. geotech. J.* **11**, 620–3.
Mesri, G. and Olson, R. E. (1970), 'Shear strength of montmorillonite', *Géotechnique,* **20**, 261–70.
Norman, J. W., Lievowitz, T. H. and Fookes, P. G. (1975), 'Factors affecting the detection of slope instability with air photographs in an area near Sevenoaks,

Kent', *Q. J. Engng Geol.* **8**, 159–76.

Prior, D. B. and Graham, J. (1974), 'Landslides in the Magho District of Fermanagh, N. Ireland', *Engng Geol.* **8**, 341–59.

—— Stephens, N. (1971), 'A method of monitoring mudflow movements', *Engng Geol.* **5**, 239–46.

—— and Stephens, N. (1972), 'Some movement patterns of temperate mudflows: examples from north-eastern Ireland', *Bull. geol. Soc. Am.* **33**, 2533–44.

——, Stephens, N. and Archer, D. R. (1968), 'Composite mudflows on the Antrim coast of north-east Ireland', *Geogr. Annlr* **50**, 65–78.

Rapp, A. (1960), 'Recent development of mountain slopes in Kärkevagge and surroundings, northern Scandinavia', *Geogr. Annlr* **42**, 71–200.

Rouse, W. C. (1975), 'Engineering properties and slope form in granular soils', *Engng Geol.* **9**, 221–35.

Savigear, R. A. G. (1965), 'A technique of morphological mapping', *Ann. Ass. Am. Geogr.* **53**, 514–38.

Scheidegger, A. E. (1975), *Physical aspects of natural catastrophes* (Elsevier, New York), 289 pp.

Sharpe, C. F. S. (1938), *Landslides and related phenomena—a study of mass-movements of soil and rock* (Columbia Univ. Press, New York), 137 pp.

Skempton, A. W. and Chandler, R. J. (1974), 'The design of permanent cutting slopes in stiff fissured clays', *Géotechnique,* **24**, 457–66.

—— and DeLory, F. A. (1957), 'Stability of natural slopes and embankment sections', *Proc. 4th int. Conf. Soil Mech. Foundation Engng* **2**, 378–81.

—— and Hutchinson, J. N. (1969), 'Stability of natural slopes and embankment sections', *Proc. 7th int. Conf. Soil Mech. Foundation Engng* (State-of-the-art Volume), 291–340.

Smalley, I. J. (1972), 'Boundary conditions for flowslides in fine-particle mine waste tips', *Trans. Instn Min. Metall.* Sect. A, 31–7.

Stacey, T. R. (1974), 'The behaviour of two- and three-dimensional model rock slopes', *Q. J. Engng Geol.* **8**, 67–72.

Temple, P. (1976), 'Studies of mass wasting in the British Isles 1970–1975', A provisional bibliography for the *I.G.U. Commission on Geomorphological Processes* (presented to the I.G.U. Conference, Moscow, 1976).

Varnes, D. J. (1958), 'Landslide types and processes', in E. B. Eckel (ed.), *Landslides and engineering practice* (Highway Res. Board, Washington, Spec. Rep. 29; NAS–NRC Publ. 544), 232 pp.

Yatsu, E. (1966), *Rock control and geomorphology* (Sozosha, Tokyo), 135 pp.

7

GLACIAL GEOMORPHOLOGY: PRESENT PROBLEMS AND FUTURE PROSPECTS

JUST GJESSING
(*University of Oslo*)

Glacial geomorphology is concerned with the bedrock forms and superficial deposits produced by glacial and fluvio-glacial processes in areas of present glaciers as well as in areas covered by glaciers during the Quaternary. The interpretation of forms created during the Ice Ages is based essentially on studies of present-day glacierised areas. Since glaciers are the main agents, glaciological studies are of vital importance. If we had been left without glaciers at the present day, the forms created by ancient glaciers would have been very difficult to interpret. We still have the problem, however, that glacial erosion, transport and deposition are difficult to observe directly, and difficult to quantify (G. S. Boulton, 1975). Most glacial geomorphologists work in areas far from present glaciers, and most conclusions relating to processes must be based on studies of the forms left after the glaciers have vanished. Studies of processes under present glaciers are difficult to carry out and the fieldwork involved is hard, dangerous and expensive. Access to subglacial areas is severely limited, and is often not possible to the very places where we need observations. If a tunnel is made for access, the conditions that we would like to observe may be changed by its construction.

To understand subglacial processes we must know the physical conditions of pressure and temperature existing under glaciers. We must know the physical properties of the materials involved, bedrock, ice, debris and water, and of the compounds of water and debris, ice and debris. The interaction of the materials and how they behave under different physical conditions are equally important. Of particular significance are the differences between temperatures below freezing and the melting-point temperature. Some of the research can be done in the field, but laboratory investigations on the physical properties of the different materials and simulation of subglacial processes are vital.

A presentation of present problems in glacial geomorphology might have been based upon an analysis of the topics dealt with in recent issues of different journals or in *Geo-Abstracts*. I have not used this method, however, for it is impossible to cover the whole field. I shall therefore deal with those topics with which I am most familiar, thereby demonstrating some types of problem in glacial geomorphology.

ICE EROSION IN LOWLANDS AND UPLANDS

As an introduction to the problems of glacial erosion, let us look at the effects of ice action on lowlands such as the Baltic and Canadian Shields, and on upland areas.

Overdeepened lake basins cut in bedrock are undeniable proof of ice erosion. Generally we must assume that depressions in all landscapes have been formed in less resistant bedrock, the hills in more resistant rock. A connection between depressions and areas of relatively closely jointed rock has often been demonstrated, for example in Finland by H. Niini (1968). In Norway, difficulties with jointed rock have frequently been reported when tunnel works have passed under valleys.

A common hypothesis is that preglacial chemical weathering penetrated deeply into the closely jointed rocks below what are now depressions. With the advance of glaciers, the weathered material was readily removed to form the basins. Further deepening has probably been effected by glacial plucking in the jointed rocks. Lake basins of this type are not very deep, mostly of the order of a few tens of metres.

Interpretation of the landscape surrounding the lakes is more difficult. One view, put forward, for instance, by D. L. Linton (1963), is that hills of this type have been 'moulded' by the ice. This may be true to a certain extent in some types of rocks, but mostly there seems to be no specific indication that the general forms of the hill-and-basin landscape are of glacial origin. It seems more likely that the main effect of glacial action would be plucking on the structurally determined sides, rather than rounding and smoothing by some sort of moulding process.

In many areas of Fennoscandia, and probably also in other countries, the landscape is composed of round-topped, steep-sided hills surrounded by gentle slopes. Hills of similar shape are found under arid to semi-arid conditions, in areas that have not been glaciated, for instance in the western parts of North America. If we can use the similarities in form as an indication of similarity in process and in history of development, it seems reasonable to infer that the hill-and-basin forms in the formerly glaciated areas have been produced by subaerial processes and possibly under different climatic conditions from those of the present. It seems most likely that these forms were produced in warmer Tertiary climates, but the effect of periglacial and other interglacial processes should not be excluded.

Structural control of slopes

A second example will expose some of the difficulties involved in efforts to distinguish between preglacial, interglacial or interstadial subaerially-produced slopes, and slopes produced by glacial erosion. It will also demonstrate the possible structural control of slopes in connection with ice erosion.

Escarpments controlled by resistant rocks resting upon less resistant beds (such as Caledonian thrusted rocks resting upon phyllite, or Permian lavas resting upon Cambro-Silurian sandstones, slates and limestones in the Oslo region) are often seen to have slope forms which might be expected if these forms had been produced by subaerial agents. The resistant rock forms an escarpment and a free face with a talus slope below. Except for minor irregularities, the surface of the less resistant rocks seems to have a gradient adjusted to their resistance, as in the case of slopes developed by subaerial processes. Similar conditions are found where slopes are controlled by gently dipping beds of limestone (such as the cuesta-forms and the ridge-and-valley landscape in the Cambro-Silurian sedimentary rocks of the Oslo region).

There seem to be two alternative interpretations:

(1) The preglacial, and possibly also interglacial and interstadial, subaerial processes have produced structurally controlled forms, which glacial erosion has not been active enough to transform. Removal of talus may have been significant, exposing bedrock to subaerial processes after deglaciation. A completely postglacial origin for these forms is, however, impossible.

(2) In places where structural control is pronounced, the rocks yield to ice action in the same way as to subaerial processes, and the forms produced by glacial erosion are then similar to the forms produced by subaerial processes, resulting in escarpments on resistant rocks, gentler slopes on less resistant rocks, cuesta-forms or ridge-and-valley landscapes on suitable structures, and so on.

VALLEYS

Many valleys show clear signs of glacial activity, but the exact importance of ice erosion for the development of specific forms may be difficult to define. In any study of the effects of ice erosion, we must also take into account the effects of other processes acting before, between and after the glaciations. The general cooling of climate in the Tertiary and the rapid and severe changes of climate that took place many times during the Quaternary, have been important factors. The same is true of the uplift in Tertiary to Quaternary times to which many areas, later glaciated, were subjected.

The decrease in temperature in the Tertiary meant that frost weathering became a major agent in the denudation systems, while the importance of chemical weathering decreased. There were also associated changes in vegetation and in hydrological conditions. Uplift and frost weathering probably resulted in vigorous stream incision before glaciation. Thus, in uplifted areas, glaciers would continue the work of powerful preglacial erosion, and would work in alternation with interglacial and interstadial subaerial processes. Owing to the rapid environmental changes, an equilibrium between agents and forms would rarely be established. A particular landform having started to develop under the influence of one agent, would be changed by another. The present valley landscape, therefore, is polygenetic, composed of elements produced by different agents or combinations of agents.

In theory we might find remnants of forms produced during different stages of valley development:

(1) Remnants of the oldest valley forms, belonging to those still remaining on upland surfaces and in mountain areas outside the incised valleys. In Norway this type of land surface has been called the 'paleic' surface (or the old surface). It is composed of rounded, smooth forms that are more or less well preserved. This land surface was probably mainly produced before Tertiary cooling of climate and possible uplift started more vigorous valley cutting.

(2) Remnants of valley forms produced in preglacial times as a result of up-

lift and/or change of climate initiating frost weathering with mass-wasting and stream incision.

(3) Forms produced directly by ice erosion.

(4) Forms produced in interglacial and interstadial periods and in postglacial time by subaerial processes such as frost weathering, mass-wasting and stream erosion, being a continuation of preglacial erosion and an 'indirect' effect of the glacial production of forms.

An important analysis of the development of valley 'generations' was carried out in the Alps by H. Annaheim (1946). In contrast to former interpretations, he found that a considerable part of the volume of the valleys below the valley shoulders, or edges of the entrenched valleys, was of preglacial origin. Only the deeper parts of the valleys appeared to have been cut during glacial and interglacial periods. No conspicuous differences could be detected between the preglacial and glacial parts of the valleys (Fig. 7.1). The most important change of erosional effect seems to have taken place when incision of the young valleys started. The reason for this was probably uplift of the Alps or a change in the denudation system, or both. The introduction of glacial activity seems in some respects to have been of less importance. Similar analyses need to be carried out in other areas, but in most cases genetic and chronological data are lacking.

Fjords and fjord valleys

Fjords and fjord valleys possess forms mostly produced as a result of the direct and indirect effects of glacial erosion. At the coast and along the outer parts of the fjords, it is common to see gentle and rounded ('paleic') landforms and lowland areas near sea level. Where fjord basins have been cut, remnants of the lowland and the old valley floors are left as benches along the fjord. Such a situation is found, for instance, in the lake region of northern Italy. Overdeepened fjord and lake basins can only have been produced by ice erosion. Since they have not been exposed to subsequent subaerial erosion, their present forms depend entirely upon glacial action and structural control. Most probably the deep fjord basins have been formed in zones of more densely-jointed rocks (*cf.* J. P. Bakker, 1965; J. Gjessing, 1966a) where preglacial weathered material may have been removed and jointed rocks quarried by glaciers.

Farther inland, narrow sections of the valleys have ice-modified sides; in wider sections, old slopes still exist. Where the old landforms are found at higher levels, young preglacial valleys have been cut into the surface below the floors of the basins. The first part of this incision was probably preglacial, but later glacial erosion has also played a part. Benches along the inner parts of the fjords are probably remnants of the gently-sloping floors of entrenched valleys of glacial type that existed before the fjord basins were produced. At the head of the fjords, gently-sloping valley sections are often found. Then follow more steeply-sloping sections with signs of glacial modification, where fluvial erosion is vigorous at the present day.

Steps in the longitudinal profile

High steps or valley heads are prominent features in many glacial valleys. Massive rocks, possibly with vertical and horizontal structures, and rocks with structures dipping up-valley and giving rise to lee sides where ice plucking can take place,

FIG. 7.1. Glaciated valley. Valley shoulder indicates a change in denudation processes, but not necessarily a change from stream incision to glacial erosion. The landscape beyond the valley is comparable with the 'paleic' landforms of Norway, altered by cirque glaciers and frost shattering.

FIG. 7.2. A glacial valley with prominent steps and waterfalls, in massive granite. Yosemite valley, California.

seem to facilitate the formation of steps. It is probable that basal plucking will increase the steepness and height of steps (Fig. 7.2). At present, steps are places undergoing vigorous attack by stream erosion, leading to the cutting of canyons of adjustment. This in itself indicates that such steps are of glacial origin. By alternating ice and stream erosion, great volumes of rock are removed, steps and steep valley sections retreat up-valley and the deeper valley sections below the step are extended inland.

Hanging fjords and hanging valleys bear witness to the glacial deepening and transformation of the main fjords and valleys. The problem is to reconstruct the preglacial valley, to determine the part played by ice erosion and to estimate the amount of erosion during glacial and interglacial periods. The most deeply entrenched valleys are found where the overall land surface rises most rapidly, from coast to upland or from upland to mountains. In some valleys headward erosion by ice and streams has been so vigorous that watersheds have been breached and tributaries of streams running in the opposite direction have been captured.

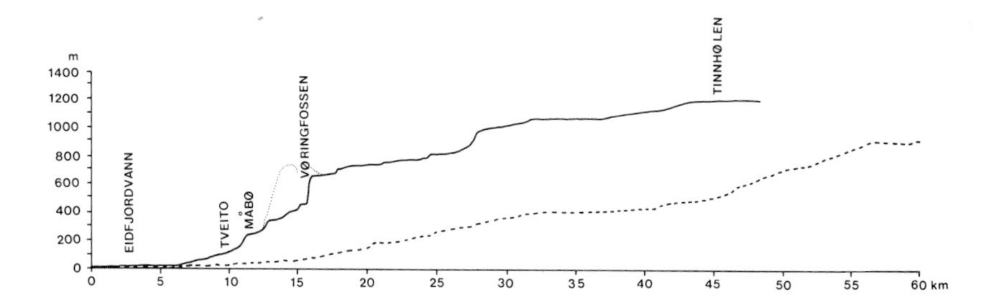

FIG. 7.3. Longitudinal profiles of two valleys in Western Norway. The upper profile has a gently sloping section, a steeper section and a valley head (with the waterfall Vøringfossen). The general trend of the lower profile is determined by stream erosion (from Gjessing, 1956).

In glaciated areas there are also valleys where stream erosion seems to have been more important than ice erosion (Fig. 7.3). This may be due to duration and intensity of ice erosion in relation to stream erosion but lithology is probably the main factor, the streams cutting deeply in more erodible bedrock.

Basins and narrow sections

Basin or trough-shaped sections with U-shaped cross profiles and gently sloping floors, sometimes with a lake or a floodplain, are distinctive features of glaciated valleys. Often, sections of this type alternate with narrow, more steeply sloping, sometimes zigzag-patterned sections with interlocking spurs and V-shaped or canyon-like cross profiles. It seems natural to think that the basins or troughs have been formed by ice erosion, while the narrow reaches are mainly products of stream incision (Fig. 7.4). The alternation between troughs and canyons may to some extent represent an inheritance from the preglacial relief, but structural control has probably been of vital importance. Where a valley has been cut

FIG. 7.4. Basin and zigzag-pattern section with interlocking spurs. The longitudinal profile of the river has only minor steps. Hanging valley in the background. Jostedalen, western Norway.

across structures of resistant rocks, a narrow section has most likely evolved.

A flood plain will protect the floor of a basin, but weathering and lateral stream erosion will attack the rocks along the foot of the valley sides, a process which tends to widen the basin even under subaerial conditions. As the interglacial and interstadial periods taken together have been of considerable duration, it is necessary to take this into account in explaining the development of the basins.

The narrow and more steeply sloping valley sections are at present attacked by vertical stream erosion. The question is, however, what will happen to the narrow, zigzag-patterned and V-shaped sections during glaciation? Will they remain untouched, will they be modified by ice erosion, or will they be further developed by ice plucking? A subglacial fluvial hypothesis has also been suggested, but evidence is often difficult to find.

U-shaped cross profiles

A U-shaped cross profile is often taken to be a classical sign of glacial erosion. The valley sides are steep, regular and parallel. The valley floor is wide and flat or curving, sometimes overdeepened to form a lake or fjord basin, or filled with debris forming a flood plain. A pre-requisite for the formation of a regular U-shaped valley is homogeneous, massive rock, with mainly vertical and horizontal structures. A U-shaped cross profile is likely to be produced if ice erosion increases with the depth of the glacier, owing to greater pressure and higher velocity (Gjessing, 1956). U-shaped valleys have often been regarded as 'mature' forms of glacial erosion, representing a stage when the ice can flow without further changing the valley form.

Mathematical functions. It has been suggested by several writers that a U-shaped cross profile can be described by a mathematical function, for instance a power or exponential function suitable for correlation and regression after transcription to logarithmic or semi-logarithmic form. Such a procedure is useful for comparison between different cross profiles, but a function is of real value only if a physical explanation can also be given. For instance, if it can be shown that a glacier, under given conditions, will produce a cross profile of a certain mathematical form in homogeneous rocks of certain types, the particular glacial conditions and rock properties might be expressed by a particular set of constants and exponents. Deviations from a function of this type would appear as noise which might be explained by the influence of other agents or by a particular structural control.

The possible significance of *subaerial processes* during interglacial and interstadial times should not be overlooked. The sides of some U-shaped valleys seem to bear traces of subaerial slope development, which have not been obliterated by glacial erosion. A steep valley side constitutes a free face exposed to weathering. Under subaerial conditions, talus will build up on the lower part of the slope, causing a subtalus slope to develop and the rock wall to retreat. On many valley sides a veneer of talus can be seen on a pronounced subtalus slope, which seems to be too well developed to have been formed in postglacial time. It seems reasonable to conclude that the subtalus slope results from mass-wasting and talus formation before the last glaciation, and that the form has been preserved without much glacial change except the removal of talus during the last (or most recent) glaciations. When the last ice melted, subaerial development began again.

Valley-side benches

It is quite common for glaciated valleys to show complex cross profiles. In certain cases, the irregularities appear to be benches which can be followed for varying distances along the valley side. There seem to be two possible explanations of such stepped profiles:

(A) In rocks with pronounced horizontal or gently sloping structures, valley-side benches may be entirely controlled by structure, cut out by differential glacial erosion.

(B) The benches may be remnants of old valley floors, of different ages and produced by different agents:

- (1) Remnants of wide valley floors belonging to the old 'paleic' surface.
- (2) Remnants of valley floors dating from preglacial times.
- (3) Remnants of a series of old valley floors of alternating
  - (a) glacial and
  - (b) interglacial and interstadial origin.

In most cases there is insufficient evidence for a genetic and chronological analysis of elements belonging to these stages.

If benches are remnants of old valley floors, there is the question of why they have been preserved and why the valley floor has become narrower as the valley was incised. The reason for incision along the axis of a valley may be a zone of weakness, most probably a zone of closely-jointed rock. Incision may have been effected by streams cutting into valley steps or into valley floors. However, certain conditions seem to point to glacial incision as in the case of fjord and lake basins. It can often be seen that tributary streams are about to cut 'canyons of adjustment' (Gjessing, 1956) through valley benches, probably because of recent lowering of local base level, which is the valley floor or the talweg of the main valley.

Tributary valleys

Hanging tributary valleys have been regarded as good indications of ice erosion (Fig. 7.4), and tributary streams cascading down the side of a main valley are a common feature in glaciated valleys. The main valley floor is the local base-level of the tributary streams. Lowering of this floor by ice erosion will increase the erosional power of the tributaries. If the bedrock is not too resistant, a notch or canyon of adjustment will be cut in the opening of the tributary valley (Fig. 7.5). Glacial plucking may increase the depth of hanging valleys, but glacial erosion has obviously not widened the deeper parts enough to make it U-shaped. In this connection we must bear in mind that even in areas that have been glaciated, V-shaped or canyon-shaped valleys are in the process of being formed by stream erosion and mass-wasting.

Forms reminiscent of rock fans are probably produced in periods before and between glaciations in connection with the formation of alluvial fans at the mouths of some tributary valleys.

FIG. 7.5. Tributary river in the process of cutting a canyon of adjustment at the mouth of a hanging valley. Leirdøla, Jostedalen, western Norway.

CIRQUES

The movement and erosional processes associated with cirque glaciers have been much investigated, but there are still several questions remaining, particularly in relation to the overdeepening of cirque floors, and to the combined action of weathering and ice erosion. When the climate changed during the Quaternary, the firn line must have moved up and down many times. With a low firn line one might expect the lower hills to provide the sites of cirque glaciers, but this is not always true. This suggests that amounts of snow great enough to produce cirques are caught only by the highest mountains. The development of cirques with a preferred lee-side orientation is related to the great snow accumulation by wind drift on the lee sides of hills, combined with the fact that snow is blowing off the windward side, making a pronounced difference in snow accumulation at the same altitude. Exposure to radiation may increase the difference. The snow cover on the lower hills has probably been more even. A study of wind conditions, snow drift and snow accumulation at the present day would shed light on these problems.

THE STRANDFLAT

The so-called strandflat is found on the coasts of some areas that have glaciers today or have been glaciated in the past, and therefore in some respects it should be regarded as a part of the glacial complex.

The strandflat is generally a low and uneven rocky plain with a shallow submarine continuation, ranging between 50 m above and 50 m below sea level. There is a variable cover of superficial deposits. A typical feature is the pronounced knick at its inner margin and a steep slope up to higher land which may have rounded and smoothed forms.

Rock type and initial relief along the coast have probably been of great significance in the development of the strandflat. Glacial incision of fjords and sounds has been emphasized (F. Nansen, 1922). The existence of a preglacial 'paleic' landscape of rounded hills and basins, with valley floors close to or at sea level, would constitute an excellent initial form for the development of a strandflat (Gjessing, 1967a).

Norwegian geomorphologists seem to agree that frost weathering and wave action have been the main agents in the development of the strandflat. A problem is what part glaciers have played. One theory is that cirque glaciers have eroded headwards into the highland rising above the strandflat, the coalescent cirque floors becoming parts of the strandflat (E. Dahl, 1946; H. Holtedahl, 1960). A similar theory was adopted by K. M. Strøm (1948) for bevelled surfaces surrounding certain high mountains. Another theory is that foreland (piedmont) glaciers have levelled the surface (O. Holtedahl, 1929). As general experience seems to be that ice more readily attacks less resistant rocks in the depressions, increasing the irregularities, a primary origin by other agents seems more likely. However, glaciers will remove debris and expose bedrock to the action of marine and subaerial processes.

Glacial erosion acting along steep slopes may make them retreat, and thus also help to expand the area of the strandflat.

GLACIAL DEPOSITION

The glaciated areas in mountains and areas of ice-sheet glaciation can be divided in two, a central area where erosion prevailed, and a peripheral area where accumulation was dominant. Both areas possess distinctive forms, in both bedrock and superficial deposits. During deglaciation, frontal deposits and fluvioglacial dead-ice deposits often form in the central area where erosion earlier prevailed. The volume of deposits in the peripheral areas will give some indication of the volume of debris removed from the area of prevailing erosion.

In 1879, A. Helland estimated the volume of Quaternary deposits in the south-eastern part of the North European lowland, and compared this volume with the area of erosion, which he assumed had been the Baltic Sea, Finland and Sweden. His estimate was that on average 30 m of Quaternary deposits cover an area of 2×10^6 km^2, which gives a volume of about 60 000 km^3. This volume would fill the Baltic Sea plus all the lakes in Finland and Sweden, and after that 20 000 km^3 would still be left. Spread evenly over Finland and Sweden, it would form a cover 25 m thick. The debris does not have as high a density as that of rock, but this is probably compensated by material exported by streams. The valleys have probably contributed much more material than the rest of the land but as the areas of the valleys are quite small they might easily be filled and there would still be a considerable volume remaining to represent general 'denudation' of the land surface.

Better calculations might be made today. Estimates of erosion in fjords and fjord valleys would be possible if information on the distribution, thickness and types of the Quaternary deposits on the continental shelf were available. In this case one might be able to see what has been derived from the different fjords and the different sections of the coast.

The origin of the Norwegian Channel, which is 700 m deep in its deepest part, remains a problem. It has been claimed that it is a graben (O. Holtedahl, 1956), but the evidence also seems to suggest ice erosion. In this case a considerable volume of rock, which has contributed more to the deposits in the North Sea and on the continental shelf than any single fjord or fjord valley, has been removed.

FLUVIOGLACIAL EROSION AND DEPOSITION

The production of fluvioglacial forms in the preglacial zone was determined by the source of debris and the drainage which partly followed courses determined by the ice. Sandur plains, eskers, kames and other related features reflect the abundant transport and accumulation of debris by glacial rivers. There are, however, problems related to the reconstruction of the courses of glacial drainage and to fluvial erosion before accumulation took place.

Englacial or supraglacial drainage. An ice sheet probably does not normally contain much englacial moraine. Judging from the prolific fluvioglacial deposits in the peripheral zones of glaciated areas, a hypothesis of englacial or supraglacial drainage would imply an abundant supply of debris from the lowest to the higher parts of the ice, a distribution which might have arisen because of internal transport along shear planes near the ice front, where debris could be picked up by the englacial or supraglacial streams. Proofs of such a hypothesis may, however, be difficult to find.

Subglacial drainage. The existence of eskers with undisturbed bedding indicates subglacial drainage. Water running under the ice would normally have plenty of till from which to pick up material. It is of interest to find out how far behind the ice front water descended from the ice surface to the bottom, and how it flowed under the ice. It has proved difficult to find evidence on locations where the water descended, to trace water courses, and to find signs of fluvial erosion in areas behind the former positions of the ice front.

Tunnel valleys. Closely connected to the problem of glacial drainage is the problem of 'tunnel valleys'. The classical explanation has been that they were formed by subglacial streams. Some writers think that this kind of valley has been formed by ice erosion and later occupied by subglacial drainage (P. Woldstedt, 1952). In places where the head of a sandur plain is found at the terminus of a tunnel valley, it seems reasonable to infer that the feeding stream followed the tunnel valley under the ice, and that it ascended, carrying debris with it, to flow out on the sandur plain. This suggests that a considerable part of the glacial drainage followed valleys and that this drainage was probably subglacial, a conclusion which is in accordance with the existence of eskers in the area immediately behind the ice border.

In cases where the apices of sandur plains are not situated at the termini of tunnel valleys, reconstruction of the courses of the feeding streams is more difficult. One can not exclude the possibility that part of the glacial drainage was less dependent upon the forms of the subglacial land surface. However, later events might also have obliterated possible connections which may have existed at earlier stages of development.

Fluvioglacial activity during recession

Problems similar to those in the peripheral zones of glacial areas are related to fluvioglacial activity during the period of deglaciation. Owing to glacio-isostatic depression, much lowland in south-eastern Norway was inundated by the sea as the ice front receded. Re-advance moraines are mainly found on higher ground and in minor valleys. Glacial drainage was generally directed in accordance with the slope of the ice surface (Gjessing, 1960), and great deposits of fluvioglacial material in the major valleys indicate that glacial drainage, to a great extent, followed the courses of the present streams. The bedload was deposited at the ice front but sedimentation of suspended material took place in the sea, forming large areas of clay and silt.

The great amounts of fluvioglacial debris can only have been derived from ground moraine and therefore it seems most natural to infer that the feeding streams were subglacial. However, it has been difficult to find traces of such drainage, of the places where water descended from the surface to the bottom of the ice, and where the areas of erosion have been. Water might have penetrated through the ice, or between the ice and the ground, when the hills and mountains were freed from ice by downmelting.

Dead-ice wastage

During deglaciation of some areas the last parts of the ice sheet became inactive and melted down as dead ice. Great drainage systems came into being and during deglaciation prominent forms were produced by fluvioglacial erosion and deposition. The drainage divide was at the highest part of the ice surface, and the drainage upon the surface, along the ice margins, and within and under the ice, ran in the directions of the ice-surface slope.

Drainage started upon the ice from ice-melt, snow-melt and rain, and drainage also came from ice-free mountain areas. The water followed the ice margins and tended to pass more or less obliquely down the valley sides between the ice and the bedrock, as sub-marginal and subglacial drainage. The water continued in tunnels along the valley floor, either up-valley or down-valley, in a direction determined by the slope of the ice surface. By the melting effected by the subglacial rivers, room was provided for deposition of debris.

Fluvioglacial erosional forms are found at various levels on valley sides. Deep canyons, bedrock stripped of its till cover, meltwater channels and other signs of vigorous erosion may be found. Forms of fluvioglacial deposition, such as kame terraces and eskers, are mainly found in the lower parts of the terrain. Altogether, the fluvioglacial forms indicate the courses of the drainage and the transport of sediment from the proximal to the distal parts of a drainage system (Fig. 7.6). In some places drainage courses can be followed from high levels in the mountain areas down to some hundreds of metres below what must have been the level of the ice surface, continuing farther on along valley floors, to escape through cols in the present watersheds (Gjessing, 1966b). Where the ice surface sloped up-valley, special conditions were presented. The supraglacial streams crossed watersheds through cols at successively lower altitudes as the ice surface melted down. The subglacial drainage, however, following the valley floor, had to ascend to cols to escape from the area.

Friction in the tunnels, transport of debris and the ascent of water and debris to cols must have required considerable energy. This can only have been produced by a sufficiently great head, or difference in altitude between the inlet and outlet of the tunnel system. The tunnels were probably kept open by the melting and by the pressure of the water, because it has a greater density than ice. One problem is how the tunnels first developed. Subglacial water flow or ground-water flow, determined by pressure differences according to the slope of the ice surface, must have existed before deglaciation started. This subglacial drainage must have developed into systems which could take great amounts of meltwater and debris.

During the later stages the drainage seems to have been less powerful. There are some problems particularly in connection with the existence of benches or 'parallel roads' at the levels of cols, and the origin of fine-grained sediments interpreted as glacio-lacustrine. It seems clear, however, that many of the benches that have been called 'parallel roads' ('sete' in Norwegian), are built of fluvioglacial debris. Even kettle-holes are found in them and, from the sides and distal parts of the benches, eskers run obliquely downhill to the bottom of the valley, in the direction of the general fluvioglacial drainage. In such places benches represent the upper limits of fluvioglacial deposition. At the same time, benches have been formed at the downward limits of fluvioglacial erosion on

FIG. 7.6. Accumulations of fluvioglacial material with knob-and-kettle relief and terraces. Water escaped through the canyon which starts in the lower right-hand corner of the picture, and later moved along the valley towards the left. Døràlen, Rondane, central-south Norway.

valley sides. (A good example is found at the outlet of Loch Treig in Scotland). Many of the former type of bench have been formed at the level of a col where the drainage escaped, the level controlling both fluvioglacial deposition and erosion (Fig. 7.7). A col blocked by dead-ice forms, eskers and terraces with kettle holes, indicates that no open lake drained directly through the col.

Parallel roads or benches at the levels of cols have generally been interpreted as strand lines (beaches) formed in ice-dammed lakes. The idea was first applied to the 'parallel roads' of Glen Roy in Scotland (T. F. Jamieson, 1863, *cf.* also J. B. Sissons, 1967). However, it is rare for the deposits or forms to provide conclusive evidence that a beach was produced by wave action under lacustrine conditions. Solifluction processes down to the water level or ice margin might explain some benches.

On the other hand, kame-terraces with foreset bedding indicate deposition in standing water. Such ice-margin lakes seem to have formed near places where water carrying sediment entered the ice. If the melting effect of the water made more room than the volume of sediments supplied, a lake would form; thus, the extent of standing water depended on the ratio between melting and the rate of debris supply.

Fine-grained sediments have been regarded as good indications of ice-dammed lakes, but fine-grained sediments may be deposited in running water when no coarser sediments are left after a long sorting process. The problem is, therefore, to find indications able to determine whether fine-grained sediments have been sorted by stream transport or in standing water. In some places, fine-grained sediment with undisturbed stratification is found in ridges, or nets of ridges, with kettle holes, like esker nets. Sediments of the fine-grained type are also found with ablation moraine on top, evidence that points to subglacial deposition.

The debris budget. The size of the erosion and deposition systems indicates that enormous amounts of debris have been transported and sorted by fluvioglacial activity. Only a fraction of the debris appears to have come from fluvioglacial erosion in bedrock: most is derived from ground moraine stripped off by the fluvioglacial drainage. In some areas, sediments have been carried out during deglaciation from tributary valleys which must at one time have been filled with debris. We do not know if the material now seen as fluvioglacial sediments was originally produced from bedrock during the last glaciation, or if much of it originated in earlier glaciations and remained on slopes or in valleys during the last interglacial or interstadial times. The high degree of sorting of some of the sediments also raises the question whether some of the debris was already sorted before the last glaciation. In a few places, sediments have been found which have not been moved during the last glaciation (for instance, in the Alps and Fennoscandia: J. Mangerud, 1965; O. F. Bergersen & K. Garnes, 1971; A. Heintz, 1971).

THE PRODUCTION AND TRANSPORT OF DEBRIS FROM THE CENTRAL AREAS OF GLACIATION WITH PREDOMINANT EROSION TO THE PERIPHERAL AREAS OF DEPOSITION

Most areas where glacial erosion prevailed during the Quaternary were already areas of prevailing subaerial erosion in the Tertiary (such as Fennoscandia,

FIG. 7.7. ‘Parallel road’ built of fluvioglacial material with ridges trending obliquely downhill towards the right. Fluvially eroded and washed rock above. Water came across the ridge and went obliquely down from left to right. Grimsdalen, central-south Norway.

Scotland, the Alps and Canada). Therefore, in most glaciated areas, ice erosion took part in and probably increased the rate of denudation.

We must assume that weathered material and alluvial deposits were present in these areas when glaciers started to work, and that this debris was the first to be re-worked and exported. During the glaciations, debris was produced by ice quarrying and scouring, and the interglacials and interstadials gave ample time for frost weathering, non-glacial slope processes and fluvial action to sort and re-distribute the material.

It seems unlikely that conditions since the last deglaciation are unique. On most areas between the valleys, only a relatively thin and partly discontinuous till cover has been left. This till has been slightly weathered chemically, but we must assume that, as in postglacial time, there was little transport of material in these areas during interglacial and interstadial periods. It is of interest to know how far till has been transported during each glaciation, where new till has been produced, and if it was produced at the same rate as it was removed.

Judging from the present, it is reasonable to assume that earlier glaciations also left fluvioglacial debris in the valleys and marine deposits on the lowlands. In interglacial and interstadial periods, till and sorted material must have been moved and re-deposited, by mass movement on valley sides and by fluvial action along valley floors, just as at present. It is of interest to know how far the material has been transported during an interglacial or an interstadial period.

The ice is thicker and erodes more vigorously in valleys, resulting in greater production of debris as well as greater transport. Fluvial material in the valleys and marine deposits on low land will become sources of new till. Different sediments will be mixed and transported towards areas of prevailing deposition. It seems certain that the valleys have been the major sources of production of sediment, and the major routes of glacial and subaerial transport, but the relative importance of preglacial, glacial, interglacial or interstadial and postglacial activity is not known.

SUBGLACIAL EROSION: PLASTIC SCOURING FORMS

Finally I wish to consider some problems relating to the production of so-called 'plastic' scouring forms. Several partly contradictory theories have been advanced. This serves to illustrate the situation when theories have to be formulated without support from observations of the actual processes involved. Formation by one substance or a mixture of substances has been suggested, as well as the action of several substances one after the other. The substances that might be involved are:

Ice containing abrasive material
Running water with abrasive material
Cavitation by rapidly flowing water
Running water first, then ice (with abrasive material)
Ice-water mixture (with abrasive material)
Water-soaked till
Ice impregnated with till

An interesting problem is associated with the formation of *sichelwannen* (sickle-shaped troughs (Fig. 7.8)) and their varieties. Formation by cavitation, caused by water flowing at high velocity between the ice and the rock surface,

FIG. 7.8. *Sichelwanne*. Ice moved along the axis of the trough, obliquely upwards in the picture.

has been suggested (E. Ljungner, 1930; F. Hjulström, 1935). Depressions of similar form can be produced by flowing water in loose sand or by air over snow. The *sichelwannen* and other plastic scouring forms, however, lack conclusive signs of cavitation.

The polished and scoured stoss sides of roches moutonnées can be explained by pressure-melting and re-freezing as the basal ice armed with abrasive material moves over the rock, fitting closely to the surface. Lateral squeezing of ice, due to increased pressure at steep stoss sides, has been suggested by J. L. Andersen and J. L. Sollid (1971). The apparent lack of scouring on lee sides can be explained by the slow plastic reaction of ice, reducing the pressure upon the rock surface or making it bridge over the lee side.

More difficult is to explain sinuous grooves, with fine parallel striations, curving both in vertical and horizontal directions. The fine parallel striae are a sign of a laminar flow in the moulding substance. It has been suggested that such sinuous grooves have been produced by running water (R. Dahl, 1965). On the other hand, rock surfaces in present streams do not show continuous parallel striae. Only the characteristic facets produced by turbulent flow with abrasive material are seen. Water can exhibit laminar flow at very low velocities. However, such velocities would not be capable of transporting abrasive material such as sand grains. Grooves or striations like those seen in the field cannot, therefore, be produced in this way.

E. Ebers and Stefaniak (in H. Flohn (ed.), 1961) have postulated that the grooves have first been produced by running water and then striated by glacier ice flowing plastically along their winding courses. An objection to this theory is that no signs of erosion by turbulent water are seen. If, however, we could find a substance able to follow the bends closely, moving in laminar flow, and to produce the striations as the grooves were being made, it would then be unnecessary to invoke a hypothesis of a 'double' history, involving first grooving and then striation. A single-stage hypothesis seems to fit the field evidence better.

Potholes are often seen in connection with plastic scouring forms. We know from present rivers that potholes can be formed by running water. But if another scouring substance were able to flow in the same manner, it would be reasonable to conclude that such a substance could produce similar forms (Gjessing, 1967b).

On rock surfaces with plastic scouring, lee sides showing striae and other scouring forms deflecting to directions nearly at right angles to the general ice movement over the area (Fig. 7.9) may also be seen. As mentioned, plastic deformation of ice by increased pressure can explain ice movement and scouring on stoss sides. But an explanation of plastic scouring on lee sides by a theory involving plastic deformation of ice due to increased pressure seems implausible. It seems more reasonable to look for an abrasive substance that can move separately under the glacier, being plastic or viscous enough to exhibit laminar flow. Under a glacier, the direction of the general pressure gradient will coincide with the direction of slope of the ice surface. Exposed to subglacial differences in pressure, a substance between the ice and the rock surface will tend to move. In addition to the general pressure field the actual flow will be determined by pressure components resulting from differences in pressure exerted by the ice upon the rock (Fig. 7.10). As a consequence, decreased pressure at the lee sides and over depressions in the rock surface will give local deviations in the general

FIG. 7.9. Lee-side deflection of ice scouring owing to pressure release. Deflection in a direction away from the observer. The main scouring direction over the area is seen in the upper right-hand part of the picture. Coast of south Norway.

pressure field and will provide passages for a subglacial substance. In this way the courses of the sinuous grooves and related features can be explained (Gjessing, 1966c).

We see that some theories depend on an increase in ice pressure to explain plastic scouring forms, while the last-mentioned theory assumes a decrease in ice pressure. Observations at present glaciers may provide some clue. The field geomorphologist needs help to find the eroding substance and the conditions of flow. Physicists understand the properties of different materials, and in a laboratory the properties of different substances can be tested, and the flow conditions and erosional processes can be simulated.

We also face another problem. Literature on the properties of different materials and their behaviour under stress is written for specialists in these fields, not for geomorphologists and not for solution of geomorphological problems. Therefore, to solve the field problems, glacial geomorphologists need to cooperate with specialists in other disciplines.

CONCLUSION

I hope that this discussion illustrates some of the kinds of problem with which glacial geomorphology is working. For further progress, organisation of knowledge is important. Classification of glacial forms, or definition of form types, as well as the formulation of general principles or 'laws', call for international cooperation, bringing together experience from different glaciated areas. Textbooks on glacial geomorphology, such as C. Embleton and C. A. M. King (1975),

FIG. 7.10. Sinuous grooves and related forms on plastically sculptured steep, oblique lee-side, probably owing to pressure release. Main direction of ice movement from lower left to upper right in the picture. Fischbach am Inn, Germany.

help, but in such books the terminology depends fundamentally on the language used.

In a study of the effects of glacial activity in a landscape, we must also understand the parts played by other agents. An analysis of the whole landscape is therefore necessary. This requires a considerable knowledge of other branches of geomorphology.

Most conclusions on the genesis of glacial forms have to be based on a study of the forms after the glaciers have retreated. There are few chances of testing the conclusions in places where the processes are actually at work. Observations of conditions and processes at present glaciers, combined with glaciological studies, are therefore of vital importance.

Glacial processes are difficult to quantify. Measurements of the loads carried by glacial streams give estimates of rates of erosion (G. Østrem, 1973; K. Nordseth, 1974), but do not indicate the genesis of particular forms.

A 'field geomorphologist' knows how to collect field evidence for his conclusions, he understands the techniques of geomorphological mapping, air photo interpretation and so on. In addition, we need specialisation in, or cooperation with specialists in, glaciology, sedimentology, fluvial geomorphology, physics, hydrodynamics and hydraulics. Testing the properties of relevant substances and the simulation of processes in the laboratory will be of vital importance.

Theoretical considerations are also necessary, but the most beautiful models are of limited value if we fail to understand their significance for the field problem.

REFERENCES

Andersen, J. L. and Sollid, J. L. (1971), 'Glacial chronology and glacial geomorphology in the marginal zones of the glaciers Midtdalsbreen and Nigardsbreen, South Norway', *Norsk geogr. Tidsskr.* **25**, 1–38.

Annaheim, H. (1946), 'Studien zur Geomorphogenese der Südalpen zwischen St Gotthard und Alpenrand', *Geogr. helv.* **1**, 65–149.

Bakker, J. P. (1965), 'A forgotten factor in the interpretation of glacial stairways', *Z. Geomorph. N.F.* **9**, 18–34.

Bergersen, O. F. and Garnes, K. (1971), 'Evidence of sub-till sediments from a Weichselian interstadial in the Gudbrandsdalen valley, Central East Norway', *Norsk geogr. Tidsskr.* **25**, 99–108.

Boulton, G. S. (1975), 'Processes and patterns of subglacial sedimentation: a theoretical approach' *in* Wright, A. E. and Moseley F. (eds.), *Ice ages: ancient and modern* (Seel House Press, Liverpool), 7–43.

Dahl, E. (1946), 'On the origin of the strand flat', *Norsk geogr. Tidsskr.* **11**, 159–72.

Dahl, R. (1965), 'Plastically sculptured detail forms on rock surfaces in northern Nordland, Norway', *Geogr. Annlr, Ser. A.* **47**, 83–140.

Embleton, C. and King, C. A. M. (1975), *Glacial Geomorphology* (Edward Arnold, London), 573 pp.

Flohn, H. (ed.) (1961), 'Der Gletscherschliff von Fischbach am Inn', *Landesk. Forsch.* **40**, 43–85.

Geo-Abstracts, ed. by K. M. Clayton, University of East Anglia, England.

Gjessing, J. (1956), 'Om iserosjon, fjorddal- og dalendedannelse (On ice erosion, fjord-valley and valley-end formation'), *Norsk. geogr. Tidsskr.* **15**, 243–69.

—— (1960), 'Isavsmeltningstidens drenering (The drainage of the deglaciation period)', *Ad Novas* 3, 492 pp.

—— (1966a), 'Some effects of ice erosion on the development of Norwegian valleys and fjords', *Norsk. geogr. Tidsskr.* **20**, 273–99.
—— (1966b), 'Deglaciation of southeast and east-central south Norway', *Norsk geogr. Tidsskr.* **20**, 133–49.
—— (1966c), 'On "plastic scouring" and "subglacial erosion"', *Norsk. geogr. Tidsskr.* **20**, 1–37.
—— (1967a), 'Norway's Paleic surface', *Norsk. geogr. Tidsskr.* **21**, 69–132.
—— (1967b), 'Potholes in connection with plastic scouring forms', *Geogr. Annlr Ser. A*, **49**, 178–87.
Heintz, A. (1971), 'Mammut-funn fra Norge', *Fauna* **24**, 173–86.
Helland, A. (1879), 'Uber die glacialen Bildungen der nord-europäischen Ebene', *Z. dt. geol. Gesell.* (1879), 63–106.
Hjulström, F. (1935), 'Studies of the morphological activity of rivers as illustrated by the river Fyris', *Bull. geol. Instn. Univ. Upsala* **25**, 221–527.
Holtedahl, H. (1960), 'The Strandflat of the Møre-Romsdal coast, west Norway', *Skr. Norg. Handelsh. Geogr. Avh.* **7**, 35–43.
Holtedahl, O. (1929), *On the geology and physiography of some Antarctic and sub-Antarctic islands* (Det Norske Videnskaps-Akademi, Oslo), 172 pp.
—— (1956), 'Junge Blockverschiebungstektonik in den Randgebieten Norwegens' *Geotektonisches Symposium zu Ehren von Hans Stille,* 55–63.
Jamieson, T. F. (1863), 'On the parallel roads of Glen Roy, and their place in the history of the glacial period', *Q. J. geol. Soc. Lond.* **19**, 235–59.
Linton, D. L. (1963), 'The forms of glacial erosion', *Trans. Inst. Br. Geogr.* **33**, 1–28.
Ljungner, E. (1930), 'Spaltentektonik und Morphologie der schwedischen Skagerrak-Küste', *Bull. geol. Instn. Univ. Upsala* **21**, 478 pp.
Mangerud, J. (1965), 'Dalfyllinger i noen sidedaler til Gudbrandsdalen, med bemerkninger om norske mammutfunn', *Norsk geol. Tidsskr.* **45**, 199–226.
Nansen, F. (1922), 'The strandflat and isostasy', *Vidensk. Skr. I. Math.-naturv. Kl.* **11**, 313 pp.
Niini, H. (1968), 'A study of rock fracturing in valleys of Pre-cambrian bedrock', *Fennia* **97**, 5–60.
Nordseth, K. (1974), 'Sedimenttransport i norske vassdrag', *Dept. Geography, Univ. of Oslo.*
Östrem, G. (1973), 'Sediment transport in glacial meltwater streams', *Glaciofluvial and glaciolacustrine sedimentation* **23**, 101–22.
Sissons, J. B. (1967), *The evolution of Scotland's scenery* (Oliver & Boyd, Edinburgh and London), 259 pp.
Strøm, K. M. (1945), 'Geomorphology of the Rondane area', *Norsk. geol. Tidsskr.* **25**, 360–78.
Woldstedt, P. (1952), 'Die Entstehung der Seen in den ehemals vergletscherten Gebieten', *Eiszeit. Gegenw.* **2**, 146–53.

8

BRITISH GLACIAL GEOMORPHOLOGY: PRESENT PROBLEMS AND FUTURE PROSPECTS

R. J. PRICE
(*University of Glasgow*)

In order to establish the present status of any field of study it is necessary to look back over its recent development as well as to examine present activities within the field. The past decade has seen a great expansion in all aspects of geomorphology in Britain, and glacial geomorphology has been one of the major areas of development.

It is not easy to define the scope of glacial geomorphology. If it is defined as the description and explanation of the origins of landforms produced by glacial and fluvioglacial processes it includes two quite distinct realms of research activity. British geomorphologists have been active in studying both the landforms produced by the Pleistocene glaciations in the British Isles as well as the landforms currently being created by glaciers and ice sheets in polar and alpine environments. In these two types of work there is a great deal of overlap with glacial geology, glaciology and Pleistocene stratigraphy and chronology. In the discussion that follows, every attempt will be made to concentrate on the work that has been done on landforms, although it is inevitable that work in the allied fields already mentioned will have to be included from time to time.

Two publications are of great value in any attempt to review the activities undertaken by British geomorphologists over the period 1965–75. First, *Current Research in Geomorphology* published by the British Geomorphological Research Group provides a great deal of information about research projects being carried out by research students and senior research workers. Secondly, *Geo-Abstracts* provides an indication of the end-products resulting from these various research projects in the form of abstracts of published papers. Although an extremely useful publication, *Current Research in Geomorphology* has to be treated with some caution. Its contents depend on the successful completion and return of questionnaires and we all know how efficient academics are in this activity! Moreover, the research projects listed are the hopes and aspirations of the research workers rather than the end-product. The same project, too, may appear in several editions and therefore total figures for projects in any particular field of activity may be exaggerated. Bearing these reservations in mind, analysis of *Current Research in Geomorphology* showed certain trends for the period 1965–75 (Table I).

There has been steadily growing activity in glacial geomorphology over the

TABLE I

Analysis of research in glacial geomorphology, 1965–75 (based on Current Research in Geomorphology)

	Glacial geomorphological studies in the U.K.	Glacial geomorphology: existing glaciers	Glacial geology	Totals
		1965 (Total entries: 304)		
Master's theses	7	0	0	7
Doctoral theses	13	3	3	19
Other projects	18	6	11	35
Totals	38	9	14	61
		1970 (Total entries: 338)		
Master's theses	1	0	0	1
Doctoral theses	14	1	2	17
Other projects	30	12	17	59
Totals	45	13	19	77
		1975 (total entries: 420)		
Master's theses	5	0	0	5
Doctoral theses	15	8	4	27
Other projects	30	12	8	50
Totals	50	20	12	82

decade. In particular, there has been a considerable increase of interest in areas of existing glaciers by British-based glacial geomorphologists. Such work has been undertaken in the Alps, Norway, Iceland, Greenland, Baffin Island, Alaska, Spitzbergen and the Antarctic.

Several University departments, mainly geography departments, have built up strong research groups over this period. The most productive group in the early part of the decade was based in Edinburgh. J. B. Sissons and his group of research students have made numerous and important contributions to the glacial geomorphology of Scotland. C. M. Clapperton and D. E. Sugden in Aberdeen have been responsible for work in Scotland, Spitzbergen, Greenland, Iceland and the Antarctic. In Glasgow, work on the glaciation of west-central Scotland was paralleled by a research project in Iceland. In England, the University of East Anglia has developed a research group concerned with the processes of glacial erosion and deposition in association with glaciers in Spitzbergen and Iceland. Another long-term project has been organised by P. Worsley of Reading University, concentrating efforts in the Okstindan area of Norway. B. S. John has been the leader of a group of workers based in Durham with interests in north-east Iceland. It is perhaps invidious to list all the Universities that have undertaken research on the glacial landforms of Britain over the last decade. No one department could be described as the centre of excellence in this field of research. About a dozen departments have had small groups of workers concerned with glacial geomorphology.

The type of entry covered by Table I reveals two trends in British glacial geomorphology over the last 10 years. First, there has been increased activity in the detailed study of landforms and deposits created by the Devensian ice sheet. Large parts of Britain have now been mapped at a large scale to reveal the patterns of glacial landforms. Only a limited amount of this work has added much to our understanding of the origins of individual glacial landforms. Secondly, there has been a great increase in the amount of work done by British glacial geomorphologists in areas of existing glaciers. Perhaps one of the biggest advantages of these projects has been that they have enabled many British glacial geomorphologists to obtain experience of existing glacierised environments and therefore to be better equipped to interpret British Pleistocene landforms.

Table I also reveals that in each of the 3 years selected there were between twenty and thirty postgraduate students working in the field of glacial geomorphology. I do not have an accurate figure of the number of dissertations concerned with glacial geomorphology which were started during the decade nor do I have any information about the number actually completed. The Natural Environment Research Council of the U.K. (NERC) awarded nineteen research studentships in this research field between 1970 and 1975. My guess would be that well over 100 research students started work on theses on some aspect of glacial geomorphology between 1965 and 1975. The implications of this expansion in postgraduate research will be considered later.

Perhaps a more reliable measure of the research activity in a particular field lies in its publications. In this context, *Geo-Abstracts* provides the necessary information (Table II). Once again the decade 1965 to 1975 is the period covered but, of course, the time-lag between a piece of work being completed

TABLE II

Publications by British glacial geomorphologists

	Glacial geomorphological studies in the U.K.	*Glacial geomorphology: existing glaciers*	*Glacial geology*	*Total*	*Books*
1965	10	5	10	25	
1966	9	7	7	23	
1967	9	2	5	16	Sissons
1968	11	4	8	23	Embleton & King West
1969	5	3	4	12	
1970	6	3	2	11	Lewis
1971	6	2	5	13	
1972	9	2	1	12	
1973	6	0	0	6	Price
1974	10	5	5	20	Embleton
1975	11	3	8	22	Embleton & King Andrews
				183	papers + 7 books

and its subsequent publication makes comparisons with the information obtained from Table I difficult.

The output by British glacial geomorphologists over the decade has been considerable. The period 1969 to 1973 was rather lean but at present we seem to be benefitting from the expansion that took place in the numbers of research students in the late 1960s. It is perhaps rather invidious to pick out the more important publications during this period. Nevertheless, this was the period when modern textbooks relating to glacial geomorphology were produced by British authors. Sissons's book on *The evolution of Scotland's scenery* appeared in 1967 and contained a great deal of systematic glacial geomorphology. It was followed in 1968 by the first modern British textbook on our subject by C. Embleton and C. A. M. King, a new edition of which appeared in 1975. *The glaciations of Wales* (1970), edited by C. A. Lewis, added to the regional picture. *Glacial and fluvioglacial landforms* (1973) by Price concentrated on work carried out in areas of existing glaciers. *Glacial systems* by J. T. Andrews (1975) deals with both existing and Quaternary glacial environments, considering glaciers as open systems. D. E. Sugden and B. S. John in 1976 added another book to the list of those referred to in Table II, falling just outside the period of review. The research papers published between 1965 and 1975 either fell into the class of routine interpretations of glacial landforms in various parts of Britain or made some important contribution to our understanding of the origins of glacial landforms. Among the latter I would place Clapperton's papers on meltwater channels, V. Haynes's articles on cirques and troughs, the work of I. J. Smalley, D. J. Unwin and A. R. Hill on drumlins, and G. S. Boulton's papers on moraines and flow tills. As a by-product of some of the regional investigations, the chronology of the wastage of the Devensian ice sheet in Britain is now much better understood.

The sources of the funds used to undertake the above research were numerous. NERC, over the period 1970–75, provided £5000 for research in glacial geomorphology, £22 000 for the University of East Anglia for work in Spitzbergen and Iceland on various aspects of glacial geology, and over £77 000 for research on radio-echo sounding of glaciers. Various research foundations, the Royal Society, and the Universities also provided funds of varying amount.

The question of funding of research leads us to consider the present status of our subject, an analysis of some of the outstanding problems and the presentation of some further questions.

In terms of the number of persons actively involved in glacial geomorphology in Britain, there can be no doubt that the last 10 years have seen a big expansion of activity. This partly reflects the general expansion in research students and staff that occurred in the Universities and Polytechnics during the mid-1960s and early 1970s. Although I cannot substantiate my next statement with accurate statistics, I am sure we have over-produced graduates with higher degrees in glacial geomorphology during recent years. The main job opportunities for these graduates have been in academic appointments in this country and in the United States, Canada, Australia and New Zealand. These opportunities no longer exist to the same extent and there may even be a contraction of existing positions. A few graduates have entered the Soil Survey, the Institute of Geological Sciences, the Nature Conservancy, or industry. Again accurate statistics do not exist but the number of such cases is probably few. At present, there is a considerable excess of qualified glacial geomorphologists in this country. What then of the future?

The strength of glacial geomorphology in the British Universities both as undergraduate courses and as a subject of staff research is considerable. This inevitably leads to a strong interest in our subject by potential research students. The main problems that remain to be tackled are:

(1) An extension of studies of Pleistocene glacial landforms, applying the results of the recently published work on areas of existing glaciers.

(2) A continuation of the work on the processes of glacial and fluvioglacial erosion and deposition in areas of existing glaciers.

University staff can undertake further work on the Pleistocene landforms of Britain without the expenditure of large sums of money. Research students can also participate in this type of work so long as they realise that the chance of an academic post in Higher education on the completion of their postgraduate degree is unlikely. Opportunities for employment outside the Universities but in situations where their scientific expertise will be utilised are also likely to be few and far between.

The question of carrying out research in areas of existing glaciers from a British base needs some consideration. I have already demonstrated that British glacial geomorphologists have been active over the last decade in the Alps, Norway, Spitzbergen, Iceland, Greenland, Baffin Island, Alaska and the Antarctic. We have, indeed, been fortunate in the reception we have received in these areas and the financial support we have received from various sources in Britain. However, the cost of these projects has grown enormously in recent years and perhaps the time has now come to rationalise our research efforts in these far-flung places. It might be argued that one of the benefits of research in widely different environments such as the French Alps, Iceland and Antarctica is that the whole range of glacial environments can be examined. On the other hand, it could be argued that spreading our personnel and financial resources as widely as we have done over the last decade has led to costly duplication of equipment and excessive travel costs.

The only British research group that receives continuous financial support from the Government to undertake research in areas of existing glaciers is the British Antarctic Survey. A limited amount of glacial geomorphological work has already been carried out under their auspices. It could be argued that this should be expanded at the expense of work in northern latitudes, although the cost per person of fieldwork in the Antarctic is obviously much higher. It would seem more reasonable for British glacial geomorphologists to select an area somewhere in the northern hemisphere and establish a permanent field centre which could be used by numerous research workers over a period of years.

At a time when research funds are increasingly difficult to obtain it would seem sensible for us to be thinking about priorities. Our relationship with NERC has been largely on an individual basis, apart from the meeting between the officers of the BGRG and NERC which took place on 30 April 1975. The report of this meeting, which appeared in *Geophemera*, was very informative and referred to the establishment of the Working Party on the Geomorphology of Water-produced Landforms. If a similar working party were set up on glacial geomorphology to identify research priorities, what would we wish to submit in the light of some of the comments I have already made? A table published by NERC shows that a total of thirty studentships were awarded for glacial geomorphology between 1970 and 1975, but this figure is misleading in that it includes studies in periglacial processes, glaciology and radio-echo sounding. Only about nineteen of these studentships were for work in glacial geomorphology. Remembering that NERC has not been the only source of funding for research students, could a case be made to NERC for the continued support of research students working in glacial geomorphology, at the same level as that of the last 5 years? There are certainly many problems worthy of attention but the career opportunities following a research training in glacial geomorphology are difficult to envisage.

Support by means of Research Grants from NERC is the other issue we should consider. Do we want to continue with the every-man-for-himself approach or is some element of co-operation feasible? Two possibilities should be considered. First, the question of expensive overseas fieldwork could be rationalised by suggesting to NERC that one field centre could be established in a particular area. Secondly, some attempt could be made to improve the relationship between University research workers and those members of the Institute of Geological Sciences (IGS) concerned with the mapping of glacial deposits. I see no reason why specific projects could not be undertaken on a contract basis for the IGS. I know of only one example of such an arrangement in which a member of the Aberdeen geography department undertook work, the results of which are to be used in the publication of a drift edition of an IGS map.

Maybe it is the view of those present that Universities should remain free to engage in research in areas of their own choice, and to compete for the limited funds available purely on the basis of the academic quality of the research proposal. On the other hand, it could be argued that some rationalisation of research effort is required and that some cognisance should be given to the employment opportunities for research students. I believe it would be more satisfactory if we, as research workers, faced up to these questions and made some attempt to inform NERC of our priorities, rather than continue in the haphazard way in which we at present operate.

REFERENCES

Andrews, J. T. (1975), *Glacial systems: an approach to glaciers and their environments* (Duxbury Press, North Scituate, Mass.), 191 pp.

Embleton, C. (ed.) (1972), *Glaciers and glacial erosion* (Macmillan, London), 287 pp.

—— and King, C. A. M. (1968), *Glacial and periglacial geomorphology* (Arnold, London), 608 pp.

—— and King, C. A. M. (1975), *Glacial geomorphology* (Arnold, London), 573 pp.

Lewis, C. A. (ed.) (1970), *The glaciations of Wales and adjoining regions* (Longman, London), 378 pp.

Price, R. J. (1973), *Glacial and fluvioglacial landforms* (Oliver and Boyd, Edinburgh), 242 pp.

Sissons, J. B. (1967), *The evolution of Scotland's scenery* (Oliver and Boyd, Edinburgh), 259 pp.

Sugden, D. E. and John, B. S. (1976), *Glaciers and landscape* (Arnold, London), 376 pp.
West, R. G. (1968), *Pleistocene geology and biology* (Longmans, London), 377 pp.

9

PERIGLACIAL GEOMORPHOLOGY: PRESENT PROBLEMS AND FUTURE PROSPECTS

JAROMIR DEMEK

(*Institute of Geography, Czechoslovak Academy of Sciences, Brno*)

Cryogenic phenomena and forms have been known for a relatively long time. As early as 1640, the Yakutian dukes reported that in Yakutia 'ground does not thaw completely even in summer', and knew about the characteristics of permafrost. In 1828, F. Shergin, a worker of the Russian-American Company, began to sink a well on the bank of a branch of the Lena River in the town of Yakutsk; he finished his work 9 years later in frozen deposits at a depth of 116·4 m. In 1843–46 A. F. Middendorf measured the temperature of the frozen deposits in this well and scientifically demonstrated the presence of permafrost.

More than 100 years ago, geologists and geographers mapping in Southern England and Wales noticed superficial non-stratified and coarse deposits which could not have originated by processes active at that time. This unsorted heterogeneous debris has been referred to as 'head', 'coombe rock', 'warp', and other terms. It was obvious that these deposits had developed in a much colder climate than that of the present time. This knowledge became the basis of a new branch of geomorphology dealing with the forms and processes of a cold environment.

However, the main development of periglacial geomorphology took place only after the Second World War. According to J. Tricart (1970, p. xv), W. M. Davis's geomorphological school regarded cryogenic forms as rarities. Prior to 1945 the attention of geomorphologists in most countries was restricted mainly to small cryogenic forms (patterned ground, stone streams). Only in the USSR since the 1930s has considerable attention been paid to the study of permafrost. Since 1945, geomorphologists engaged in the study of processes and relationships in cold environments all over the world have concentrated their attention on the investigation of permafrost and incidental cryogenic phenomena. Extensive economic utilization of the natural resources of polar regions, as well as the resulting intensification of investigations and the increasing number of expeditions in these often relatively inaccessible regions, contributed greatly to the development of periglacial geomorphology.

The problems of periglacial geomorphology may be divided into two basic groups:

1. problems of phenomena linked to processes and forms conditioned by permafrost

2. problems connected with cold climates affecting the operation of geomorphological processes.

TERMINOLOGY

It is first of all necessary to consider a terminological question. The term 'periglacial' was introduced by the Polish explorer Walery Loziński in 1909 and is now firmly entrenched. In the last 3 years it has appeared in the names of significant monographs on periglacial geomorphology, namely those by A. L. Washburn (*Periglacial processes and environments*, 1973), C. Embleton and C. A. M. King (*Periglacial geomorphology*, 1975), and most recently by A. Jahn, (*Problems of the periglacial zone*, 1975). But since the term 'periglacial' still tends to evoke the idea that we are dealing with a region, a zone, at the margin of a glacier, it is quite often wrongly thought that the periglacial zone is the result of the climatic effects of the glacier. The term 'periglacial zone' also creates the impression that the climate of this zone is a more temperate variant of a glacial climate and that in the case of climatic cooling, a glacier or ice sheet will develop in the present periglacial zone.

But most typical periglacial zones with characteristic landforms and the greatest thicknesses of permafrost occur in the interiors of continents or in high mountains where either there was no glaciation at all or the glaciers were insignificant. A typical periglacial climate differs substantially from a glacial climate mainly by the aridity and the negative heat balance of the ground surface. Unfortunately, the periglacial zone has not been well-defined climatically, being delimited so far rather on the basis of the occurrence of cryogenic phenomena. The development of cryogenic phenomena is connected with the freezing of soil and the phase transformation of water. Ground freezing can be perennial (ground temperature below freezing point for more than 2 years), seasonal (several months) and short-period (from several hours up to days).

The core of the periglacial zone comprises the region of perennial ground freezing with permafrost.

SPATIAL DIFFERENTIATION OF THE PERIGLACIAL ZONE

In the spatial differentiation of the periglacial zone, the difference between the Arctic and Antarctic periglacial zone on the one hand, and the periglacial zone in mountain ranges on the other, is of the greatest significance. In the Arctic and Antarctic periglacial zone, there is also a significant difference between maritime and continental climatic conditions. In the maritime zone (for example, Iceland, Spitzbergen) the temperature of the permafrost approaches 0°C, higher humidity causing cryogenic processes to be active over much of the year. The active layer is of considerable thickness and therefore solifluction is rather intensive. In the continental zone, however, the permafrost exhibits much lower temperatures. The activity of cryogenic processes is concentrated in a short period in spring only when the area becomes dry, and cryogenic processes become less active. In autumn, the temperature decreases rapidly, the active layer freezes, and the effects of geomorphological processes are small. A typical feature of this continental periglacial zone is the occurrence of extensive cryogenic planation surfaces (cryopediments, cryoplanation terraces).

PERMAFROST AS A SYSTEM

Over about 24 per cent of the surface of continents and in some marginal parts of the oceans, the rocks exhibit a negative thermal balance and a temperature below 0°C for more than 2 years. If these rocks contain water then this water usually freezes and is transformed into the solid state–ice. The freezing of water in rocks substantially changes their permeability and their physical-mechanical, thermal and other properties. Loose sediments become consolidated. Owing to the freezing of rocks and the transformation of water into the solid state, a complicated system of solid (mineral or organo-mineral skeleton ice), liquid and gaseous phases develops. This system is called permafrost. The origin of permafrost and the related changes of rock properties manifest themselves naturally during the operation of geomorphological processes and the development of specific cryogenic forms.

Most permafrost is linked to present conditions, mainly climate, relief, vegetation, hydrological conditions and the activities of man. It is accordingly a dynamic system adapting itself to the changing conditions of the environment. Unfortunately, relatively little is known about the dynamics of permafrost in spite of the fact that research has made considerable progress, especially in the U.S.S.R., the U.S.A. and Canada.

In any consideration of permafrost dynamics, and also, therefore, of the development of landforms, the temperature, thickness and origin of the permafrost are of greatest importance. In low temperatures permafrost is stable and the development of landforms is slow. Around 0°C, permafrost dynamics are, on the contrary, much more active and the development of specific cryogenic forms takes place. Permafrost thickness also plays an important part. The greatest permafrost thicknesses known are to be found in the U.S.S.R.: in the Udokan Ridge, where a thickness of 1300 m has been recorded, and in the basin of the Mạrcha River (a tributary of the Viljuj River) where a thickness of 1500 m has been established (P. I. Melnikov and N. I. Tolstikhin, 1974, p. 96).

The origin of permafrost is of considerable geomorphological significance, mainly because it affects the distribution of ground-ice in frozen rocks, and especially in sediments. In epigenetic permafrost, ground-ice develops mainly in the upper parts. On the other hand, in syngenetic permafrost, ground-ice is fairly regularly distributed through the whole permafrost thickness. The significance of this for periglacial geomorphology follows from the fact that ground-ice forms from 50 to 80 per cent by volume of the fine-grained sediments of polar regions. The dynamics of permafrost, that is, its aggradation and degradation, are of unusual significance for the development of landforms in the periglacial zone.

Increase in the negative thermal soil balance results in a decrease of permafrost temperature. Because of this, permafrost aggradation also begins to result in a decrease of the thickness of the active layer and this in turn affects the dynamics of the geomorphological processes. But the aggradation of permafrost also results in the freezing of taliks and in this way some typical forms, such as pingos, originate. An acceleration of the growth of ice wedges and the development of landforms, such as low-centred polygons, also takes place. In some places in polar regions, ice-wedge polygons occupy as much as 60 per cent of the surface. In the deposits at the foot of the Kular Ridge, ice wedges up to depths of 6 m amount to 71 per cent by volume, and in the lowland of Yano-Indigirskaya-nizmennost as much as 76 per cent by volume, of all deposits (T. Czudek and Demek, 1973, p. 10). The ice wedges in exposures on the Indigirka River bank near the community of Sypnyy Yar are as much as 40 to 50 m deep.

However, the growth of ice wedges can result simultaneously in local degradation of the permafrost, and the pushing-out of soils on the margins and the development of ramparts. Low-centred polygons with a concave core develop. The growth of the ramparts breaks the turf cover, thus exposing the soils to the direct effects of insolation. In places where polygon boundaries cross, and in the concave cores, water accumulates. The temperature of the water in the depressions is always higher than that of permafrost. Owing to the thermal and chemical effects of the water, ice wedges begin to thaw, even in very cold climates where permafrost temperatures range between −9° and −11°C.

Forest fires in the taiga may also be a cause of local permafrost degradation. In eastern Siberia, disturbance of the vegetation cover changes the thermal conditions of the ground to depths of 10 to 15 m and can result in the onset of thermokarst processes which considerably affect the development of the relief.

Regional permafrost degradation leads to marked changes in the relief of polar lowlands where the permafrost is often characterised by a high ground-ice content. Permafrost degradation by thermokarst processes results in local subsidence of the soil, owing to the thawing of ground-ice, and the development of depressions (Fig. 9.1). When flying over the Central Yakutian Lowland between the Lena and Aldan Rivers, numerous circular or oval-shaped thermokarst depressions called *alases* occupying 40 to 50 per cent of the surface of the lowland can be seen. Most are closed, often containing a lake, but some merge into elongated dry depressions with flat grassy floors. The light-green grass on the floor of the *alas* contrasts sharply with the surrounding dark green of the larch taiga. Many thermokarst depressions in the Central Yakutian Lowland developed as a result of regional permafrost degradation during the Holocene climatic optimum, 4000–5000 years ago.

Geomorphologists have so far paid relatively little attention to thermokarst phenomena in spite of the fact that they are of considerable geomorphological significance. An important result of thermokarst processes is the degradation of the surface of lowlands in the periglacial zone. This process is designated 'thermo-planation', which can take place either laterally or vertically.

Lateral permafrost degradation and thermo-planation connected with this phenomenon are more significant in areas of dissected relief. These processes are usually linked to the thermo-erosion of streams or to thermo-abrasion. Characteristic forms developing by lateral thermo-planation are arcuate depressions on slopes, termed 'thermocirques'. The sides of the thermocirques usually exhibit parallel retreat. The thermocirque floor becomes larger, and adjacent thermocirques thus often coalesce with each other. In this way a terrace develops on the slope, increasing in size by backwearing of the thermocirque walls. Gradually a lower level of the polar lowland develops.

Vertical permafrost degradation occurs mainly in flat undissected terrain, often on watersheds. The extent of relief change owing to permafrost downwearing depends on the amount of ground-ice in the deposits, and on its type.

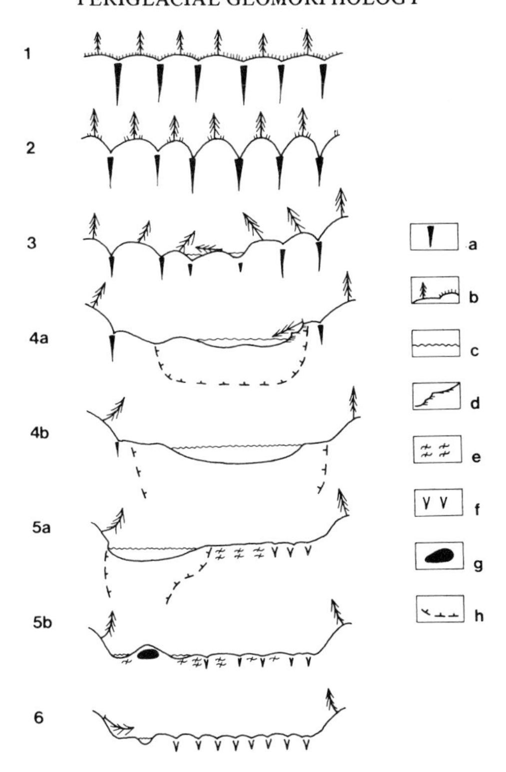

FIG. 9.1. Schematic representation of alas development as a result of permafrost degradation from above (T. Czudek and J. Demek, 1973). **Key**: 1. original surface with syngenetic ice wedges, 2. beginning of permafrost degradation and origin of baydjarakhs, 3. duyoda, 4a and b, alas, 5a and b, khonu with pingo, 6. thermokarst valley.

Considerable changes will take place if there is a high ground-ice content in the deposits, especially if the ice-wedge network is well-developed. The resulting landforms are designated by Yakutian names. In the first stage, at the beginning of downwearing, the ice wedges begin to thaw and high-centred polygons develop. The vegetation usually remains undisturbed. In the second phase, the depression above the ice wedges becomes deeper, the humus layer breaks and slips down into the depressions. The polygon cores change into conical elevations called *baydzharakhs*. The vegetation cover is disturbed and deposits are eroded. In the third stage, the *baydzharakhs* are gradually disturbed and a depression called a *duyoda* develops, in whose centre water accumulates. If water remains in the depression for the greater part of the year, thermokarst processes begin to accelerate. On the slopes of the *duyodas*, *baydzharakhs* develop, and sliding and solifluction take place owing to oversaturation of the ground with water from the melting ground-ice.

In the fourth stage, a distinct depression called an *alas* develops, with steep slopes and a flat floor. In the Yakutian language this term means an oval or circular depression with steep slopes and a flat floor covered with green grass and a lake. When the water in the lake attains a certain depth, freezing of the water to the bottom in winter ceases and a lake talik develops. The ground-ice thaws out slowly and the deposits settle, so that the *alas* becomes deeper. On the shores of the lake, thermo-abrasion takes place. Due to wind action, the shores of lakes and *alases* shift.

In the fifth phase, the lake progressively disappears and renewed permafrost aggradation takes place. Owing to freezing of the talik, a pingo usually develops. The depression with secondary ice wedges and a pingo is called a *khonu*.

The coalescence of *alases* and *khonu* results in thermokarst valleys. In ground-plan, the thermokarst valleys consist of wider sections (*alases, khonu*) and narrower sections marking the position of former watershed ridges. The thermokarst valleys have numerous meanders, blind projections, and may run counter to the general dip of the surface in places.

Permafrost degradation thus causes substantial changes in the relief of the polar lowlands and the thermokarst processes lower the surface of the terrain to create a lower level.

PERMAFROST STRUCTURE AND CRYOGENIC TEXTURES

Depending upon its mode of origin and development, permafrost usually contains various kinds of ground-ice. 'Texture ice', involving ice crystals from some tens of millimetres up to a few centimetres in size, forms a typical cryogenic texture in the case of rock freezing. This texture is defined by its form, dimensions and the distribution of ground-ice crystals in the frozen rock. The kind of texture depends on the genesis of the permafrost and the conditions under which it developed.

The main differences in cryogenic textures are related to the different origins of syngenetic or epigenetic permafrost (Fig. 9.2). In the case of syngenetic permafrost, the cryogenic textures develop as the permafrost thickens and grows upwards and as, at the same time, sediments accumulate under subaerial or subaqueous conditions (Fig. 9.3). Under subaerial conditions the lower part of the active layer of colluvial, eolian, proluvial and fluvial deposits changes gradually with accumulation into permafrost. Under subaqueous conditions, the deposits of river beds, ox-bow lakes, lacustrine and marine deposits, freeze. Freezing often takes place not only in the vertical but also in the horizontal direction. Texture-ice in the case of this type of permafrost develops from the water above the upper permafrost line which freely migrates towards the freezing-point. In clays, loams and silts, the water of adhesion migrates. The growing ice crystals push off rock particles and form lenses and bands of texture ice. In sands, gravels and coarser debris, water penetrates as far as the freezing-front only by gravity or precipitation from vapour. The ice fills the free spaces among the mineral particles. This is why, in syngenetic permafrost, rather complicated textures of segregation type prevail. The principal types of texture are shown in Figure 9.2.

In the case of the development of epigenetic permafrost, vertical freezing dominates (Fig. 9.4). Most cryogenic textures therefore develop under conditions of water migrating under pressure towards the freezing-front. Ice develops from water under the permafrost. The water migrating under pressure occurs not only in water-bearing rocks but even in rocks normally considered impermeable in hydrogeology. Epigenetic freezing of rocks with ground ice development leads

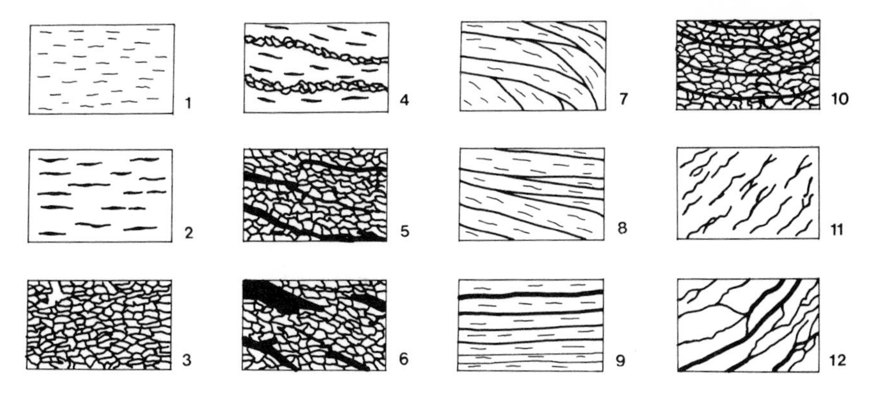

FIG. 9.2. Main cryogenic textures caused by syngenetic freezing of unconsolidated deposits (P. I. Melnikov and N. I. Tolstikhin, 1974). **Key**: *cryogenic textures of subaerially freezing deposits. Simple textures*: 1. texture of fine ice lenses (typical of solifluction deposits), 2. lenticular texture (typical of dry slope deposits), 3. reticulate texture (typical of solifluction terraces and streams). *Complicated textures*: 4. texture of undulating bands (typical of humid slope deposits). 5. reticulate texture, tilted (typical of rather humid slope deposits), 6. texture of interrupted bands and nets (typical of creep deposits), 7. strongly undulating lenticular texture (typical of mobile solifluction terraces and streams), 8. fan-like lenticular texture (typical of alluvial cones), 9. horizontal stratified texture (typical of floodplains), 10. stratified texture with concave curvature (typical of higher levels of floodplain with ice-wedges).
Cryogenic textures of subaqueously freezing deposits: 11. Texture of oblique ice lenses (typical of ox-bow deposits), 12. obliquely stratified texture (typical of ox-bows).

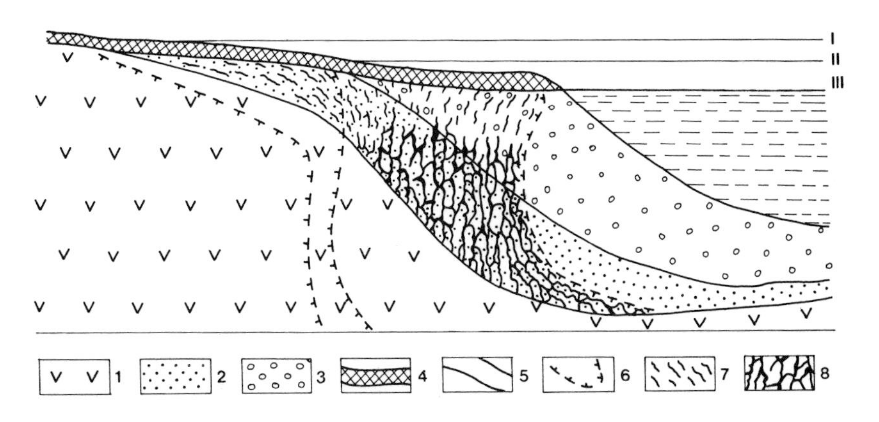

FIG. 9.3. Syngenetic freezing of lake-floor deposits (P. I. Melnikov and N. I. Tolstikhin, 1974). **Key**: 1. bedrock, 2. deposits of first generation on lake floor, 3. deposits of second generation on lake floor, 4. active layer, 5. lake floor in various stages of filling with deposits, 6. upper permafrost table in various stages of lake filling with deposits, 7. lake-floor deposits with a dense pattern of relatively fine ice lenses, 8. deposits on lake floor with a widely-spaced pattern of relatively broad ice bands. I–III: water level in the lake in various stages of development.

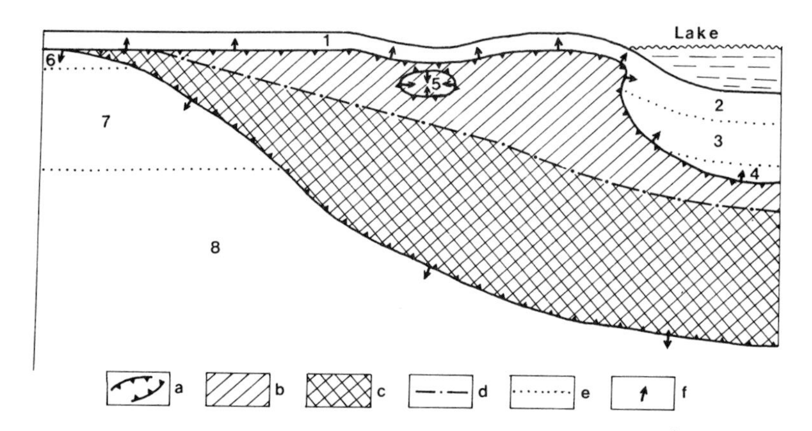

FIG. 9.4. Diagram to show freezing of loose deposits (P. I. Melnikov and N. I. Tolstikhin, 1974). **Key:** *Syngenetic type.* a. upper and lower permafrost tables, b. region of syngenetically freezing deposits, c. region of epigenetically freezing deposits, d. borders between regions b. and c., e. borders of horizons and lenses, f. direction of migration of permafrost tables, 1. Horizon of subaerially freezing deposits with superimposed cryogenic textures of segregation type, 2. Horizon of subaqueously freezing deposits with superimposed cryogenic textures of segregation type, 3. Horizon of subaqueously freezing deposits with syngenetic cryogenic textures of segregation type, 4. Horizon of passive ice formation with lithogenic cryotextures, 5. Lense with cryogenic textures of injection type.
Epigenetic type: 6. Horizon of ground-ice formation under conditions of free migration of water towards the border of freezing with cryogenic textures of segregation type, 7. Horizon of ice formation under conditions of water migration under pressure towards the freezing border with cryogenic textures of injection type, 8. Horizon of passive ice formation with lithogenic cryotextures.

to an increase in their volume. Freezing rocks therefore exert intense differential pressures on the underlying unfrozen rocks. Since the compression and consolidation of underlying rocks is restricted, uplift and compression of the permafrost take place. In the case of thin permafrost, deformation can easily take place and the pressure on the substratum is small. The conditions of water migration therefore resemble those of syngenetic permafrost. But with increasing permafrost thickness, the pressure on unfrozen rocks increases. Clays, loams and silts become consolidated under this pressure. Some of the groundwater is freed and is forced out into the less consolidated part. This water then gradually freezes in fissures, larger pores or in less compressed rocks. This is why in epigenetic permafrost, both segregation and injection textures can be found, but pressure migration of water predominates. In epigenetic permafrost, three texture types therefore occur in a vertical direction (from above):

a) a horizon of ground-ice originating under conditions of free-water migration towards the freezing-front and showing textures of segregation type

b) a horizon of ground-ice developing under conditions of water migration under pressure towards the freezing-front, showing cryogenic textures of injection type

c) a horizon of passive ground-ice with textures depending on the rock properties (Melnikov and Tolstikhin, 1974, p. 151).

The cryogenic textures of epigenetic permafrost thus differ substantially from those of syngenetic permafrost. Cryogenic textures can therefore be utilized for an analysis of permafrost facies and to reconstruct the geomorphological conditions of permafrost development. In geomorphology, most attention has been paid to phenomena showing disturbance of the original rock structures, such as ice wedges, frost wedges and cryoturbation. Little attention has so far been paid to analysis of permafrost facies on the basis of cryogenic textures.

POST-CRYOGENIC TEXTURES AND THEIR PALAEOGEOGRAPHICAL SIGNIFICANCE

During permafrost degradation, ground-ice thawing and changes of rock properties take place. On steeper slopes, solifluction and mudflows develop owing to oversaturation of soils with water from melting ground ice. These movements naturally result in the disturbance of cryogenic textures. On gentle slopes or in the case of a lower degree of saturation of soils with water, rock consolidation takes place without any disturbance of the textures. In place of ice crystals, fine cracks are preserved, forming a so-called post-cryogenic texture. The walls of the cracks are often covered with a fine film of precipitated ferrous oxide or clay. Study of post-cryogenic textures in deposits allows the presence of permafrost in the Pleistocene periglacial zone, and its type, to be established.

An analysis of airphotos of various parts of the Pleistocene periglacial zone and detailed geomorphological investigations, for instance in Sweden and the U.S.S.R., shows that, in the case of permafrost degradation at the end of the Pleistocene, numerous landforms were preserved in the former periglacial zone, such as pseudomorphs of ice wedges and thermokarst phenomena (H. Svensson, 1964; A. A. Velichko, 1969, 1972, 1973). A study of exposures has shown that even post-cryogenic textures have been preserved, enabling us to analyse the palaeogeographical conditions of the origin of these surface forms under permafrost conditions.

In Central Europe, post-cryogenic textures have been found mainly in slope deposits and loess. In loess especially, these investigations raise several questions. The thickness of loess deposits separated by buried soils in some locations in Central Europe reaches several tens of metres. The loess in the present-day periglacial zone in Siberia and Alaska (T. L. Péwé, 1968) exhibits a considerable ground-ice content. Ice wedges in the loess deposits of the polar lowlands in the north-eastern U.S.S.R. attain vertical thicknesses between 40 and 50 m (S. V. Tomirdiaro *et al.*, 1974, p. 47). In Central Europe, pseudomorphs of ice wedges of considerable size have also been found in loess. Simultaneously, however, the post-cryogenic textures point to a significant content of texture-ice. During permafrost thawing, considerable changes in volume, consolidation of sediments and relief changes must have taken place (Demek and J. Kukla, 1969) which have not yet been sufficiently evaluated by geomorphologists and Quaternary geologists.

TALIKS AND THEIR GEOMORPHOLOGICAL SIGNIFICANCE

Taliks occur even in northern regions where temperatures in the permafrost are very low. Their number and size naturally increase towards the southern permafrost border. Taliks most often occur below rivers and lakes. Open taliks exist under large rivers such as the Ob, Yenisey, Lena, Indigirka and Kolyma. Beneath smaller rivers and lakes, closed taliks can be found. Sometimes an interesting phenomenon can be observed: in the valley-head below the stream there may be no talik; farther downstream a closed talik may appear; while in the lower reaches, an open talik appears. Taliks can also be found where strong groundwater springs, mainly salt-water, emerge.

Taliks are of considerable importance as regards the geomorphological activity of streams in the periglacial zone. The thermal effects of the water accelerate erosion, evoking the phenomenon termed thermo-erosion, that is, the combined thermal and mechanical activity of running water. Thermo-erosion is the cause of the wide valleys of Siberian rivers.

Taliks also determine the origin of other landforms such as icings and various types of ground-ice mounds. Little attention has so far been paid to the geomorphological significance of taliks in the development of periglacial geomorphology.

EROSION AND PLANATION IN THE PERIGLACIAL ZONE

In his geomorphological division of continents J. Büdel (1963, 1969a) used the designation 'zone of intensive linear erosion and pronounced valley formation'. This specification has appeared even in some significant textbooks and in the *Encyclopaedia of Geomorphology,* edited by R. W. Fairbridge (1968). Büdel claimed that in the periglacial zone linear erosion is the dominant activity, while lateral planation takes place only to a small extent or even not at all. This idea is wrong.

Intensive linear erosion undoubtedly takes place in the periglacial zone. And Büdel's (1969b) idea of the mechanism of linear erosion in the periglacial zone taking place along ice crusts (*Eisrindeneffekt*) is valid provided it is restricted to the uppermost valley sections without a talik. Nevertheless, linear erosion in the periglacial zone is favoured by the common mechanical and thermal effects of running water (so-called 'thermo-erosion'). Especially in regions with a well-developed ice-wedge system, linear erosion is affected and controlled by these ice wedges. As ice wedges on slopes thaw out, thermo-erosion gullies develop, not only on syngenetic ice wedges in loose deposits but also on epigenetic ice wedges in solid rocks. On the steep banks of the Lena River incised in the limestones and mudstones of the Vysokoe Prilenskoe Plateau, gullies as much as 30 m in depth have developed at regular intervals, undoubtedly as a result of thermo-erosion on a system of epigenetic ice wedges. The extent of lateral thermo-erosion is also considerable, sapping the valley sides and leading to the development of broad trough-shaped valleys.

Another phenomenon connected with thermo-erosion is the development of one of the most characteristic forms of the periglacial zone—dells. Dells are small dry valleys shallowly cut in large numbers along the slopes of the periglacial zone, running straight down the direction of slope. They have flat floors, gently inclined sides and reach lengths of several hundreds of metres, widths of several tens of metres and depths of several metres. Often they began to develop as a result of local permafrost degradation and the thawing of ice wedges. Their further modelling is controlled by solifluction, sheet wash and other cryogenic processes. With respect to the greater humidity along their floor lines, the dells act as the main denudation and transport lines on periglacial slopes. Owing to their frequency on slopes, dells are an important aspect of subaerial modelling.

As well as linear erosion in the periglacial zone, extensive planation takes place. This has been stressed in the last 25 years in the papers of numerous geomorphologists. In some regions, mainly with extremely continental periglacial climates, linear erosion is less active than planation. Sheet denudation and pediplanation owing to cryogenic processes such as frost creep, frost heaving, nivation, solifluction, suffosion and sheet wash, attain considerable importance in the periglacial zone.

Planation takes place here both at the foot of slopes, and on summits and watershed ridges. At the foot of slopes, piedmont planation surfaces designated 'cryopediments' develop owing to cryogenic processes in the periglacial zone. Extensive pediments on the margins of mountain ranges in eastern Siberia are often linked to Tertiary pediments. In valleys, valley cryopediments occur, these being continuations of Quaternary terraces. The valley cryopediments in the flysch Carpathians in Czechoslovakia are usually linked to Upper Pleistocene river terraces, sometimes even to the floodplain. The cryopediments have developed in various rocks and cut across rocks of varying resistance (H. M. French, 1973; Czudek and Demek, 1973; P. Macar, 1969).

The principal process leading to the development of cryopediments is the retreat of steep slopes caused by frost weathering and the sapping of slopes by nivation. In the modelling of the gentle foot-slope, many different processes play a part, especially solifluction, suffosion and sheet wash. Of great significance are the dells along whose axes there is greater moisture content in the floor deposits, and consequently the active layer above the permafrost is thicker. The gentle foot-slope is protected by permafrost from further denudation, the slope serving only as a surface of transport of the material that has descended the scarp.

The cryopediments extend from the main valleys into lateral valleys, even joining by pediment passes over watersheds. As the cryopediments coalesce, a regional planation surface develops called a cryopediplain.

Planation surfaces in the periglacial zone also develop on summits and watershed ridges. The summits and watershed ridges of the periglacial zone often exhibit a distinct stepped topography. The planation process usually begins by the formation of nivation hollows. As they widen and merge, and because of the retreat of rock steps termed 'frost-riven cliffs' and 'frost-riven scarps', cryoplanation terraces develop. On the summits, several steps of successive cryoplanation terraces are often to be seen. As they merge, a cryogenic planation surface termed a 'cryoplain' develops.

APPLIED PERIGLACIAL GEOMORPHOLOGY

For several decades, periglacial geomorphology has been considered a science dealing with phenomena and processes without any special practical significance. As late as the Second World War, and especially in the last two decades, the practical aspects of periglacial geomorphology have come to the forefront. This is connected above all with the economic utilisation of polar regions, mainly the Arctic. When man utilises the natural resources of extensive regions of Canada and Siberia, he encounters the characteristic natural complexes linked to permafrost along with the widespread occurrence of ground-ice and typical natural processes. The economic utilization of the periglacial zone requires careful planning of construction, the use of mineral resources, and the economic utilisation of the whole region. Permafrost and the associated natural complexes are very sensitive systems. Small mistakes which might have only negligible effects on construction in other climatic zones may result in catastrophic effects in the periglacial zone where permafrost occurs. The disturbance of the permafrost equilibrium by man leads to an irreparable disturbance of the surface of the terrain owing to thermokarst, thermo-erosion or thermo-abrasion processes. Under such conditions, the prediction of the development of the surface during construction work is of great significance.

Geomorphologists are able nowadays, with the aid of geomorphological mapping based mainly on air and space photographs, to establish the extent of permafrost, delimit taliks, establish the course of cryogenic processes and make predictions about possible changes in natural complexes resulting from the development of certain projects. Most constructional projects (for instance, gas and oil pipelines) are designed with conservation of the permafrost in mind. From the evidence of landforms, the disturbance of the thermal balance of the permafrost can be monitored, the beginning of its degradation can be established, and protective measures can be recommended. In the case of major structures (railways, roads, long oil and gas pipelines), the basic principles governing the periglacial zone must be studied and small-scale geomorphological mapping of extensive areas is necessary. As well as general geomorphological maps, special geomorphological maps for particular purposes must be compiled.

Geomorphology also plays an important part in prospecting for mineral deposits in the periglacial zone. To establish the occurrences of gold, tin and other deposits, geomorphological analysis aimed at distinguishing regions with different tectonic régimes is important, that is, regions in tectonic equilibrium or regions of minor uplift in which cryopediments develop and gold placers are lacking, or subsiding regions where talus deposits and extensive gold placers may develop, or regions of strong uplift with small placers in river beds.

In mining from placers in slope or fluvial sediments, economically favourable modes of permafrost thawing for systematic mining must be sought. Geomorphologists can give valuable recommendations on the basis of their knowledge of thermokarst phenomena. When mining on the margins of taliks, artificial freezing is often necessary to prevent water from flowing from the talik into the deposit.

Geomorphologists are facing special problems resulting from the development of modern forms of agriculture in the periglacial zone. In the continental zone especially, agriculture often requires irrigation which leads to permafrost degradation. Owing to the increase in the thickness of the active layer when fields are irrigated, suffosion takes place, and suffosion wells, sinks and sub-surface hollows develop.

Recently, the construction of large water reservoirs and dams has been started in the periglacial zone. The accumulation of water causes degradation of the permafrost and disturbance of the bottom and shores built of frozen rocks. Prediction of the rate of permafrost thawing on the bottom and the shores, and the related extent of seepage and loss of water from the reservoir are of great practical significance. Permafrost thawing on the shores can result in dangerous slides and mudflows. Filling of major reservoirs is usually accompanied by earth tremors not only in the solid rocks but also in the frozen deposits because frozen unconsolidated rocks possess a stability modulus closely resembling that of unfrozen solid rock. Special problems occur in the construction of dams in frozen calcareous rocks with karst phenomena. Ground-ice in karst spaces may prevent their being filled by injections of concrete and later, in the case of permafrost degradation, leakage of water can take place.

The protection of the environment in the periglacial zone is therefore of immediate economic significance, and geomorphologists have great possibilities of helping in the rational utilisation of the natural resources of polar regions, by the study of cryogenic forms and processes.

CONCLUSION

In connection with the economic utilisation of arctic and sub-arctic areas, better information has been obtained in recent decades concerning cryogenic processes and forms in the periglacial zone. Simultaneously, the practical significance of such knowledge in the utilisation of these regions has become apparent. But the amount of information is still insufficient in comparison with knowledge of processes operating in other climatic zones. Of principal significance for the future development of periglacial geomorphology will be a more detailed knowledge of the processes acting in the permafrost as in a dynamic natural system. Quantitative measurements of cryogenic processes using field equipment are still relatively undeveloped, encountering many difficulties because of the unfavourable climatic conditions in the periglacial zone. Qualitative and quantitative measurements of the processes will, however, become still more important because further progress will require the development of mathematical models of frozen rocks and the processes active in them. The increase in the amount of information will also require other ways of data processing, with the use of statistical analysis and computers. In this connection the immediate practical significance of periglacial geomorphology will further increase.

REFERENCES

Büdel, J. (1963), 'Klimagenetische Geomorphologie', *Geogr. Rdsch.* **15**, 269–85.

—— (1969a), 'Das System der klima-genetischen Geomorphologie', *Erdkunde* **23**, 175–83.

—— (1969b), 'Der Eisrinden-Effekt als Motor der Tiefenerosion in der excessiven Talbildungszone', *Würzb. geogr. Arb.* **25**, 1–41.

Czudek, T. and Demek, J. (1970), 'Thermokarst in Siberia and its influence on the development of lowland relief', *Quatern. Res.* **1 (1)**, 103-20.

—— and —— (1971), 'Pleistocene cryoplanation in the Ceská Vysočina Highlands, Czechoslovakia', *Trans. Inst. Br. Geogr.* **52**, 95–112.

—— and —— (1973), 'The valley cryopediments in eastern Siberia', *Biul. peryglac.* **22**, 117–30.

—— and —— (1973), 'Die Reliefentwicklung während der Dauerfrostbodendegradation', *Rozpr. čsl. Akad. Věd,* řada MPV **83** (2), 1–69.

Davies, J. L. (1969), *Landforms of cold climates* (M.I.T. Press, Cambridge, Massachusetts and London), 200 pp.

Dedkov, A. (1965), 'Das Problem der Oberflächenverebnungen', *Petermanns geogr. Mitt.* **109**, 258–64.

Demek, J. (1968), 'Cryoplanation terraces in Yakutia', *Biul. peryglac.* **17**, 91–116.

—— (1969), 'Cryoplanation terraces, their geographical distribution, genesis and development', *Rozpr. čsl. Akad. Věd,* řada MPV **79(4)**, 1–80.

—— (1972), 'Die Pedimentation im subnivalen Bereich', *Göttinger geogr. Abh.* **60**, 145–54.

—— and Kukla, J. (eds.) (1969), *Periglazialzone, Löss und Paläolithikum der Tschechoslowakei* (Czechoslovak Academy of Sciences, Institute of Geography, Brno), 155 pp.

Dostovalov, B. N. and Kudrjavcev, V. A. (1967). *Obshchee merzlotovedenie* (Izdatel stvo Moskovskogo universiteta, Moscow), 403 pp.

Embleton, C. and King, C. A. M. (1975), *Periglacial geomorphology* (Arnold, London), 203 pp.

Fairbridge, R. W. (ed.) (1968), *The encyclopedia of geomorphology* (Reinhold, New York-Amsterdam-London), 1295 pp.

Ferrians, O. J., Kachadoorian, R. and Greene, G. W. (1969), 'Permafrost and related engineering problems in Alaska', *U.S. geol. Surv. Prof. Pap.* **678**, 37 pp.

French, H. M. (1973), 'Cryopediments in the chalk of Southern England', *Biul. peryglac.* **23**, 149–56.

Jahn, A. (1965), *Problems of the periglacial zone* (PWN, Warszawa), 223 pp.

Karrasch, H. (1972), 'Flächenbildung unter periglazialen Klimabedingungen?' *Göttinger geogr. Abh.* **60**, 155–68.

Macar, P. (1969), 'Actions périglaciaires et évolution des pentes en Belgique', *Biul. peryglac.* **19**, 137–52.

Melnikov, P. I. and Tolstikhin, N. I. (1974), *Obshchee merzlotovedenie* (Nauka, Novosibirsk), 291 pp.

—— (1973), 'Permafrost: North American contribution', *2nd int. Permafrost Conf. Yakutsk* (1973), National Academy of Sciences, Washington D.C., 782 pp.

Péwé, T. L. (1968), 'Loess deposits of Alaska', *23rd int. geol. Congr. Czechoslovakia* (1968), Proceedings of Section 8 (Academia, Prague), 297–309.

—— (ed.) (1969), *The periglacial environment* (McGill-Queen's Univ. Press, Montreal), 487 pp.

Schunke, E. (1975), 'Die Periglazialerscheinungen Islands in Abhängigkeit von Klima und Substrat', *Abh. Akad. Wiss. Göttingen* (Mathematisch-physikalische Klasse), Dritte Folge **30**, 1–273.

Shvecov, P. F. and Dostovalov, B. N. (1959), *Osnovy geokriologii* (*merzlotovedenie*) (Izdatelstvo Akad. nauk S.S.S.R., Moscow), 459 pp.

Svensson, H. (1964), 'Aerial photographs for tracing and investigating fossil tundra ground in Scandinavia', *Biul. peryglac.* **14**, 321–5.

Tomirdiaro, S. V. *et al.* (1974), 'Fiziko-geograficheskaja obstanavka i osobennosti formirovanija lessovo-ledovogo pokrova na ravninakh severovostoka Azii'. (Physico-geographical environment and peculiarities in formation of loess-glacial cap, north-east Asiatic Plains), *Geologiya geof.* **7**, 47–61 (Novosibirsk).

Tricart, J. (1970), *Geomorphology of cold environments* (Macmillan, London), 320 pp.

Velitchko, A. A. (1969), 'Milieu géologique et géomorphologique de la zone périglaciaire de la plaine est-européenne', *Biul. peryglac.* **18**, 183–93.

—— (1972), 'La morphologie cryogène relicte: caractères fondamentaux et cartographie', *Z. Geomorph., Suppl.* **13**, 59–72.

—— (1973), 'Osnovnye osobennosti reliktovou kriogennoi morfoskulptury i obschchiie principy ee kartirovanya' (The main pecularities of relict cryogenic morphosculpture and general principles of its mapping) in *Paleokriologya v chetvertichnoi stratigrafii i paleografii* (Nauka, Moscow), 121–34.

Washburn, A. L. (1973), *Periglacial processes and environments* (Arnold, London), 320 pp.

IO

PERIGLACIAL GEOMORPHOLOGY IN BRITAIN

R. S. WATERS
(*University of Sheffield*)

ANTECEDENTS

The geological consequences of periglacial, or cryergic, activity have been recognised in Britain for more than 100 years. Indeed, as early as 1758, W. Borlase described a superficial deposit in Cornwall of the kind to which H. T. De la Beche subsequently applied the term 'head' (Borlase, 1758; De la Beche, 1839). It was not until 1851, however, that R. A. C. Godwin-Austen perceived that such accumulations were the result of processes no longer active and not until 1855 that O. Fisher attributed the warp of eastern England to Pleistocene frost action (Godwin-Austen, 1851; Fisher, 1866). In 1882 S. V. Wood understood the significance of involutions and five years later C. Reid explained the origin of coombe rock (Wood, 1882; Reid, 1887). But this early recognition by geologists of specific examples of periglacial deposits would appear to have stimulated little or no research into the geomorphological effects of cryonival processes. Apart from a few studies of individual landforms such as dry valleys, and superficial deposits and structures, little attention was given to periglacial geomorphology as such. Even H. Breuil's prescient and well-illustrated account of the importance of solifluction in northern France and southern England seems to have gone unnoticed (Breuil, 1934). Until the early 1950s British geomorphologist were concerned almost exclusively with denudation chronology, glaciation and coastal forms.

But 25 years ago, following the immediate post-war years of academic austerity and relative isolation, the hitherto necessarily introverted geomorphologists of Britain were particularly receptive to new ideas and were stimulated by renewed international contacts. At that time three events or circumstances appear to have encouraged the development of periglacial geomorphology in the U.K.

The symposium in the U.S.A. in honour of the 100th anniversary of the birth of W. M. Davis had a major impact, especially the contribution by L. C. Peltier: 'The geographic cycle in periglacial regions' (Peltier, 1950). Peltier's paper revived interest in earlier British publications dealing with evidence or periglacial fashioning of landforms and reminded us of European work in the field of *Klimamorphologie*.

Further stimulus came from the direct contacts which were now possible with geomorphologists in continental Europe where, from the time of W. Lozinski and B. Högbom, interest had been maintained both in contemporary frost phenomena and solifluction in high latitudes, and in the morphological legacies of Pleistocene cryergic processes (Lozinski, 1909, 1912; Högbom, 1914). The remarkable development of interest among our European colleagues in all aspects of periglacial geomorphology culminated in the publication in Poland of the *Biuletyn Peryglacjalny*. Since 1954 this premier vehicle for research papers has served as a powerful stimulus to further work.

The third event which encouraged interest in the periglacial theme was the visit of M. T. Te Punga to Britain in the mid-1950s. He reviewed the distribution of relict periglacial features and evaluated the extent of cryergic morphogenesis in southern England (Te Punga, 1957). From Kent in the east to Cornwall in the west he noted the occurrence of phenomena which are held to be diagnostic of Pleistocene frost action: he showed that the last periglacial phase ended comparatively recently, certainly less than 10 000 years ago; and he concluded that the subdued forms of this 'typical periglacial landscape' were moulded during a succession of cryergic episodes, it being 'unlikely that inter-periglacial erosion, seeing that it was restricted essentially to linear processes, was competent to obscure or obliterate earlier developed periglacial landscape form' (Te Punga, 1957, p. 410). This fresh evaluation of the extent to which the land form had been modified by 'vigorous downwearing as a result of mass wasting in which solifluction was particularly active' represented such a radical departure from traditional interpretations that it could hardly be ignored.

ACHIEVEMENTS

These and doubtless other external stimuli encouraged research in periglacial geology and geomorphology at a time of rapidly growing interest in all aspects of the British Pleistocene, and it was this that led to the founding of the Quaternary Research Association in 1964. So it is hardly surprising that the periglacial theme has developed primarily as an integral part of Quaternary studies.

The last 25 years have seen an ever-increasing amount of attention being given, first, to relict periglacial phenomena themselves, their stratigraphic relations and the information they are presumed to provide about environmental conditions, and, secondly, to the nature and extent of landform creation and landscape modification by cryonival processes.

Relict periglacial phenomena. A large amount of research since 1950 has been in the field of periglacial geology, much of it involving detailed, regional surveys (e.g. D. F. Ball and R. Goodier, 1970) and systematic investigations of the kinds of fossil cryergic phenomena that were noted by earlier observers. Many of these studies have been concerned primarily with the stratigraphy and/or the climatic indications of relict phenomena supposedly diagnostic of (a) frozen ground and (b) periglacial mass movements. The results of more than two decades of field investigation by geologists, geomorphologists and pedologists have demonstrated beyond doubt the wide distribution of fossil features whose identification with permafrost and ground-ice segregations has been made more secure by work on contemporary analogues in high latitudes. Involutions, pingo relics, fossil thermokarst basins, ice-wedge pseudomorphs and the ground patterns of former ice-wedge polygons have been reported and described from many widely distributed sites in the U.K. and Ireland; and they have been used by R. B. G. Williams to define permafrost areas lying south of the Devensian (Weichselian) ice-limit (Williams, 1969). Of particular significance for the reconstruction of Devensian and Late-glacial permafrost and hydrological conditions are the recent investigations by E. and S. Watson of the remarkable suites of pingo remains in south-west Wales (Watson, 1974) and by B. W. Sparks and others of the Breckland depressions in East Anglia (Sparks *et al.*, 1972). It is concluded that the Welsh pingos were comparable with the open-system type of pingo, of central Alaska and southern Yukon, whereas the Breckland meres represent Late-glacial thermokarst features.

The importance of periglacial mass movements is apparent from the ubiquitous gelifluction deposits and from the bedded screes and other products of frost creep and slope wash in more accidented terrain (A. S. Potts, 1971). Their diagnostic value as climatic indicators is more problematical, as is that of other sorted patterns and disturbed regoliths (including some features called involutions) which occur widely beyond the limits of the frozen ground phenomena. But the nature of the possible mechanisms that are involved in gelifluction and related movements has been elucidated by many recent studies in engineering geology (e.g. R. J. Chandler, 1970; J. N. Hutchinson, 1974).

The coastal head deposits that are so characteristic of south Wales or south-west England, where they attain their greatest thicknesses, have of course been commented upon by field workers for many years. More recently they have been investigated more systematically, but they still present problems of interpretation (G. E. Groom, 1956; D. N. Mottershead, 1971). Less attention has been directed to the inland heads of southern, unglaciated, Britain, in spite of the lead given by H. G. Dines and others in 1940 (Dines *et al.*, 1940) and the studies of Late-glacial deposits in Cornwall by H. Godwin and his collaborators in 1950 (A. P. Connolly *et al.*, 1950). Farther north, within the generally accepted limit of glaciation, gelifluction deposits together with other periglacial phenomena have been identified and accommodated within the established Pleistocene sequences. It is, of course, in the formerly glaciated areas that the value of identifying and investigating gelifluction spreads and distinguishing them from tills has often been demonstrated. One example from many must suffice.

On the western Pennine slopes in the neighbourhood of the meltwater channels of north-east Cheshire, R. H. Johnson was able to obtain important evidence relating to details of the Devensian deglaciation from four temporarily exposed sections showing the stratigraphic and altitudinal relations of glacial and periglacial deposits, and the presence of gelifluction materials on the side slopes and floors of the meltwater channels (Johnson, 1968). The most satisfactory hypothesis for explaining the observed relationships implies that the channels were cut by meltwater flowing sub-glacially or marginally to the Main Devensian ice sheet when only its relatively drift-free, upper surface was melting. After reaching its maximum limit the ice began downwasting, thus leaving the upper slopes to be attacked by frost while ice stagnated in the valleys. The intense cryergic denudation of the upper hillslopes led to the infilling of the abandoned meltwater channels at the same time as boulder clays and fluvioglacial deposits were accreting beneath or at the edge of the rapidly down-wasting ice. Still later, in Zone III, further gelifluction caused the frost-weathered mantle and its

Alleröd soil to move down over the till surface, and exposed the bedrock slopes to further frost weathering. This example not only provides details of local Devensian events; as Johnson notes, the interpretation also calls into question the criteria that are commonly used for determining the limits of a former ice sheet. 'Clearly it is not permissible to use the till margin, the upper altitudinal limit reached by erratics, or the position of meltwater channels to fix precise boundaries'. He also deplores, as do many periglacial geologists, the use of the term 'local drift' for what in many parts of the U.K. 'are unequivocably periglacially and not glacially formed deposits' (Johnson, 1968, p. 386). Clearly recent work of this kind on periglacial mass movements and frozen ground phenomena has contributed to our understanding of Pleistocene history and environmental conditions even in those parts of glaciated Britain in which there is no lack of glacial and fluvioglacial deposits. The testimony of periglacial features is of still greater importance in unglaciated Britain where other geological indicators of Quaternary cold stages are absent.

Cryonival morphogenesis. It was southern England that Te Punga described as a 'typical periglacial landscape' possessed not only of an abundance of features indicative of Pleistocene frost action but also of a morphology which was moulded by that frost action and by gelifluction during a succession of cryergic episodes. Therefore the basic question is this: to what extent are the effects of the well-documented cold stages on the superficial geology of unglaciated Britain reflected in its land form? Although periglacialists have been rather less concerned with geomorphological than with geological studies, recent work is beginning to suggest the answer (Williams, 1968; Waters, 1971).

The landscape elements, or individual landforms, which have been interpreted as wholly or mainly periglacial in origin, include such erosional forms as dry valleys, asymmetrical valleys, cryopediments, cryoplanation terraces and tors, and the primarily constructional gelifluction flats or terraces, both coastal and valley-floor.

It is hardly surprising that the chalklands have been found to exhibit the effects of periglaciation more clearly than other southern English landscapes. Chalk is particularly susceptible to frost weathering and other cryonival processes, and the changes in hydrology induced by its conversion from cold stage impermeability to mild stage permeability have ensured the retention of periglacial forms.

Recent studies have corroborated Reid's perceptive interpretation of the coombe rock and its association with the formation of small dry valleys on the Chalk escarpments. Analyses of the morphometry of these valleys and of the nature and age of the deposits which often mask their floors and extend fan-like beyond their outlets have established them as products of Late-glacial cryergic episodes (M. P. Kerney *et al.*, 1964). Investigations of asymmetrical valleys both on the Chiltern Hills (C. D. Ollier and A. J. Thomasson, 1957) and in Wessex (H. M. French, 1972) have revealed a regional consistency in the asymmetry which matches that found in many parts of western and central Europe. Other detailed surveys have identified cryopediments as characteristic elements of the asymmetrical valleys (French, 1973). Cryopediments have been shown to exist as concave slopes of low gradient at the foot of the steeper south-west-facing, valley-side slopes; and are interpreted as replacement slopes of transport which extended outwards as the upper steeper slope retreated. Deposits indicative of surface wash or frost creep are absent and it is apparent that the conveyance of frost-weathered, chalky rubble across the cryopediment to the valley floor was effected almost entirely by gelifluction. In this respect, as in its position on the hillslope, this distinctive periglacial landform of the chalklands differs from the more widely recognised cryoplanation terraces which occur on the middle and upper parts of slopes on more resistant bedrock. The cryopediment is, in effect, a periglacial *glacis* or footslope which must be presumed to develop relatively rapidly on lithologies particularly susceptible to cryergic weathering and mass movements.

Cryoplanation terraces and their relations with frost-riven cliffs and tors have also been identified and examined in many parts of upland Britain (A. Guilcher, 1950; Te Punga, 1956; Waters, 1962). They are characteristic of the upper slopes and summit areas on the Palaeozoic rocks of south-west England, where their distribution, geometry and degree of development reflect the influence of strong lithological and structural controls. They are developed most fully on the metamorphosed sediments and metadolerites which encircle the Dartmoor granite outcrop and on indurated arenaceous sediments on Exmoor, i.e., on well-jointed and closely-bedded rocks which were highly susceptible to frost weathering and which yielded a sufficiency of fines for gelifluction to operate particularly effectively. In the absence of any absolute dating they can only be presumed to represent the cumulative effects of the several Pleistocene episodes of selective gelifraction, nivation and gelifluction.

These and other erosional forms that have been investigated during the last few years exemplify the creation of relief and the diversification of pre-existing slopes by cryergic processes. Evidence from the granite uplands of the South-West also suggests that, at higher elevations at least, the pre-existing land form was diversified rather than smoothed by periglaciation. This was effected by a large-scale, downslope transfer of pre-existing regoliths by slope wash, frost creep and, above all, by gelifluction, which lowered the broad interfluve surfaces and laid bare the tors (Waters, 1965). Indeed, from studies of debris layers in numerous pits it has been possible to reconstruct the history of the progressive emergence of the tors and their subsequent modification as surface features during, at most, two periglacial phases. During the first identifiable cold phase, pre-existing soil and regolith were removed from the upper parts of slopes and deposited lower down as a lower or main head. The succeeding intercryergic phase appears to have been too short or perhaps bioclimatically unfavourable for a new weathering profile to develop on the stripped portions of the slopes, though the main head commonly shows evidence of truncation. The second cold phase was characterised by the downslope transfer of boulders and blocks of sound bedrock, detached from the newly exposed tors, and their deposition as the upper head. Thus as the result of the substitution of a cryergic morphogenetic system, in which the rate of removal of waste greatly exceeded its rate of production by weathering, for the pre-existing rain-and-rivers system (under which weathering rates exceeded those of removal), interfluve surfaces on the granite uplands were lowered by an amount at least equal to the vertical extent of the tors that differentiate them. Beyond the immediate environs of the tors with their encircling masses of boulders and blocks (the upper head), some of which display rudimentary patterns of nets and stripes, the downslope movement of the lower head merely preserved the pre-existing slope form. Consequently the development of cryoplanation terraces is minimal on the granite, not only because the predominantly coarse-grained, massive granite is much less susceptible to frost weathering than are the aureole rocks but also because the thickness of the pre-existing regolith available for cryergic redistribution precluded the exposure of bedrock slopes.

The downslope transfers of waste were also responsible for the constructional landforms, the flats and terraces built of geliflual material. In south-west England, for example, they occur in sheltered coastal locations, where relatively thick accumulations of head remain on the raised beach platform, occupy the floors of first- and second-order basins and form the so-called 'rubble-drift' terraces along third-order streams on the granite outcrops (Waters, 1971).

Recent work on different aspects of the periglacial geomorphology of southern England certainly suggests that the land form was modified by cryonival processes. The extent of the periglacial metamorphosis varied spatially and altitudinally and was clearly related to lithology and structure. Landforms were created and the surface diversified in some areas while in others the amplitude of relief was reduced by mass transfers of both pre-existing regoliths and material provided by frost weathering of exposed bedrock.

Where it has been possible to suggest dates for these periglacial modifications they have been attributed to the Devensian. The implication of this is, to say the least, far-reaching. For instance, field evidence suggests that none of the details of the relief on the granite areas in south-west England is older than the periglacial phase to which the lower (or main) head testifies. But the preceding mild phase of selective bedrock decomposition, together with the production of deeply weathered regoliths, was a prerequisite for the extensive periglacial metamorphosis, after which virtually no trace of the pre-existing landform remained. Broad, 'flat'-topped interfluves on the granite uplands can no longer be interpreted as the terminal surfaces of pre-Pleistocene planations; they are, rather, surfaces at the base of the zone of weathering, stripped of their mantles of waste during the Devensian cold stage; their negligible gradients are indicative only of the minimal slopes across which the geliflual transfers were effected. Indeed, if a similar wholesale stripping of regoliths occurred during each of the preceding cold phases, these uplands may well have been lowered by some 40–60 m during the Quaternary. It is not possible to estimate the magnitude of the denudation in other parts of south-west England, where pre-periglacial weathering was less, but frost weathering was more, important than it was on the granite outcrops (Waters, 1971, p. 30).

In spite of the steady, albeit unspectacular, progress of periglacial studies during the last 25 years, the course of Quaternary landform development in southern Britain is far from clear; further investigations of relict cryergic features might be expected to provide additional illumination.

REFERENCES

Ball, D. F. and Goodier, R. (1970), 'Morphology and distribution of features resulting from frost-action in Snowdonia', *Fld Stud.* **3**, 192–217.

Borlase, W. (1758), *The natural history of Cornwall* (Printed for the author, Oxford), 326 pp.

Breuil, H. (1934), 'De l'importance de la solifluxion dans l'étude des terrains quaternaires de la France et des pays voisins', *Rev. Géogr. phys. Géol. dyn.* **7**, 269–331.

Chandler, R. J. (1970), 'Solifluction on low-angled slopes in Northamptonshire', *Q.J. Engng Geol.* **3**, 65–9.

Connolly, A. P., Godwin, H. and Megaw, E. M. (1950), 'Studies in the post-glacial history of British vegetation: XI. Late glacial deposits in Cornwall', *Phil. Trans. R. Soc.* **234B**, 397–469.

De la Beche, H. T. (1939), 'Report on the geology of Cornwall, Devon and West Somerset', *Mem. geol. Surv. Gt Br.* 648 pp.

Dines, H. G., Hollingworth, S. E., Edwards, W., Beecham, S. and Welch, F. B. A. (1940), 'The mapping of head deposits', *Geol. Mag.* **77**, 198–226.

Fisher, O. (1866), 'On the warp, its age and probable connection with post geological events', *Q. J. geol. Soc. Lond.* **22**, 553–65.

French, H. M. (1972), 'Asymmetrical slope development in the Chiltern Hills', *Biul. peryglac.* **21**, 51–73.

—— (1973), 'Crypediments on the chalk of southern England', *Biul. peryglac.* **22**, 149–56.

Groom, G. E. (1956), 'The development of the Dale valley', *Ann. Rep. Fld Stud. Council,* 34–42.

Godwin-Austen, R. A. C. (1851), 'Superficial accumulations of the coasts of the English Channel', *Q. J. geol. Soc. Lond.* **7**, 118–36.

Guilcher, A. (1950), 'Nivation, cryoplanation et solifluction quaternaires dans les collines de Bretagne occidentale et du Nord de Devonshire', *Rev. Géomorph. dyn.* **1**, 53–78.

Högbom, B. (1914), 'Über die geologische Bedeutung des Frostes', *Bull.geol. Instn Univ. Upsala* **12**, 257–389.

Hutchinson, J. N. (1974), 'Periglacial solifluxion: an approximate mechanism for clayey soils', *Géotechnique* **24**, 438–43.

Johnson, R. H. (1968), 'Four temporary exposures of solifluction deposits on Pennine hillslopes in north-east Cheshire', *Mercian Geol.* **2**, 379–87.

Kerney, M. P., Brown, E. H. and Chandler, T. J. (1964), 'The Late-glacial and Post-glacial history of the chalk escarpment near Brook, Kent', *Phil. Trans. R. Soc.* **248B**, 135–204.

Lozinski, W. (1909), 'Über die mechanische Verwitterung der Sandstein in gemassigten Klima', *Bull. Acad. Sci. Cracovie* (*Cl. sci. math. et nat.*) **1**, 1–25.

—— (1912), 'Die periglaziale Fazies der mechanischen Verwitterung', *C.r. 11th int. geol. Congr.* (*Stockholm, 1910*) **2**, 1039–53.

Mottershead, S. N. (1971), 'Coastal head deposits between Start Point and Hope Cove, Devon', *Fld Stud.* **3**, 433–53.

Ollier, C. D. and Thomasson, A. J. (1957), 'Asymmetrical valleys of the Chiltern Hills', *Geogr. J.* **122**, 71–80.

Peltier, L. C. (1950), 'The geographic cycle in periglacial regions as it is related to climatic geomorphology', *Ann. Ass. Am. Geogr.* **40**, 214–36.

Potts, A. S. (1971), 'Fossil cryonival features in central Wales', *Geogr. Annlr* **53A**, 39–51.

Reid, C. (1887), 'On the origin of the dry chalk valleys and of the coombe rock', *Q. J. geol. Soc. Lond.* **43**, 364–73.

Sparks, B. W., Williams, R. B. G. and Bell, F. G. (1972), 'Presumed ground-ice depressions in East Anglia', *Proc. R. Soc.* **327A**, 229–43.

Te Punga, M. T. (1956), 'Altiplanation terraces in southern England', *Biul. peryglac.* **4**, 331–8.

—— (1957), 'Periglaciation in southern England', *Tijdschr. K. ned. aardrijksk. Genoot.* **64**, 401–12.

Waters, R. S. (1962), 'Altiplanation terraces and slope developments in Vest-Spitsbergen and south-west England', *Biul. peryglac.* **11**, 89–101.

—— (1965), 'The geomorphological significance of Pleistocene frost action in south-west England', in Whittow, J. B. and Wood, P. D. (eds.) *Essays in geography for Austin Miller* (Univ. of Reading), 39–57.

—— (1971), 'The significance of Quaternary events for the landform of south-west England', in Gregory, K. J. and Ravenhill, W. L. D. (eds.) *Exeter Essays in Geography* (Univ. of Exeter), 23–31.

Watson, E. and S. (1974), 'Remains of pingos in the Cletwr basin, southwest Wales', *Geogr. Annlr* **56A**, 213–25.

Williams, R. B. G. (1968), 'Some estimates of periglacial erosion in southern and eastern England', *Biul. peryglac.* **17**, 311–35.

—— (1969), 'Permafrost and temperature conditions in England during the last glacial period' in Péwé, T. L. (ed.) *The periglacial environment* (Montreal), 399–410.

Wood, S. V. (1882), 'The Newer Pliocene Period in England', *Q. J. geol. Soc. Lond.* **38**, 667–745.

II

TROPICAL GEOMORPHOLOGY: PRESENT PROBLEMS AND FUTURE PROSPECTS

IAN DOUGLAS
(*University of New England, Australia*)

Tropical geomorphology is slowly emerging from a somewhat dilettante occupation for expatriate, peripatetic, physical geographers and Quaternary geologists, and is becoming a dynamic contributor to inter-disciplinary studies of the potential, response and limitations of the tropical environment as the habitat of the world's most rapidly growing populations. The great naturalists and travellers of the nineteenth century set the scene for recording the earliest impressions of the character of tropical landform evolution. While the first European explorers of the interiors of Africa and Australia were finding their concepts of perennial river systems upset by the inland deltas of the Niger, Congo, Nile, Warrego and Coopers Creek, the colonial administrators of the Indian sub-continent were coping with the practical problems of managing irrigation schemes and using local building materials. Buchanan's writings on laterite (1807) and early studies of fluvial dynamics were among the products of this colonial phase. Yet the exploration and administration of colonial territories led to the first collections of fundamental geomorphological observations on tropical landforms. Perhaps the Germans, leaders in scientific investigation at the end of the nineteenth century, with explorers in the tradition of Alexander von Humboldt, produced the most useful early descriptions and syntheses of tropical landforms. In New Guinea, K. Sapper (1910, 1914) and W. Behrmann (1917, 1921) explored, observed, recorded and later produced syntheses on landform development under humid tropical conditions. S. Passarge (1904, 1923), E. Obst (1915) and W. Bornhardt (1900) described the level surfaces and isolated hills of east African inselberg landscapes, while in the wetter, forested Cameroons, C. Guillemain (1908, 1914), Passarge (1910) and F. Thorbecke (1911, 1914) examined the interrelationships of recent volcanics, older intrusive rocks and inselberg landscapes associated with a short but marked dry season. Passarge also travelled extensively in South America, but the major German contribution to tropical geomorphology from that continent was made later by F. W. Freise (1930, 1932, 1934, 1936) in a series of investigations of processes of denudation and climatic conditions under rain forest. With this background, it was not suprising that the first general discussion of tropical geomorphology occurred at a meeting chaired by Thorbecke (1927) in Düsseldorf. At this conference, the contrast between equatorial continuously humid climates and tropical climates with a dry season was emphasised and the contributors were asked to examine the extent to which present-day landforms were dependent on present-day climates. For 50 years the question has remained unanswered, and more recent work has merely demonstrated the likelihood that most tropical landforms have evolved through a series of climatic oscillations.

Even though the early German tropical geomorphology was impressive, perhaps the less prolific but equally observant American travellers to the Pacific (J. D. Dana, 1850) and South America (J. C. Branner, 1896) made process observations which gave rise to some of the generalisations about tropical conditions that have persisted to the present day. R. J. Chorley, A. J. Dunn and R. P. Beckinsale (1964) give Dana credit for being the first to publicise to Europeans the influence of tropical climate on landforms. Dana (1850) illustrated the fundamental roles of lithology and aspect in the landforms of Oahu and Hawaii, showing how the flutings of the windward side of the Koolau Range of Oahu were deep, winding, narrow valleys cut by running water during heavy rainstorms. Quoting W. Hopkins (1844) on the 'sixth power law of traction', Dana comments:

> There is every thing favorable for degradation which can exist in a land of perpetual summer: and there is a full balance against the frosts of colder regions in the exuberance of vegetable life, since it occasions rapid decomposition of the surface, covering even the face of a precipice with a thick layer of altered rock, and with spots of soil wherever there is a chink or shelf for its lodgement.

The seasonal contrast between the relative calm of base-flow conditions and the rapid rise of streams following intense rainfall was shown to be responsible for the changing of 'a lofty volcanic dome. . .to a skeleton island like Tahiti'.

Although Branner's evidence (1896) that, in the tropics, rain has four or more times as much nitric acid as in the temperate zone has been questioned, his comments that the process of exfoliation in Brazil is not essentially different from that common in other parts of the world, and that a given quantity of rainfall evenly distributed throughout the year would do less work than if it were concentrated into a few months, have been substantiated by more recent observations, such as those of C. D. Ollier (1965) on granite weathering and F. Fournier (1960) and I. Douglas (1967) on denudation rates.

From these beginnings the period between the two world wars saw an evolution of regional case studies, particularly within the context of colonial administration and education, with a few attempts at international comparisons and synthesis. Many of the early German impressions of tropical geomorphology are summarised in Sapper's *Geomorphologie der feuchten Tropen* (1935) while P. D. Krynine's review (1936) of Sapper's work puts the American experience in contact with European views. Despite the syntheses, preoccupations with specific types of landforms, perhaps those strangest to European eyes, in particular colonies, led to an often specialised type of discussion by a national group of tropical geomorphologists. Thus their contrasting experiences in east Africa and New Guinea gave German geomorphologists clear notions that the selva landscapes of continually wet (*immerfeuchten*) tropics have a characteristic feral relief (C. A. Cotton, 1957) with V-shaped valleys (*Kerbtäler*) as against the broad, gentle valleys (*Flachmuldentäler*) of the seasonally wet (*wechselfeuchten*) savanna climates (H. Louis, 1961).

Most geomorphological work in British colonies was a by-product of geological or soil surveys, but such themes as drainage evolution, erosion surfaces and the geomorphological significance of inselbergs and laterites recurred and were subsequently re-examined by academic geomorphologists after 1945. The drainage of Uganda (E. J. Wayland, 1921), peninsular India (D. N. Wadia, 1966) and peninsular Malaysia (J. A. Richardson, 1947) were all interpreted in Davisian terms, while erosion surfaces in Africa (F. Dixey, 1942; Wayland, 1934), Sri Lanka (Wadia, 1941, 1945) and India (A. M. Heron, 1938) were later found to have much in common. While both inselbergs and laterites were mentioned incidentally in geological surveys, such as J. B. Scrivenor's *Geology of Malaya* (1931), two particularly significant pedological works were J. B. Harrison's study of weathering (1933) and G. Milne's development of the catena concept (1935).

In Indonesia, Dutch scientists pursued detailed studies of Java and Sumatra (L. M. R. Rutten, 1928) some of which benefited from excellent topographic maps (A. J. Pannekoek, 1939). The pressure for commercial development in the Belgian Congo saw a steady growth of pedological and geomorphological studies (J. Cornet, 1896; J. Bayens, 1938; M. Robert, 1927) while French pedological studies (L. Aufrère, 1932, 1936; F. Blondel, 1929; J. de Lapperent, 1939) led eventually to the integration of pedogenesis and landscape evolution in H. Erhart's concept (1956, 1961) of alternating periods of stability and landscape disruption. By 1940, E. de Martonne (1939, 1946) had highlighted the importance of tropical geomorphology but he did not argue that tropical landform evolution was inherently different from that in the humid temperate zone (de Martonne, 1951).

Post-1945 trends have confirmed rather than eliminated these indications of national specialisations, but a counter-current to their generally academic trends, a great growth of practical applications of geomorphology to tropical regions, began. In their colonial period the Dutch had problems of both soil and water management that led to fundamental enquiries (E. C. J. Mohr and F. A. Van Baren, 1954; H. Boissevain, 1942) of soil genesis and fluvial geomorphology. The problems encountered by irrigation schemes in India led to enquiries into the stability of natural channels which are still basic to any examination of meandering (G. Lacey, 1930, 1935). Studies of soil erosion in French African territories eventually provided the basis for global syntheses of the relationship between climate and erosion (Fournier, 1960) while in the post-war period the need to have basic information on the land saw geomorphologists working as members of land evaluation teams, particularly with C.S.I.R.O. in northern Australia and New Guinea (J. Mabbutt and G. A. Stewart, 1963) where work culminated in the production of a geomorphological map of Papua New Guinea in time for the country's independence (E. Löffler, 1974). Fundamental contributions to tropical weathering and denudation (B. P. Ruxton, 1967, 1969; Ruxton and J. McDougall, 1967), karst morphology (M. J. J. Bik, 1967; J. N. Jennings and Bik, 1962), fluvial geomorphology (J. G. Speight, 1965a, 1965b, 1967) and tropical high mountain geomorphology (Löffler, 1970, 1971, 1972) were the direct result of this programme. The Overseas Surveys of Britain developed similar land evaluation programmes in Africa and eventually such work was sponsored by UNESCO (R. L. Wright, 1971).

TROPICAL GEOMORPHOLOGY AS A BRANCH OF CLIMATIC GEOMORPHOLOGY

French (de Martonne and P. Birot, 1944), German (Louis, 1957) and American (L. C. Peltier, 1950) geomorphologists produced arguments that special tropical geomorphological conditions give rise to unique landform assemblages whose occurrence in extra-tropical zones must indicate the former existence of tropical climates. In certain cases, such as karst (H. Lehmann, 1953; Birot, 1960; M. M. Sweeting, 1958), the forms are strikingly distinct, but in other cases the contrasts among forms in different humid climates are but differences of degree rather than of kind. Notions of weathering and chemical action in the humid tropics prevalent in the post-war period are typified by the following statement:

> In the humid tropics, under the influences of high temperature and humidity, weathering takes place rapidly, leaching is very effective, and organic matter, although produced in greater abundance, is generally much more rapidly and completely decomposed than in cooler and drier climates. Leaching under some tropical conditions removes a wider, or at least a different, range of substances. Even silica, which is the leaching residue in many northern cold regions, disappears from many tropical soils, leaving a residue of iron and aluminium oxides (R. R. Fosberg, 1962).

It is held that rock decomposition is more rapid than the transport of waste down slopes, which in turn is more rapid than the evacuation of material by streams (Birot, 1960). Generalisations on weathering in rain forests say that it yields a waste mantle often tens of metres thick (J. Büdel, 1957; Birot, 1968). A double planation process, through the inward and downward migration of the weathering front which accompanies the surface stripping of the weathered mantle, is sometimes recognised (Büdel, 1957; J. J. Nossin, 1964). Weathering is so effective that the upper regolith rarely has individual particles exceeding 3-4cm in diameter (Birot, 1966; J. P. Bakker, 1967). This and the continued chemical action on particles in rivers means that coarse particles for the abrasion of streambeds are lacking; hence, vertical erosion is weak and waterfalls persist well downstream in tropical rivers (J. Tricart, 1959).

The generalisations on tropical landforms suggest that avalanching, landslides, slumping and underground earthflows are important on steep slopes (Freise, 1938; Sapper, 1935), while on gentler slopes, rainwash, soil creep, subsurface lateral mechanical eluviation and removal in solution are said to be of roughly equal importance (Birot, 1966). Slopes are held to retreat (Nossin, 1964), particularly where they are steep (K. Bryan, 1940; Freise, 1938), flattening from below upwards (Birot, 1966; Bryan, 1940). The characteristic humid tropical landscape has often been taken to be one of wide plains with scattered inselbergs (M. F. Thomas, 1965; J. C. Pugh, 1966), although this probably reflects the emphasis on African research by most post-1945 British tropical geomorphologists.

From these generalisations emerge the limitations of experience of the expatriate Europeans and the lack of flow of information between workers in different tropical regions. However, the most serious deficiency was the lack of direct observations of process and the reliance on using form and depth of regolith to infer efficiency of process. Yet right from the start of tropical landform studies, local and regional climatic and vegetative contrasts were shown to be significant for the efficiency of processes. Examining the Nicaragua canal route, C. W. Hayes (1899) found that the contrast between the humid east and the seasonally dry west of the country give rise directly to 'very striking differences in vegetation, and either directly or indirectly, to differences in the appearance and structure of the soils, in the topographic forms of the land surface, and in the effectiveness of various physiographic processes'. Hayes went on to argue that:

After a careful study of the region it was considered that the absence of frost more than counterbalances the enormous rainfall, and that degradation of the surface is, on the whole, slower than in temperate regions, where the rainfall is less than a quarter of that in Nicaragua, but where the surface soil is thoroughly loosened by the action of frost.

As soon as detailed process measurements were made, such statements could be tested. J. Corbel (1957a, 1957b), for example, finds that the rate of removal of silica in solution is higher in equatorial regions than in other areas; G. Rougerie (1961) shows that there is little difference in the rate of loss of silica from crystalline rocks in the central-west of the humid tropical Ivory Coast and from similar rocks in Limousin in humid temperate France. Rougerie, however, notes that silica concentrations are highest in the driest periods of the year. J. P. Carbonnel (1964a) confirms this, suggesting that:

l'érosion chimique est un phénomène lent qui n'a pas le temps de se produire quand les débits sont très élevés (1964b, p. 154).

On the other hand, only small variations of silica concentrations with discharge are noted by S. N. Davis (1964) who finds that stream water appears to acquire most of its silica within a relatively short time, and that silica concentrations are not primarily controlled by temperature. This question of the mobility of silica serves to illustrate both the lack of general knowledge of the geochemistry of weathering and denudation in the tropics, and the impossibility of making a general statement on denudation processes in the humid tropics from the results of a few isolated quantitative studies in widely differing areas. No wonder, then, that Tricart (1965) complained of lack of information when he wrote the first edition of his tropical geomorphology:

Nos connaissances sur les particularités morphogénétiques des régions chaudes restent. . .encore trés insuffisantes, . . .il nous manque encore beaucoup de données précises, beaucoup d'éléments quantitatifs notamment (Tricart, 1965, p. 7).

INTERDISCIPLINARY STIMULI FOR TROPICAL GEOMORPHOLOGY

In the last decade, process studies in tropical geomorphology have moved rapidly, but much of the momentum has come from outside the traditional bounds of the discipline, particularly from ecosystem studies in both rainforest and savanna, but also from hydrological projects. The International Biological Programme and International Hydrological Decade have both helped to encourage projects in developing countries.

The definitive study of weathering and erosion rates is of necessity a biogeochemical study, taking into account the flow of elements from the abiotic to the biotic and back to the abiotic components of the ecosystem (E. P. Odum,

1971; N. D. Turvey, 1974). Since most plant nutrients and other elements are transported in aqueous solutions through the ecosystem, it is necessary and convenient to study the processes through the hydrological cycle, with the drainage basin as a minimum ecosystem unit for such an investigation. The first conclusions on such rainforest ecosystem dynamics came from the work of P. H. Nye (1961), Nye and D. J. Greenland (1960) and Rougerie (1960) in West Africa, but more recently the concept of the drainage basin as an ecosystem has drawn geochemists (Y. Tardy, 1969), biologists (H. T. Odum *et al.*, 1970; J. B. Kenworthy, 1971) and hydrologists (Turvey, 1974, 1975) to examine nutrient cycling in rainforest catchments.

The role of water in the rainforest is important in maintaining a flow of nutrients through the system. H. T. Odum *et al.* (1970) measured the hydrogen budget (essentially the rate of water movement) for the rainforest. They found that approximately 50 per cent of the inflow was used in evapotranspiration, and that hydrogen in most compartments, except for wood, turned over in a few days. Most of the hydrogen maintained in the forest was as water. Water in the litter and soil accounted for 5803 $g H m^2$, compared with 366 $g H m^2$ stored in wood, 248 $g H m^2$ in litter and 57 $g H m^2$ in live leaves. Hydrogen was incorporated into the organic state at a rate of 2·2 $g m^2$ day^{-1}. Taken with the above storage rates, this demonstrates the rapid passage of large amounts of water from the soil storage compartment through the system and out through evapotranspiration.

This internal cycle of nutrients in the forest is complex, but H. T. Odum (1970) suggests that a euphotic upper zone can be compartmentalised and separated from a lower regenerative zone, the two being coupled by the mineral cycle and by feedback regulating energy and nutrient-flow mechanisms. In this system, losses to drainage water are small. At the El Verde site, the mean concentration of dissolved solids in stream water is 27ppm, in rainfall 22ppm, and in throughflow 53ppm (P. Sollins and G. Drewry, 1970). Turvey (1974) regards this as evidence of the highly efficient nutrient uptake and return by the root systems of the regenerative zone.

Such a nutrient cycling approach has also been adopted in studies of dry evergreen forest, woodland and savanna communities in upper Shaba, Zaire (R. Freson, G. Goffinet and F. Malaisse, 1974). The total annual litterfall is 3·5 times lower in *miombo* (fire and hatchet-induced woodland) than in the *muhulu* (dry evergreen forest) climax vegetation. The creepers, mosses and epiphytes of *muhulu* create a hydrological situation quite different from that of the *miombo* with its sparse shrub layer. The rate of average abundance of soil intercalcic macrofauna extracted during successive rainy and dry seasons decreases from about 4:1 in savanna, 3:1 in *miombo*, to 1·5:1 in dry evergreen forest. Such a variation with vegetation cover under the same regional climate illustrates the significance of local site variations for the organic activity and consequently for geomorphological processes in the tropics.

Instrumented catchment experiments in the tropics have also increased in number in the last decade. The excellent series of studies from Tanzania (A. Rapp, L. Berry and P. H. Temple, 1972) have been paralleled in West Africa by the Korhogo experimental study in savanna (Tardy, 1969) and the Amitoro catchment study in forest (P. Mathieu, 1971). Turvey (1975) in New Guinea and

Kenworthy (1971) in Malaysia examined rainforest catchments, while in the Cameroons, savanna catchments ranging in size from 77 000 to 4·3 km^2 have been studied by J. F. Nouvelot (1969). The growing body of information from these and similar studies is beginning to indicate how processes vary from one tropical environment to another.

A further type of interdisciplinary stimulus to tropical geomorphology has come from the limnological and ecological studies of Amazonia (H. Sioli, 1956, 1961; Sioli, G. H. Schwabe and H. Klinge, 1969) which have reinforced forest nutrient cycling concepts. E. J. Fittkau (1973) sums up the stability of the Amazonian ecosystem:

The greatest diversity of species known, in a restricted area of forest or in a jungle watercourse, occurs in central Amazonia, a region offering extremely poor food supply. Here forest stands on weathered and washed-out soils, from which practically no nutrients can be freed. Unavoidable losses in the cycling of materials can thus be compensated only by capture of allochthonous nutrients, particularly those brought to the forest by rain. If the number of plant species in such a forest increases, it is to be expected that this greater differentiation of the vegetation will favour not only energy uptake, but also the retention of both allochthonous nutrients and those bred by the continual decomposition of dead organisms; as a result the nutrient cycle is more effectively closed. Thus we can speak of a biological litter for autochthonous and allochthonous nutrients in the ecosystem, the performance of which increases with increased species diversity, contributing to the stabilization of the system.

However, the same group of workers have highlighted the problem of climatic change in the Amazon region, raising the questions of alternating arid-humid systems discussed by H. F. Garner (1974) and shown by Tricart (1973, 1974a, 1974b) to have affected the geomorphological evolution of the whole area. Reviewing geomorphological and other evidence, Fittkau (1974) concluded that during periods of Quaternary low sea level, lowered watertables and more rapid surface runoff could have caused rainforest to retreat from many parts of the Basin. Forest refuges as described by Garner (1974) probably remained. Similar questions about forest evolution and species distribution have concerned botanists in south-east Asia. While evidence of climatic change in the lowlands of Malaysia and Borneo is lacking, it is highly probable that, during times of low sea level a heterogeneous patchwork of forest and savanna existed in Sundaland, the general pattern being somewhat analogous to that existing in equatorial South America today (P. Ashton and M. Ashton, 1972). H. Th. Verstappen (1975), however, considers the extent of forest patches in Quaternary dry periods to have been much more restricted than that suggested by biologists.

While these interdisciplinary experiences have changed the thinking of some geomorphologists, others still keep their eyes on the major forms and the stratigraphic details of sections through colluvial and valley-fill deposits. Thus many geomorphologists are still primarily concerned with the fundamental questions of slope evolution, laterite development and inselberg morphology with which their late nineteenth-century predecessors were so involved.

WIDELY RECOGNISED PROBLEMS OF TROPICAL GEOMORPHOLOGY

The types of investigation currently pursued by tropical geomorphologists range from those concerned with fundamental measures of the rates of operation of

TABLE I

Problems for Tropical Geomorphology

Type of problem	*Authors recognising problem*
Facts about landforms	Swan (1970)
Nature of slope and valley development	Louis (1961); Zonneveld (1975)
Earth movements and slope form	Simonett (1967)
Relative efficiency of soil creep, slope work and subsurface eluviation	Ruxton (1967); Simonett (1968)
Role of mudflows in feral selva relief	Simonett (1968)
Controls of streamheads and drainage density	Morgan (1972)
Influence of structure on tropical landforms	Michel (1973)
Influence of lithology on tropical landforms	Mainguet (1972), Swan (1970)
Variability of depth of weathering	Simonett (1968)
Karst evolution, especially formation of karst plains	Simonett (1968)
Weathering of granite rocks	Ollier (1965), Tardy (1969)
Geomorphological role of termites	De Ploey (1964), Alexandre (1966), M. A. J. Williams (1968)
Rates of denudation	Paton and M. A. J. Williams (1972), Zonneveld (1975)
Fluvial erosion	Tricart (1959), Michel (1973)
Lateritisation	Maingien (1964), Paton and M. A. J. Williams (1972)
Stone-lines	Tricart (1961), Collinet (1969)
Relation of landform assemblages to morphogenetic systems	Stoddart (1969), Mainguet (1972)
Scale in the analysis of tropical landforms	Stoddart (1969), Morgan (1976)
Erosion surface development	Swan (1970), Demangeot (1969)
Inselberg landscapes	Thomas (1965), Pugh (1966)
Quaternary climatic changes	Michel (1973), M. A. J. Williams (1975), Zonneveld (1975)
Anastomosing and deranged channels	Garner (1966, 1974)
Relative significance of actual and inherited forms	Thomas (1965), Demangeot (1969), Twidale (1976)
Neotectonics	Blake and Ollier (1969)
Depositional landforms	Sternberg (1975)

processes and the role of scale in relationships between landform, lithology and climate, to the detailed study of specific features such as inselbergs and stone-lines (Table I). Some problems, such as the role of earth movements, can be seen as more likely to arise in one tectonic or physiographic situation than in others. Concern with the role of termites, for example, has come from the seasonally wet environments of the Northern Territory of Australia (M. A. J. Williams, 1968) and Shaba, Zaire (J. Alexandre, 1966). Most significant perhaps are the questions about climatic geomorphology raised by D. R. Stoddart (1969):

(i) Are there unique tropical landforms?

(ii) Are there tropical climatic regions corresponding to morphological regions?

(iii) Have changes of climate produced recognisable sequences of tropical landforms?

(iv) How does tropical geomorphology differ from other sorts of geomorphology?

In our present state of knowledge the answers to these questions tend to be negative. With all that is now known about weathering, pedological and denudation processes, it is hard to argue that there cannot be similar humid landform genesis in both temperate and tropical regions, given similar rock types, similar precipitation and similar length of time. Tardy (1969), for example, calculates that in humid climates, cold, temperate or tropical (with base flow greater than 500 mm), less than 200 000 years are needed to alter completely 1 m^3 of rock. About 30 000 years are necessary, in either tropical or temperate environments, seriously to decompose 1 m^3 of granite weathering into kaolinite (Table II). M. Mainguet (1972) felt that while no particular sandstone landform type could be said to occur under every climate, a series of landform characteristics which are peculiar to sandstones and which vary to a certain extent with climate, especially under the influence of chemical weathering but less so under mechanical attack, could be recognised. The lithological influence on landform is modified by climate.

TABLE II

Silica denudation based on SiO_2 content of waters (after Tardy, 1969)

Rock	*Base flow (mm)*	*siO_2 ppm*	*Years required to transform 1m³ into*	
			Kaolinite	*Gibbsite*
Norwegian granite	1250	3	85 000	225 000
Vosges granite	850	9·2	52 000	105 000
Alrance (S.E. Massif Central) granite	680	11·5	41 000	100 000
Alrance migmatite	680	5·9	100 000	200 000
Korhogo (savanna, Ivory Coast) migmatite	540	20·0	65 000	100 000
Alrance amphibolite	640	14·0	68 000	110 000
Madagascar basalts	1500	16·0	40 000	60 000

While Büdel (1969) has claimed that climatic geomorphology systematically investigates the variations in processes and forms caused by climate and that the tropics can be divided into two zones, one of extensive planation and the other, equatorial, of partial planation, there is little reason for supposing that the action of water on the tropical landscape is any different in kind from that in a temperate area with similar rainfall and ground cover. Indeed it seems highly probable that variations in the rates of operation of processes and the types of landform they create within either the humid temperate or humid tropical zones are greater than variations between them. Any correlation between tropical climatic regions and morphological regions might arise because climate and morphology both depend on tectonic style and structure which give rise to the distribution of mountain ranges, stable Gondwanaland shield surfaces or sedimentary basin structures.

Changes in climate do appear to have produced recognisable sequences of tropical landforms as illustrated by Garner's study (1966) of the Rio Caroni and P. Michel's examination (1973) of the Senegal. The biostasy-rhexistasy concept of Erhart (1956) and the periodicity notions of Mabbutt and R. M. Scott (1966) offer frameworks for further investigation of this question.

Tropical geomorphology does not differ from other kinds of geomorphology. Recently all the tools of quantitative landform analysis have been applied to tropical drainage networks, slopes and river channels. Morphometric analysis in West Malaysia (R. J. Eyles, 1968, 1969; R. P. C. Morgan, 1970) has revealed both the influence of structure on the drainage network and how the country is deeply dissected, with much alluviation and many isolated prominent landforms. Structural and local climatic influences have been shown to affect slopes in Sri Lanka (Madduma Bandara, 1974). Morphometric analysis of karst in New Guinea (P. W. Williams, 1971, 1972) has shown that a range of karst styles can develop despite essentially similar morphogenetic conditions, but that A. Grund's universal karst cycle (1914) is equally applicable to tropical and extra-tropical karst evolution. Nor is there any evidence that the hydraulic geometry of tropical rivers is basically different from that of temperate rivers (L. A. Lewis, 1966, 1969; J. B. Thornes, 1970; A. Gupta, 1975).

QUESTIONS THAT OUGHT TO BE ASKED BY TROPICAL GEOMORPHOLOGISTS

While fundamental questions of landform genesis remain unanswered, the discipline of geomorphology may still in part shrug off the questions raised by other aspects of the environment. Yet few tropical geomorphologists in their field traverses can fail to notice the geomorphological problems that arise from man's impact on the landscape. Even so, W. C. Clarke's comment (1973) about geographers applies particularly strongly to geomorphologists:

> Geographers have always seen a connection between man and environment, even if they have sometimes seen the flow of influence going only one way. However, with few exceptions, geographer's studies have been of scholarly interest only.

As C. H. Leigh and K. S. Low (1976) point out:

> Because of their limited interest in human impact on the environment, geomorphologists remained unreceptive to, or disinclined to follow the leads suggested by naturalists, human geographers, or contemporaries in other disciplines.

Since most contributors to the literature on tropical landforms are still based in extra-tropical countries and are therefore guests in the nations where they carry out fieldwork, the remoteness of many contemporary geomorphological investigations from the people of the host nations must surely be questioned. Both Tricart (1965) and Garner (1974) conclude their textbooks by commenting on man's impact on the environment and the problems which natural feedback mechanisms bring.

> C'est dans les régions chaudes et humides que les ruptures d'équilibre anthropique déclenchent les plus violentes catastrophes, car le potentiel érosif du climat est grand et les sols sont fragiles. Raison de plus pour mieux connaitre cette dynamique du milieu naturel dont la géomorphologie fait partie afin de

sauvegarder ces richesses et de mieux utiliser chaleur et humidité d'origine cosmique pour nourrir des Hommes dont la plupart ont faim (Tricart, 1965, p. 302).

Tricart (1962, 1966; Tricart and Michel, 1965) and his colleagues from the Centre de Géographie Appliquée have had long experience of co-operative applied research in South America. Normally working in association with personel from local universities, Strasbourg geomorphologists have participated in land and water resource inventories in Venezuela, including the problems of development of the region to the south of Lake Maracaibo. In Colombia, P. Usselman has led similar land and water resource enquiries, while J. Khobzi has been associated with interdisciplinary studies of the relationship between geomorphology and pedology. Tricart (1956) has often commented on man's impact on the tropical environment, advocating the use of geomorphological mapping (Tricart and Michel, 1965) as one way of ascertaining areas of geomorphological risks.

Land evaluation has been widely adopted in resource inventories in tropical countries (A. Young, 1973a, 1973b) but some of the maps produced have not had much impact on the way in which the land is being used. In assessing the merits of various types of land-systems mapping and geomorphological mapping, R. U. Cooke and J. C. Doornkamp (1974) leave doubts about the effective use of either technique to identify and plan management programmes for geomorphological hazards in tropical countries. Perhaps the military version of the technique is the most readily applicable (MEXE, 1965).

Some geomorphologists from developing countries see the potential of land evaluation (E. A. Olofin, 1974) and both land-systems mapping and geomorphological mapping have been widely adopted by UNESCO. Mapping, however, is but the resource inventory stage of environmental management. G. E. K. Ofomata (1965, 1966), for example, advocates the participation of geomorphologists in road planning and construction teams and in the provision of guidelines for rural land-use. M. J. F. Brown (1974) has shown how geomorphologists can appraise the consequences of mining development and mining waste disposal for landforms downstream. Such studies are not without repercussions, for, in Brown's case, Bougainville Copper have produced a strong defence of their environmental management programme. The clearing of land, now proceeding more rapidly than ever before in countries such as Brazil, Indonesia and Malaysia, brings a series of environmental problems of which soil erosion (Leigh, 1973) is one of the most marked. The consequences of such erosion include changed channel geometry, altered river-flow régimes, and often increased flood risks downstream.

Despite all these rural land effects, it is in the burgeoning urban areas of the tropics that the most immediate tasks for geomorphologists lie. The steep hillsides of the inselbergs of the Rio de Janeiro district have given that city slope-stability problems that are probably greater than those of any other city of its size in the world (R. Mousinto de Meis and R. da Silva, 1968; A. Jones, 1973). Slope failures were associated with rains and early building on the hills and, in the nineteenth century, several hills were removed and others were reafforested. However, as the city has expanded upslope more recently owing to lack of space on the plains, trees have been removed and excavations for building have been placed at successively higher levels. The cuts caused by buildings at the toes of the slopes have severed the surface soil mantle at its most critical point. The inhabitants of both the tower blocks at the base of hills in the Copacabana district and the *favelas* (squatter settlements) on steep hillsides are equally at risk.

Where terrain is steep, urban soil erosion and gullying become hazards, creating flood risks downstream. Ilunga Lutumba (1975) has suggested that it ought to be possible to evaluate geomorphological risks in urban areas and to persuade urban authorities to avoid housing construction in high risk areas. However, perhaps the greater difficulty is in keeping illegal squatters off hazardous vacant land close to the city centre. Nevertheless, there is an urgent need for geomorphologists to apply their knowledge of slope stability, hydraulic geometry, soil erodibility and fluvial sedimentation to the problems of the tropical city.

Just as O. Ribeiro (1972) finds that no geographer can undertake a piece of fundamental research in an area without becoming aware of its significance for the human condition in that area, so no piece of applied geomorphological research is without some feedback for the fundamental questions that concern tropical geomorphologists. If the peripatetic expatriates do not pay attention to these questions, the young geomorphologists from Africa, Asia, Latin America and the Pacific Islands must be given every encouragement to do so. The present signs are that they are far more aware of the real issues than most of their mentors.

TOWARDS AN UNDERSTANDING OF THE GEOMORPHOLOGY OF TROPICAL LANDS

While the literature of tropical geomorphology has expanded rapidly, the answers to some of the questions raised at Düsseldorf 50 years ago are still lacking. The diversity of the tropical environment is now more apparent. The distribution of lowland forest types varies with edaphic conditions and rainfall régimes while lower montane forest occurs at varying altitudes in different mountainous areas of the tropics (P. J. Grubb, 1974). Leaf litter fall varies from one rainforest site to another (Table III). Consequently, denudation pro-

TABLE III

Rate of fall of leaf litter at tropical forest sites

Site	*Litter fall (tons ha yr^{-1})*	*Source*
Shaba, Zaire (*miombo*)	1·4	Freson, Goffinet and Malaisse, 1974
Shaba, Zaire (*muhulu*)	4·6	Freson, Goffinet and Malaisse, 1974
Puerto Rico	4·8	Odum and Pigeon, 1970
Central Amazon Basin	4·8–6·4	Klinge and Rodrigues, 1968
New Guinea (Lower Montane forest)	6·4	Grubb, 1974
Trinidad	6·8–7·0	Cornforth, 1970
Ghana	7·0	Nye, 1961
Nigeria	7·2	Hopkins, 1966
Ivory Coast	7·2–9·2	Bernhard, 1970
Colombia	8·5	Jenny *et al.* 1949

cesses vary from site to site. The detailed pattern of tropical landform evolution thus depends on the ecological conditions of particular sites.

However, tropical landform evolution needs to be looked at on a global scale where the two independent variables of geology and climate can be shown to offer considerable variety within the tropics. Seasonal climatic classifications such as those of F. H. Schmidt and J. H. A. Ferguson (1952) or C. Troll (1964), reveal the true diversity of tropical climates, distinguishing those dominated by seasonal rains, particularly from cyclones, and those with uniformly distributed precipitation. Channel forms, and possibly valley forms, are related to the volumes of water to be carried at peak flows. Thus the channels of streams in north-east Queensland are larger than those of streams on similar rocks with similar annual precipitation in West Malaysia because the rain-runoff events in Queensland come from tropical cyclonic downpours.

Geological features of structure and lithology have been under-emphasised by geomorphologists, even though pedologists (Mohr and van Baren, 1954) have constantly demonstrated the links between geology, landform, soils and vegetation. While the new global plate-tectonic approach to earth structure has had relatively little impact on geomorphology, it is probably an overriding factor in the geomorphological diversity of the humid tropics. F. Machatschek (1955) considers tropical landforms under the following headings:

Die süd- und südostasiatische Kettenbergszone
Das altweltliche Gondwanaland
Ozeanien
Mittelamerika
Das andine Südamerika
Das ausserandine Südamerika

He distinguishes the Gondwanaland remnants from the great Tertiary mountain arcs. Unlike J. Demangeot (1969) who describes tropical landforms solely in terms of the extensive Gondwanaland remnants, Machatschek is at pains to show how tectonic contrasts have produced morphological contrasts. If geomorphology is concerned with going beyond process measurement to explain and account for the landforms actually found on the earth's surface, then, in the tropics, it must follow Machatschek's lead.

In southern and south-east Asia, lands of differing geological history have been pushed close together by the northward movement of the Indian and Australian Gondwanaland plates. The erosion surfaces and duricrusts of Sri Lanka and Australia are in strong contrast with the feral relief of New Guinea and the volcanic relief of Java and Sumatra. The Tertiary rocks of Borneo and the Palaeozoic and Mesozoic formations of West Malaysia offer yet further contrasts of tectonic style. Perhaps the extremes may be seen as the rapid erosion of tectonically active areas such as New Guinea (C. F. Pain and J. M. Bowler, 1973) or fresh volcanoes (Ollier and Brown, 1971) and the extremely slow changes on old, upland surfaces as those of central Africa (S. Alexandre-Pyre, 1967). Where processes are active, slopes are steep, with a highly dissected relief; where slow, planation surfaces are maintained. However, the activity of rivers is frequently a lateral planation process reducing areas near sea level to extensive plains with isolated residuals, such as the coastal plain of Selangor, Malaysia. Dissection down to low Pleistocene sea levels and subsequent alluviation makes it difficult

TABLE IV

Comparison of some pedological and geomorphological features of erosional landscapes in Northern Australia, Sri Lanka, Peninsular–Malaysia, Borneo, New Guinea and Hawaï

Feature	*Australia*	*Sri Lanka*	*Malaysia*	*Borneo*	*New Guinea*	*Hawaii*
Dominant age of landscape	Old	Old	Moderately old	Young	Very young	Very young
Proportion of labile rocks	Low	Low	Moderate	Moderate	High	Moderate
Recent tectonism	Practically none	Practically none	Negligible	Some	Intense	Infrequent
Cainozoic vulcanism	Practically absent	Practically absent	Practically absent	Some	Abundant	Highly abundant
Tropical karst	Practically absent	Virtually absent	Residuals only	Abundant	Abundant	None
Relief	Low to moderate	Moderate	Moderate	Moderate (save for Kinabulu)	Very high	High
Rate of current soil development	Slow	Slow	Slow to moderate	Moderate	Rapid	Rapid
Rate of current erosion and deposition	Low to moderate	Low to moderate	Moderate	High	Very high	Very high
Depth of soil on slopes	Generally shallow	Generally shallow	Variable	Variable	Variable	Variable but usually high
Lateritic crusts	Widespread	Widespread	Fragments	Possible	Absent	Absent
Pallid zone of deep weathering	Widespread	Widespread	Few	Unknown	Rare	Rare

Note: The inspiration for this Table comes from Galloway and Löffler (1972), Table 2.1.

to see the original topography of these plains, save where tin mining exposes the sub-alluvial karst surface of tropical *Karstrandebene*. Eventually the extensive planation surfaces which Demangeot (1969) argues are typical of the tropics may be formed, but the real world has the diversity of tropical landforms which will continue to exist as long as the mid-oceanic ridges spew out lava and the oceanic plates crush against the Gondwanaland remnants. The regional diversity, evident on many counts (Table IV), needs further exploration, with further comparisons at this regional scale, before further examining the details of allegedly 'typical' tropical landforms.

CONCLUSIONS

Although not a special kind of geomorphology, the study of tropical landforms deserves further attention for the information it can convey on processes under various vegetation types, for the opportunity to investigate stable ecosystems, some of which may have persisted for thousands of years, and particularly for the environmental problems which beset the poor and growing populations of tropical countries. In this essay I have attempted to draw together some of the main themes that have persisted through a century of tropical landform study and to demonstrate that cross-comparisons between tropical areas are necessary to avoid the pitfalls of traditional generalisations. It has not been possible to discuss the growing knowledge of Quaternary events in tropical countries which is likely to suggest that few equatorial rainforest areas escaped from the many climatic oscillations. However, as rainforest today extends over a wide range of seasonal climates, it remains possible that, although climates became drier or more seasonal, the forests may not have been eliminated in some areas. The growing interdisciplinary involvement of tropical geomorphologists is likely to lead to a better understanding of both past environments and the future environments which will have to be managed as the pressures of population growth and land development expand. If there is one special appeal in this paper to geomorphologists working in the tropics, it is that they should ascertain the problems of most importance to the people of tropical countries and endeavour to give them priority.

REFERENCES

Alexandre, J. (1966), 'L'action des animaux fouisseurs et des feux de brousse sur l'efficacité érosive du ruissellement dans une région de savane boisée', *Cong. Coll. Univ. Liège* **40**, 43–9.

Alexandre-Pyre, S. (1967), 'Les processus d'aplanissement de piémont dans les régions marginales du Plateau des Biano', *Publs. Univ. Officielle Congo Lubumbashi* **16**, 3–52.

Ashton, P. and Ashton, M. (eds.) (1972), 'Transactions of the second Aberdeen-Hull symposium on Malesian Ecology', *Univ. Hull, Dept. Geography Misc. Ser.* **13**, 122 pp.

Aufrère, L. (1932), 'La signification de la latérite dans l'évolution climatique de la Guinée', *Bull. Ass. Geogr. Fr.* **60**, 95–7 and 457–8.

—— (1936), 'La géographie de la laterite', *C.R. Soc. Biogéogr. France* **13**, 3–11.

Bakker, J. P. (1967), 'Quelques aspects du problème de sediments correlatifs en climat tropical humide', *Z. Geomorph.*, N.F. **1**, 3–34.

Bayens, J. (1938), *Les sols d'Afriques centrale, spécialement du Congo Belge* (Publications INEAC, Hors-Série, Bruxelles).

Behrmann, W. (1917), 'Der Sepik (Kaiserin-Augusta-Fluss) und sein Strömgebiet', *Mitt. dt. Schutz-gebieten* **12**, 1–100.

—— (1921), 'Die Oberflächenformen in den feucht-warmen tropen', *Z. Gesell. Erdk. (Berlin)* **56**, 44–60.

Bernhard, F. (1970), 'Étude de la litière et de sa contribution au cycle des éléments mineraux en foret ombrophile de Cote-d'Ivoire', *Oecologia Plantarium* **5**, 247–66.

Bik, M. J. J. (1967), 'Structural geomorphology and morphoclimatic zonation in the Central Highlands, Australian New Guinea' *in* Jennings, J. N. and Mabbutt, J. A. (eds.), *Landform studies from Australia and New Guinea* (ANU Press, Canberra), 26–47.

Birot, P. (1960), *Géographie physique générale de la zone intertropicale* (Centre de documentation universitaire: Paris), 244 pp.

—— (1966), *General physical geography* (Harrap, London), 360 pp.

—— (1968), *The cycle of erosion in different climates* (Batsford, London), 144 pp.

Blake, D. H. and Ollier, C. (1969), 'Geomorphological evidence of Quaternary tectonics in Southwestern Papua', *Rev. Géomorph. dyn.* **19**, 28–32.

Blondel, F. (1929), 'Les altérations des roches en Indochine Française', *4e Congr. Sci. Pacif. 1929, Bull. Serv. gèol. Indochine* **18**, Fasc. 3, 10 pp.

Boissevain, H. (1942), 'De Riviervormen in Sedimentatie-gebieden', *Tijdschr. K. ned. Aardrijksk. Genoot.* **58**, 723–56.

Bornhardt, W. (1900), *Zur Oberflächen gestaltung und Geologie, Deutsch-Ostafrika* (Collection Deutsch-Ostafrika, Berlin), 595 pp.

Branner, J. C. (1896), 'Decomposition of rock in Brazil', *Bull. geol. Soc. Am.* **7**, 255–314.

Brown, M. J. F. (1974), 'A development consequence—disposal of mining waste on Bougainville, Papua New Guinea', *Geoforum* **18/74**, 19–27.

Bryan, K. (1940), 'The retreat of slopes', *Ann. Ass. Am. Geogr.* **30**, 254–68.

Buchanan, F. (1807), *A Journey from Madras through the countries of Mysore, Canara and Malabar, performed under the orders of the Most Noble the Marquis Wellesley, Governor-General of India, for the express purpose of investigating the state of agriculture, arts and commerce; the religions, manners and customs; the history natural and civity and antiquity in the Dominions of the Rajah of Mysore, and the countries acquired by the Honourable East India Company.*

Büdel, J. (1957), 'Die "Doppelten Einebnungsflächen" in den feuchten Tropen', *Z. Geomorph. N.F.*, **1**, 201–28.

—— (1969), 'Das System der Klima-genetischen Geomorphologie', *Erdkunde* **23**, 165–83.

Carbonnel, J. P. (1964a), 'Valeur de l'érosion au Cambodge', *C.R. Acad. Sci. Paris* **259**, 3315–18.

—— (1964b), 'Rapport d'une première année d'études sédimentologiques sur le bassin du Grand-Lac au Cambodge, *Cah. Pacif.* **6**, 143–69.

Chorley, R. J., Dunn, A. J., and Beckinsale, R. P. (1964), *The history of the study of landforms or the development of geomorphology. Volume one: Geomorphology before Davis* (Methuen, London), 678 pp.

Clarke, W. C. (1973), 'The Dilemma of Development' *in* Brookfield, H. C. (ed.) *The Pacific in Transition* (Arnold, London), 275–98.

Collinet, J. (1969), 'Contribution à l'étude des stonelines dans la région du Moyen-Ogoué (Gabon)', *Cah. ORSTOM, Pédol.*, **7(1)**, 3–42.

Cooke, R. U. and Doornkamp, J. C. (1974), *Geomorphology in environmental management* (Oxford University Press), 413 pp.
Corbel, J. (1957a), 'L'érosion chimique des granites et silicates sous climats chauds', *Rev. Géomorph. dyn.* **8**, 4–8.
—— (1957b), 'Hydrologie et morphologie du nord-ouest Américain', *Rev. Géomorph. dyn.* **8**, 97–112.
Cornet, J. (1896), 'Les dépots superficiels et l'érosion continentale dans le bassin du Congo', *Bull. Soc. belge Géol.* **10**, 44–116.
Cornforth, I. S. (1970), 'Leaf fall in a tropical rain forest', *J. appl. Ecol.* **7**, 603–8.
Cotton, C. A. (1958), 'Fine-textured erosional relief in New Zealand', *Z. Geomorph. N.F.*, **2**, 187–210.
Dana, J. D. (1850), 'On denudation in the Pacific', *Am. J. Sci. (Series 2)*, **9**, 48–62. Reprinted in Schumm, S. A. (ed.) (1972) *River morphology* (Dowden, Hutchinson and Ross, Stroudsburg), 24–39.
Davis, S. N. (1964), 'Silica in streams and ground water', *Am. J. Sci.* **262**, 870–91.
Demangeot, J. (1969), 'Nappes de gravats et couvertures argilo-sableuses au Bas-Congo. Leur genèse et l'action des termites' *in* A. Bouillin, *Etudes sur les termites africains* (Edit. Univ., Léopoldville), 399–415.
Dixey, F. (1942), 'Erosion cycles in central and Southern Africa', *Trans. geol. Soc. S. Afr.* **45**, 151–81.
Douglas, I. (1967), 'Erosion of granite terrains under tropical rain forest in Australia, Malaysia and Singapore', *Publs Ass. int. Hydrol. scient.* **75**, 31–9.
Erhart, H. (1956), *Le genèse des sols, en tant que phénomène géologique. Esquisse d'une théorie géologique et géochimique. Biostasie et Rhexistasie* (Masson, Paris), 90 pp.
—— (1961), 'Sur la genése de certaines gites miniers sedimentaires, en rapport avec le phénomène de bio-rhexistasie et avec des mouvements tectoniques de faible amplitude', *C.r. Acad. Sci. Paris.* **252**, 2904–6.
Eyles, R. J. (1968), 'Stream net ratios in West Malaysia', *Bull. geol. Soc. Am.* **79**, 701–12.
—— (1969), 'Depth of dissection of the West Malaysian landscape', *J. trop. Geogr.* **28**, 24–32.
Fittkau, E. J. (1973), 'Artenmannigfaltigkeit amazonischer Lebensräume aus ökologischer Sicht', *Amazoniana* **4**, 321–40.
—— (1974), 'Zur ökologischen Gliederung Amazoniens. I. Die erdgeschichtliche Entwicklung Amazoniens', *Amazoniana* **5**, 77–134.
Fosberg, R. R. (1962), 'The physical background of the humid tropics substratum' *in Symposium on the impact of man on humid tropics vegetation* (Administration of the Territory of Papua and New Guinea; Port Moresby), 35–7.
Fournier, F. (1960), *Climat et érosion* (Presses Universitaires de France, Paris), 201 pp.
Freson, R., Goffinet, G. and Malaisse, F. (1974), 'Ecological effects of the regressive succession Muhulu-Miombo-Savanna in Upper-Shaba (Zaire), *Proc. 1st int. Congr. Ecol. (The Hague, Netherlands)*, 365–71.
Freise, F. W. (1930), 'Beobachtungen über den Schweb einiger Flüsse des brasilianischen Staates Rio de Janeiro', *Z. Geomorph.*, Old Series, **5**, 241–4.
—— (1932), 'Beobachtungen über Erosion an Urwaldsgebirgsflüssen des brasilianischen Staates Rio de Janairo', *Z. Geomorph.*, Old Series **7**, 1–9.
—— (1934), 'Erscheinungen des Erdfliessens in Tropenurwalde', *Z. Geomorph.*, Old Series, **9**, 88–98.

—— (1936), 'Das Binnenklima von Urwäldern im subtropischen Brasilien', *Petermanns geogr. Mitt.* **82**, 301–4.

—— (1938), 'Inselberge und inselberg-landschaften im granit und greissgebiete Brasiliens', *Z. Geomorph., Old Series* **10**, 137–68.

Galloway, R. W. and Löffler, E. (1972), 'Aspects of geomorphology and soils in the Torres Strait Region' *in* Walker, D. (ed.) *Bridge and barrier: the natural and cultural history of Torres Strait* (ANU Publication BG/3), 11–28.

Garner, H. F. (1966), 'Derangement of the Rio Caroni, Venezuela', *Rev.Géomorph. dyn.* **16**, 50–83.

—— (1974), *The origin of landscapes: a synthesis of geomorphology* (Oxford University Press, New York), 734 pp.

Grubb, P. J. (1974), 'Factors controlling the distribution of forest-types on tropical mountains: new facts and perspectives', *Univ. Hull, Dept. Geogr. Misc. Ser.* **16**, 13–46.

Grund, A. (1914), 'Der geographische Zyklus in Karst', *Erdkunde* **52**, 621–40.

Guillemain, C. (1908), 'Ergebnisse geologischer Forschung im Schutzgebiet Kamerun', *Mitt. dt. Schutzgeb.* **21**, 15–35.

—— (1914), 'Geomorphologische Forschungen aus Kamerun', *Petermanns Mitt.* **60**(2), 131–5, 183–6.

Gupta, A. (1975), 'Stream characteristics in eastern Jamaica, an environment of seasonal flow and large floods', *Am. J. Sci.* **275**, 825–47.

Harrison, J. B. (1933), *The katamorphism of igneous rocks under humid tropical conditions* (Imperial Bureau of Soil Science, Harpenden), 79 pp.

Hayes, C. W. (1899), 'Physiography and geology of regions adjacent to the Nicaragua Canal route', *Bull. geol. Soc. Am.* **10**, 285–348.

Heron, A. M. (1938), 'The physiography of Rajputna', *Proc. 25th Indian Sci. Congr.* **2**, 421–2.

Hopkins, B. (1966), 'Vegetation of the Olokemiji Forest Reserve, Nigeria. *IV.* The litter and soil with special reference to their seasonal changes', *J. Ecol.* **54**, 687–703.

Hopkins, W. (1844), 'On the transport of erratic blocks', *Trans. Camb. phil. Soc.* **8**, 220–40.

Ilunga Lutumba (1975), 'Erosion dans la région de Bukavu, Zaire', Unpublished paper presented at the *Colloque International sur la Géomorphologie de l'environnement* (Lumbumbashi).

Jennings, J. N. and Bik, M. J. (1962), 'Karst morphology in Australian New Guinea', *Nature, Lond.* **194**, 1036–58.

Jenny, H., Gessel, S. P. and Bingham, F. T. (1949), 'Comparative study of decomposition rates of organic matter in temperate and tropical regions', *Soil Sci.* **68**, 419–32.

Jones, A. (1973), 'Landslides of Rio de Janeiro and the Serra das Araras Escarpment, Brazil', *U.S. geol. Surv. Prof. Pap.* **697**, 42 pp.

Kenworthy, J. B. (1971), 'Water and nutrient cycling in a tropical rain forest', *Univ. Hull, Dept. Geogr. Misc. Ser.* **11**, 49–65.

Klinge, H. and Rodrigues, W. A. (1968), 'Litter production in an area of Amazonian terra Roma forest. Part I. Litter fall, organic carbon and total nitrogen contents of litter', *Amazoniana* **1**, 287–302.

Krynine, P. D. (1936), 'Geomorphology and sedimentation in the humid Tropics', *Am. J. Sci.,* **232**, 297–306.

Lacey, G. (1930), 'Stable channels in alluvium', *Minutes Proc. Instn Civ. Engnrs* **229**, 259–92.

—— (1935), 'Uniform flow in alluvial rivers and canals', *Minutes Proc. Instn Civ. Engnrs.* **237**, 321–56.

Lapperent, J. de (1939), 'La décomposition latéritique du granite dans la région de Macenta (Guinée francaise). L'arénisation prétropicale et prédésertique en A.O.F.', *C.r. Acad. Sci. Paris* **208**, 1767–9 and **209**, 7.
Lehmann, H. (1953), 'Karst-entwicklung in den Tropen', *Die Umschau* **53**, 559–62.
Leigh, C. H. (1973), 'Land development and soil erosion in west Malaysia', *Area* **5**, 213–17.
—— and Low, K. S. (1976), 'Land development in Malaysia: the role of the geomorphologist and hydrologist', *in* Lea, D. A. M. (Ed.) *Geographical research: applications and relevance* (Armidale), 71–85.
Lewis, L. A. (1966), 'The adjustment of some hydraulic variables at discharges less than one Cfs', *Prof. Geogr.* **18**, 230–4.
—— (1969), 'Some fluvial geomorphic characteristics of the Manati Basin, Puerto Rico', *Ann. Ass. Am. Geogr.* **59**, 280–93.
Löffler, E. (1970), 'Evidence of Pleistocene glaciation in East Papua', *Austr. geogr. Stud.* **8**, 16–26.
—— (1971), 'The Pleistocene glaciation of the Saruwaged Range, Territory of New Guinea', *Austr. Geogr.* **11**, 463–72.
—— (1972), 'Pleistocene glaciation in Papua and New Guinea', *Z. Geomorph.*, N.F. *Suppl. Bd.* **13**, 32–58.
—— (1974), 'Explanatory notes to the geomorphological map of Papua New Guinea', *Land Res. Ser. C.S.I.R.O. Australia* **33**, 19 pp.
Louis, H. (1957), 'Rumpfflächenproblem, Erosionzyklus und Klimageomorphologie', *Petermanns geogr. Mitt. Ergänz.* **262**, 9–26.
—— (1961), *Allgemeine Geomorphologie* (de Gruyter, Berlin), 355 pp.
Mabbutt, J. and Scott, R. M. (1966), 'Periodicity of morphogenesis and soil formation in a savannah landscape near Port Moresby, Papua', *Z. Geomorph. N.F.*, **10**, 69–89.
—— and Stewart, G. A. (1963), 'The application of geomorphology in resources surveys in Australia and New Guinea', *Rev. Géomorph. dyn.* **14**, 97–109.
Machatschek, F. (1955), *Das Relief der Erde: Versuch einer regionalen Morphologie der Erdoberfläche* (Borntraeger, Berlin, II Band, 2nd Ed.), 594 pp.
Madduma Bandara, C. M. (1974), 'The orientation of straight slope forms on the Hatton Plateau of central Sri Lanka', *J. trop. Geogr.* **38**, 37–44.
Maingien, R. (1964), *Review of research on laterites* (UNESCO), 155 pp.
Mainguet, M. (1972), *Le modelé des gres* (Institut Géographique National, Paris), 657 pp.
Martonne, E. de (1939), 'Sur la formation des "Pains de Sucre" au Brésil', *C.r. Acad. Sci. Paris,* **208**, 1163.
—— (1946), 'Géographique zonale: la zone tropicale', *Annls Géogr.* **55**, 1–18.
—— (1951), *Traité de géographie physique. Tome II, Le relief du sol* (9th Ed. Colin, Paris), 499–1057.
—— and Birot, P. (1944), 'Sur l'évolution des versants en climat tropical humide', *C. r. Acad. Sci. Paris* **218**, 529–32.
Mathieu, P. (1971), 'Érosion et transport solide sur un bassin versant forestier tropical (Bassin de l'Amitoro, Cote d'Ivoire)', *Cah. ORSTOM, Sér. Géol.* **3** (2), 115–44.
MEXE (Military Engineering Experimental Establishment) (1965), *The classification of terrain intelligence Reports of the Combined Pool (AER) 1960 to 1964* (M.E.X.E. No. 915) (Christchurch), 90 pp.
Michel, P. (1973), 'Les bassins des fleuves Sénégal et Gambie: étude géomorphologique', *Mém. ORSTOM* **63**, 752 pp.

Milne, G. (1935), 'Some suggested units for classification of tropical and sub-tropical soils', *Soil Res.* **4**, 183–98.
Mohr, E. C. J. and Van Baren, F. A. (1954), *Tropical soils* (Van Hoeve: The Hague), 498 pp.
Morgan, R. P. C. (1970), 'An analysis of basin asymmetry in the Klang Basch, Selangor', *Bull. geol. Soc. Malaysia* **3**, 17–26.
—— (1972), 'Observations on factors affecting the behaviour of a first-order stream', *Trans. Inst. Br. Geogr.* **56**, 171–85.
—— (1976), 'The role of climate in the denudation system with reference to an analysis of drainage density and texture in West Malaysia', *in* Derbyshire, E. (Ed.) *Geomorphology and climate* (Wiley, London).
Mousinto de Meis, R. and da Silva, R. (1968), 'Mouvements de masse récents à Rio de Janeiro: une étude de géomorphologie dynamique', *Rev. Géomorph. dyn.* **18**, 145–51.
Nossin, J. J. (1964), 'Geomorphology of the surroundings of Kuantan, Eastern Malaya', *Geol. Mijnb.* **43**, 157–82.
Nouvelot, J. F. (1969), 'Mesure et étude des transports solides en suspension au Cameroun', *Cah. ORSTOM, Sér. Hydrol.* **6** (4), 43–86.
Nye, P. H. (1961), 'Organic matter and nutrient cycles under moist tropical forest', *Plant Soil* **13**, 333–46.
—— and Greenland, D. J. (1960), The soil under shifting cultivation', *Tech. Commn Commonwealth Bur. Soils* **51**, 156 pp.
Obst, E. (1915), 'Das abflusslose Rumpfschollenland im nordöstlichen Deutsch-Ostafrika', *Mitt. geogr. Gesell. Hamburg* **29**, 3–105.
Odum, E. P. (1971), *Fundamentals of ecology* (3rd Ed., Saunders, Philadelphia), 574 pp.
Odum, H. T. (1970), Summary: 'An emerging view of the ecological system at El Verde', *in* Odum, H. T. (Ed.) *A tropical rainforest: a study of irradiation and ecology at El Verde, Puerto Rico* (Office of Information Services, U.S. Atomic Energy Commission, Oak Ridge, Tennessee), **I** 191–**I** 289.
—— Moore, A. M. and Burns, L. A. (1970), 'Hydrogen budgets and comport-ments in the rain forest', *in* Odum, H. T., op cit. (1970), H 105–H 122.
—— and Pidgeon, R. F. (1970), *A tropical rain forest* (Division of Technical Information, U.S. Atomic Energy Commission, Oak Ridge, Tennessee), 1678 pp.
Ofomata, G. E. K. (1965), 'Factors of soil erosion in the Enugu area of Nigeria', *J. geogr. Ass. Nigeria* **8**, (1), 45–59.
—— (1966), 'Quelques observations sur l'éboulement d'Awgu, Nigèrie oriental', *Bull. Inst. fond. Afr. noire, sér. A,* **28**, (2), 433–43.
Ollier, C. D. (1965), 'Some features of granite weathering in Australia', *Z. Geomorph.* N.F., **9**, 285–304.
—— and Brown, M. J. F. (1971), 'Erosion of a young volcano in New Guinea', *Z. Geomorph.* N.F. **15**, 12–28.
Olofin, E. A. (1974), 'Classification of slope angles for land planning purposes', *J. trop. Geogr.* **39**, 22–77.
Pain, C. F. and Bowler, J. M. (1973), 'Denudation following the November 1970 earthquake at Madang, New Guinea', *Z. Geomorph.* N.F. Suppl. Bd. **18**, 92–104.
Pannekoek, A. J. (1941), 'Einige karsterreinen in Nederlandsch-Indie', *Ned. indische geogr. Meded.* **1**, 16–19.
Passarge, S. (1904), 'Rumpfflächen und Inselberge', *Z. dt. geol. Gesell.* **56**, 193–215.

—— (1910), 'Geomorphologische Probleme aus Kamerun', *Z. Gesell. Erdkunde, Berlin* (1910), 448–65.

—— (1923), 'Inselberglandschaft der Massaisteppe', *Petermanns geogr. Mitt.* **69**, 205–9.

Paton, T. R. and Williams, M. A. J. (1972), 'The concept of laterite', *Ann. Ass. Am. Geogr.* **62**, 42–56.

Peltier, L. C. (1950), 'The geographic cycle in periglacial regions as it is related to climatic geomorphology', *Ann. Ass. Am. Geogr.* **40**, 214–36.

Pugh, J. C. (1966), 'The landforms of low latitudes', *in* Dury, G. H. (Ed.) *Essays in geomorphology* (Heinemann, London), 121–38.

Rapp, A., Berry, L., and Temple, P. H. (1972), 'Soil erosion and sedimentation in Tanzania—The Project', *Geogr. Annlr.* **54A**, 105–9.

Ribeiro, O. (1972), 'Réflexions sur le métier de géographe' *in Études de géographie tropicale offertes à Pierre Gourou* (Mouton, Paris), 69–92.

Richardson, J. A. (1947), 'An outline of the geomorphological evolution of British Malaya', *Geol. Mag.* **84**, 129–44.

—— (1927), *Le Katanga physique* (Lamertin, Bruxelles), 282 pp.

Rougerie, G. (1960), 'Le façonnement actuel des modelés en Côte d'Ivoire forestiére', *Mém. Inst. Afr. noire* **58**, 542 pp.

—— (1961), 'Etude comparative de l'évacuation de la silice en milieux cristallins tropicale humide et tempéré,' *Ann. Géogr.* **70**, 45–50.

Rutten, L. M. R. (1928), *Science in the Netherlands East Indies* (Koningklijke Akademie van Wetenschappen, Amsterdam), 432 pp.

Ruxton, B. P. (1967), 'Slopewash under mature primary rainforest in northern Papua' *in* Jennings, J. N. and Mabbutt, T. A. (Eds.) *Landform studies from Australia and New Guinea* (ANU Press, Canberra), 85–94.

—— (1969), 'Rates of weathering of Quaternary volcanic ash in northeast Papua', *Trans. 9th int. Congr. Soil Sci., Adelaide,* 367–76.

—— and McDougall, I. (1967), 'Denudation rates in northeast Papua from potassium-argon dating of lavas', *Am. J. Sci.* **265**, 545–61.

Sapper, K. (1910), 'Wissenschaftliche Ergebnisse einer amtlichen Forschungsreise nach dem Bismarck Archipel im jahr 1908: 1. Beitrag zu Landeskunde von Neu-Mecklenburg und seiner Nachbarinseln', *Mitt. Schutzgeb. Ergänz.* **3**, 1–130.

—— (1914), 'Über Abtrangungsvorgange in den regenfeuchten Tropen und ihre morphologischen Wirkungen', *Geogr. Z.* **20**, 5–16, 81–92.

—— (1935), *Geomorphologie der feuchten Tropen* (B. G. Teubner, Berlin), 150 pp.

Schmidt, F. H. and Ferguson, J. H. A. (1952), 'Rainfall types based on wet and dry period ratios for Indonesia with Western New Guinea', *Verh. Djawatan Meteorol. Djakarta* **42**, 77pp.

Scrivenor, J. B. (1931), *The geology of Malaya* (London), 217 pp.

Simonett, D. S. (1967), 'Landslide distribution and earthquakes in the Bewani and Torricelli Mountains, New Guinea' *in* Jennings, J. N. and Mabbutt, J. A. (Eds.) *Landform studies from Australia and New Guinea* (A.N.U. Press, Canberra), 64–84.

—— (1968), 'Selva landscapes', *in* Fairbridge, R. W. (Ed.) *Encyclopaedia of geomorphology* (Reinhold, New York), 989–91.

Sioli, H. (1956), 'Über Natur und Mensch im brasilianischen Amazonasgebiet', *Erkunde, Berlin* **10** (2), 89–109.

—— (1961), 'Landschaftsökologischer Beitrag aus Amazonien', *Natur Landsch. Mainz* **36** (5), 73–7.

——, Schwabe, G. H. and Klinge, H. (1969), 'Limnological outlooks on landscape-ecology in Latin America', *Trop. Ecol.* **10**, (1), 72–82.
Sollins, P. and Drewry, G. (1970), 'Electrical conductivity and flow rates of water through the forest canopy' *in* Odum, H. T. and Pidgeon, R. F (eds.), op. cit. (1970).
Speight, J. G. (1965a), 'Meander spectra of the Angabunga River, Papua', *J. Hydrol.* **3**, 16–36.
—— (1956b), 'Flow and channel characteristics of the Angabunga River, Papua', *J. Hydrol.* **3**, 16–36.
—— (1967), 'Spectral analysis of meanders of some Australian rivers' *in* Jennings, J. N. and Mabbutt, J. A (Eds.) *Landform studies from Australia and New Guinea* (A.N.U. Press, Canberra), 48–63.
Sternberg, M. O'R. (1975), 'The Amazon river of Brazil', *Geogr. Z. Beihefte* **40**, 74 pp.
Stoddart, D. R. (1969), 'Climatic geomorphology: review and re-assessment', *Progr. Geogr.* **1**, 159–222.
Swan, S. B. St.C. (1970), 'Landforms of the humid tropics: Johor, Malaya', Unpubl. Ph.D. thesis, Univ. of Sussex, 218 pp.
Sweeting, M. M. (1958), 'The karstlands of Jamaica', *Geogr. J.* **124**, 184–99.
Tardy, Y. (1969), 'Géochimies des altérations: étude des arénes et des eaux de quelques massifs cristallins d'Europe et d'Afrique', *Mem. Serv. Carte géol. Alsace et Lorraine* **31**, 199 pp.
Thomas, M. F. (1965), 'Some aspects of the geomorphology of domes and tors in Nigeria', *Z. Geomorph. N.F.*, **9**, 63–81.
Thorbecke, F. (1911), 'Das Manenguba Hochland', *Mitt. dt. Schutzgeb.* **24**, 279–304.
—— (1914), 'Geographische Arbeiten in Tikar und Wute', *Verh. dt. Geogr., Strasburg,* 147–65.
—— (1927), 'Klima und Oberflächenformen: die Stellung des Problems', *Düsseldorfer geogr. Vorträge und Erörterungen, 3 Teil: Morphologie der Klimazonen* (Ferdinand Hirt, Breslau), 1–3.
Thornes, J. B. (1970), 'The hydraulic geometry of stream channels in the Xingu-Araguaia headwaters', *Geogr. J.* **136**, 376–82.
Tricart, J. (1956), 'Dégradation du milieu naturel et problèmes d'aménagement au Fouta-Djalon (Guinée)', *Rev. Géogr. alp.* **44**, 7–36.
—— (1959), 'Observations sur le façonnement des rapides des rivières intertropicales', *Bull. Sect. geogr. Com. Trav. Hist. Sci.* **71**, 289–313.
—— (1961), 'Les caractéristiques fondamentales du systéme morphogénétique des pays tropicaux humide', *Inf. géogr.* **25**, 155–69.
—— (1962), *Problèmes de mise en valeur des montagues tropicales et subtropicales. Fascicule II. Problèmes du développement dans les Andes Vénézuéliennes* (Centre de Documentation Universitaire. Paris). 96 pp.
—— (1965), *Le modelé des régions chaudes: forêts et savanes* (SEDES, Paris), 322 pp.
—— (1966), 'Géomorphologie et aménagement rural (exemple du Vénézuela)', *Cooperation Technique* **44-5**, 69–81.
—— (1974a), 'Existence de périodes seches au Quaternaire en Amazonie et dans les régions voisches', *Rev. Géomorph. dyn.* **23** (4), 145–58.
—— (1974b), 'Influence des osidlatars climatiques récentes au le modelé en Amazonieanakle (région de Santarém) d'aprés les images radar latéral,' *Z. Geomorph., N.F.,* **19**, 140–63.
—— (1973a), 'Etude de l'évolution morphologique récente d'un secteur de

l'Amazonie sur des mosaiques de radar latéral (Wd'Obidos, Brésil),' *Photo-interprétation,* **73**, 34–8.
—— and Michel, M. (1965), 'Monographie et carte géomorphologique de la région de Lagunillas (Andes vénézuéliennes),' *Rev. Géomorph. dyn.* **15**, 1–33.
Troll, C. (1964), 'Karte der Jahreszeiten-Klimate der Erde', *Erdkunde* **18**, 5–28.
Turvey, N. D. (1974), 'Nutrient cycling under tropical rainforest in Central Papua', *Univ. of Papua New Guinea, Department of Geography Occasional Paper* **10**, 96 pp.
—— (1975), 'Water quality in a tropical-rain-forested catchment', *J. Hydrol.* **27**, 111–25.
Twidale, C. R. (1976), *The analysis of landforms* (Wiley, London), 572 pp.
——, Bourne, J. A. and Smith, D. M. (1974), 'Reinforcement and stabilisation mechanisms in landform development', *Rev. Géomorph. dyn.* **23**, 11–125.
Verstappen, H. Th. (1975), 'On palaeo-climates and landform development in Malesia', *in* Bartstra, G-J. and Casparie, W. A. (Eds.) *Modern Quaternary research in Southeast Asia* (Balkema, Rotterdam), 3–35.
Wadia, D. N. (1941), 'The making of Ceylon', *Spolia Zeylonica* **23**, 18–20.
—— (1945), 'Three superposed peneplains of Ceylon—their physiography and geological structure', *Ceylon Dept. Mineralogy Prof. Pap.* **1**, 25–32.
—— (1966), *Geology of India* (3rd ed.) (Macmillan, London), 536 pp.
Wayland, E. J. (1921), 'Some features of the drainage of Uganda', *Geol. Surv. Uganda Ann. Rep. 1920*, 75–80.
—— (1934), 'Peneplains and some other erosional platforms', *Geol. Surv. Uganda Ann. Rep. 1934,* 77–9.
Williams, M. A. J. (1968), 'Termites and soil development near Brock's Creek, Northern Territory', *Austr. J. Sci.* **31**, 153–4.
—— (1975), 'Late Pleistocene tropical aridity synchronous on both hemispheres', *Nature, Lond.* **253**, 617–18.
Williams, P. W. (1971), 'Illustrating morphometric analysis of karst with examples from New Guinea', *Z. Geomorph.* N.F. **15**, 40–61.
—— (1972), 'Morphometric analysis of polygonal karst in New Guinea', *Bull. geol. Soc. Am.* **83**, 761–9.
Wright, R. L. (1971), 'Regional seminar on integrated surveys, range ecology and management', *Nat. Resources* **7** (2), 20–2.
Young, A. (1973a), 'Soil survey procedures in land development planning', *Geogr. J.* **139**, 53–64.
—— (1973b), 'Rural land evaluation' *in* Dawson, J. A. and Doornkamp, J. C. (Eds.) *Evaluating the human environment* (Arnold, London), 5–33.
Zonneveld, J. I. S. (1975), 'Some problems of tropical geomorphology', *Z. Geomorph.* N.F. **19**, 377–92.

12

DENUDATION IN THE TROPICS AND THE INTERPRETATION OF THE TROPICAL LEGACY IN HIGHER LATITUDES—A VIEW OF THE BRITISH EXPERIENCE

MICHAEL F. THOMAS
(*University of St. Andrews*)

British experience of the tropics has come about largely as a result of the period of rapid colonial expansion in the late nineteenth century, although it began earlier in India, and also in Australia, where subsequently a discernible 'Australian school' of geomorphology has developed an independent status. The subdivision of the tropical world into spheres of European national influence fundamentally affected the location and to some extent the course of subsequent research, not only for workers from Britain but also for Dutch, French and German colleagues in particular. Moreover, for reasons of language and tradition, the geographical bias towards former colonial areas remains prominent in the research preferences of French and British scientists.

Most of the early pioneers in the field of tropical geomorphology were geologists, employed to survey the mineral resources of newly acquired colonies. By the turn of the century these expatriate geologists carried with them the new ideas of W. M. Davis concerning the erosion cycle which sprang naturally from an evolutionary view of the landscape, but also carried with it the prejudicial concept of humid temperate 'normality'. This persistent viewpoint contributed to the emphasis commonly placed on forms and deposits which were exotic or unusual, while paradoxically such features were generally fitted into or excepted from the progress of the 'normal' cycle of erosion. But the pioneer geologists were alert to the important features of the tropical landscape, including the great depths of weathering which were commonly observed. They were forced by circumstances to acquire an eye for country and in interest in landform, because they frequently had no topographic survey to guide them. The landform was a clue to the geology, and where so much of the terrain was thickly mantled by saprolite, a parallel interest in superficial deposits, particularly laterite, became evident. From the standpoint of European geomorphology this pioneer stage came late and was prolonged. Academic geomorphologists and geologists only began to work widely in the tropics following the establishment of new universities in the two decades following the end of the second World War. Even in Australia, interest in the tropical northern areas and in New Guinea began to expand only after 1945. It is noticeable that the early independence of India in 1947 meant that less interest has been taken by British geomorphologists in this sub-continent than in Africa and Malaysia, where close links with the new universities were established in the 1950s and 1960s, the years leading up to and immediately following independence.

The scattered reports of the early writers included important contributions to knowledge: the work of J. D. Falconer in Nigeria (1911); F. Dixey in Sierra Leone (1922a, 1922b); J. B. Scrivenor in Malaya (1931); E. J. Wayland in Uganda (1933), and of course J. M. Campbell (1917), R. T. Jutson (1934) and W. G. Woolnough (1927, 1930) in Australia, and of the chemist J. B. Harrison (1934) in British Guiana (Guyana) being notable among them, though there were many more. From this short list and the writings of their contemporaries we can derive the origins of most modern concepts concerning deep weathering and lateritisation, etch-planation and the stripping of saprolite from crystalline basements. Yet their findings were neglected if not ignored. In Wayland's case his notes on etching were buried in an obscure report (Wayland, 1933), and he apparently did not feel confident enough to voice these ideas in his more widely read papers of the time. But Falconer not only published a book in 1911 that contains his reflections on the northern Nigerian inselberg landscapes, but he also addressed the British Association in the following year 'On the origins of kopjes and inselbergs'.

It would appear that the dominance of the essentially monoclimatic concepts of the Davisian cycle of erosion ensured that the only 'climatic accident' to which geomorphologists paid heed in Britain at that time was glaciation, even though evidence of former warm conditions in Britain was growing. Whether for this or for other reasons, only a passing interest was taken in work on tropical landscapes until C. A. Cotton in 1947 published his useful summaries of German work on inselbergs and the 'savanna cycle'. However, perhaps because this was part of a book entitled *Climatic accidents in landscape making*, these ideas remained isolated from the central interests of British geomorphologists. But reading Cotton today makes it clear that this was not the situation in Germany. His list of German contributors to tropical landform studies prior to 1939 is impressive, and the work of Bornhardt, Credner, Freise, Jaeger, Jessen, Krebs, Passarge, Obst, Sapper and Thorbecke (see Cotton, 1947), although in some cases still unfamiliar to the anglophone reader, has become the foundation of a corpus of European work on climatic geomorphology. In Europe the traditions set by Davis were never so strong (R. P. Beckinsale, 1976) and there was always the powerful voice of W. Penck (1924) to provide the necessary debate.

In Britain the major challenge to the thinking of Davis came after 1945, represented in the work of L. C. King, a New Zealander working in the subtropics of Natal, South Africa. King's insistence on pediplanation as the 'normal' process of landscape evolution embodied in the 'Canons of landscape evolution' (1953) and 'The uniformitarian nature of hillslopes' (1957) aroused great controversy because it was a direct challenge to established orthodoxy, whereas the work of K. Bryan (1922) on pediments or of Falconer (1912) on weathering, for example, was not. It was also in 1953 that the English translation of W. Penck's major work appeared. But the ensuing debate tended to obscure the earlier ideas, and divorced the disputes about slope evolution from a discussion of weathering processes which were relegated to a minor role in landscape evolution.

With this background it is perhaps clear how important to British geomorphology was D. L. Linton's celebrated paper on *The problem of tors* (1955) which made reference to the work of Scrivenor (1931) in Malaya (Malaysia) and drew heavily on the ideas which J. R. F. Handley (1954) had advanced to explain tor landscapes in Tanganyika (Tanzania). Some 20 years after their counterparts in Germany, British geomorphologists were now faced with a view of British landforms that required an understanding of tropical conditions of denudation as well as those of the glacial and periglacial régimes. The long-known tropical conditions of the early Tertiary were seen at least by some to be relevant to an account of contemporary British landscapes. But the full impact of Linton's statement was delayed, first because quite serious problems were raised by his interpretation of the Dartmoor tors, and this made full acceptance of his approach difficult (J. Palmer and R. A. Neilson, 1962), but secondly because British studies in the tropics at that time reflected, in significant instances, the ideas of L. C. King (J. C. Pugh, 1956), though R. W. Clayton (1956) made detailed use of current German thinking in his studies from West Africa. The important paper by J. Büdel (1957), on 'Die doppelten Einebnungsflächen in den feuchten Tropen', which also carried implications for the study of European relief, reached a wider anglophone readership quite quickly due once more to the advocacy of Cotton (1961), while independent evidence of the importance of deep weathering in granitoid terrains in the tropics was afforded in the influential paper by B. P. Ruxton and L. Berry (1957) on the 'Weathering of granite and associated erosional features in Hong Kong'. Studies of laterites were also given new impetus by the monograph from J. A. Prescott and R. L. Pendleton (1952), and their effects on landforms were discussed notably by J. W. Pallister (1956) and A. M. J. DeSwardt (1946, 1964) in Africa, and by P. E. Playford (1954) and others in Australia.

The rapid expansion of university education in tropical countries at this time brought in turn a series of papers on tropical landscapes in which geomorphologists of British origin made contributions to the study of deep weathering (C. D. Ollier, 1960; M. F. Thomas, 1966), the role of duricrusts and of etch-planation (J. A. Mabbutt, 1961, 1965; A. F. Trendall, 1962; Thomas, 1965a), and the origins of residual hills (Ollier, 1960; R. A. G. Savigear, 1960; C. R. Twidale, 1964; Thomas, 1965b). Studies of pediments came mainly from Australia (G. H. Dury and T. Langford-Smith, 1964; Mabbutt, 1966; Twidale, 1967), and revealed noticeable differences of interpretation from earlier work in America (B. A. Tator, 1952; Y. F. Tuan, 1959).

Through these papers runs the dominant theme of tropical planation, whether seen in terms of pediplanation or etch-planation, and the setting for most of the work was either Africa or Australia. Comparable studies from Guyana (R. B. McConnell, 1968; M. J. Eden, 1971) and Malaysia (S. B. St. C. Swan, 1970, 1972) came later, while work in the West Indies, New Guinea and Indonesia was concerned particularly with karst studies (M. M. Sweeting, 1958; J. N. Jennings and M. J. Bik, 1962). Quaternary climates were the basis of important studies in East Africa that cannot be detailed here, but the geomorphological implications of Quaternary climatic change in the tropics were exemplified particularly by A. T. Grove (1958; Grove and R. A. Pullan, 1963).

Geomorphologists in the tropics were, and are still, frequently working on a scale demanding the sketching of an outline or framework for more detailed studies, and the relevance of denudation chronology has seldom been questioned as it has in Britain. The broad canvases and generalised conclusions of workers such as King (1962), Mabbutt (1965), DeSwardt and Trendall (1964) seemed of little interest to a new generation of geomorphologists in Britain, concentrating on processes or detailed studies of form. As information on tropical conditions and landscapes became more generally available an interest in such work as a key to past conditions elsewhere declined.

It is perhaps also a corollary of this situation that the results of those detailed studies which were being carried out (for instance, Ruxton, 1958; Clayton, 1956; Mabbutt, 1966) were seen as relevant mainly to the environments where they were undertaken. It is not surprising, therefore, that as recently as 1971 G. H. Dury sought to review the evidence for 'Relict deep weathering and duricrusting. . .of middle latitudes'. Geomorphology was not alone in awaiting a review of past enviornments in Britain and Europe, and the discussion by H. M. Montford (1970) shows that geologists too have become more aware of the tropical inheritance in recent years, partly as new evidence has come to light concerning continental shelf sedimentation. However, it should be recalled that J. P. Bakker and Th. W. M. Levelt (1964) had provided a summary in English of a European view some years before. This study was also concerned with the manner in which landscapes had evolved, and this is a topic which this paper seeks to examine.

This is not the place to review the extensive European literature, but Dutch, French and German scientists have made major contributions to tropical studies, particularly in the fields of weathering and pedogenesis (see for example E. C. Mohr and F. A. Van Baren, 1954; R. Maignien, 1966; G. Millot, 1964; UNESCO, 1971; Y. Tardy and others, 1971, 1973), and have applied their results in Europe (Bakker, 1967; F. Touraine, 1974). Detailed regional studies in West Africa by G. Rougerie (1960) and P. Michel (1967) are outstanding examples of the French regional tradition applied to tropical geomorphology, while both German and French geomorphologists have made important contributions in India (Büdel, 1965; J. Demangeot, 1975) where British geomorphologists have seldom worked. Serious discussion of climatic geomorphology has also been current for nearly three decades in Germany (Büdel, 1948, 1963, 1970; H. Rhodenburg, 1971) and with equal interest in France (J. Tricart and A. Cailleux, 1965), while in Britain we often remain either healthily or stubbornly sceptical (D. R. Stoddart, 1969).

Awareness of much of this European work has come late; we have had to wait until the 1970s for translations of key work by Millot (1964, transl. 1971), Tricart (1965, transl. 1972) and Büdel (1948, 1963, transl. in Derbyshire 1973), while the major work by H. Wilhelmy (1958) remains unavailable in English. Now, a few years after the appearance of such works in English, we can hope that British interest in the geomorphology of the tropics will be stimulated, both for its own sake and for its relevance to conditions in Britain and Europe.

However, in the fields of slope studies (A. Young, 1972) and in fluvial geomorphology (I. Douglas, 1973), British geomorphologists have established an independent reputation that is concerned in part with tropical conditions. Outside Australia, however, commitment to research in tropical and subtropical environments is if anything dwindling, although a corresponding increase in interest in the warm, humid south-eastern areas of the U.S.A. is evident (E. T. Cleaves *et al.*, 1970, 1974; R. Kesel, 1973). The expatriate geomorphologist, while still welcome in most countries of Africa and elsewhere, is now frequently only able to work on an expeditionary basis that precludes monitoring processes over time. Meanwhile the development of separate national interests in geomorphology is patchy; certainly the number of geomorphologists now working in the tropics is few, and so far as Britain is concerned lacking an institutional framework for their encouragement. The notable achievement of the soil erosion studies in Tanzania (see A. Rapp *et al.*, 1972) stands almost alone in the field of applied studies, and seems unlikely to be repeated. It will thus be seen that geomorphological studies in the tropics present a disconnected patchwork, conducted on greatly varying scales. A predominance of studies having a basis in the cyclic denudation of the Gondwana shields, and the continuing divergence of viewpoint concerning the mode of planation in the tropics, both contribute to the reluctance of British geomorphologists in general to embrace concepts of tropical denudation and to apply them within the context of landscape evolution in higher latitudes.

Despite the well-known arguments concerning the formation of tors, the origins of the 'clay-with-flints', the occurrence of pediments in southern England and the provenance of the 'Sarsen' stones, a general appreciation of the effects of warm climatic conditions in Britain has been lacking. One further aspect of this situation is the tendency in Britain for the continuity of landform development to become obscured by the glacial events of the middle and late Pleistocene. Yet recent discussions of the effectiveness of glacial erosion in Britain and elsewhere (T. Feininger, 1971; W. A. White, 1972; D. E. Sugden, 1968; F. Gravenor, 1975; K. M. Clayton, 1974) should warn us against overstating the interruptions. The hallmark of geomorphological study in the warm climates (not only tropical) has been an awareness of the long-term development of the landform and of superficial deposits. Interruptions have been viewed within a conceptual framework concerning ground-surface stability (B. E. Butler, 1959, 1967; H. Fölster, 1969) or of 'biostasie' and 'rhexistasie' (H. Erhart, 1955). Such ideas have been developed on greatly varying scales, to be applied to local soil sequences (Van Dijk *et al.*, 1968) or to regional sedimentation (Millot, 1917; Touraine, 1974). Interpretations based on these concepts appear able to integrate changing patterns of sedimentation with bioclimatic change and crustal disturbance, and to relate both to concepts of planation. In the subsequent section some pointers with respect to British conditions will be discussed after related ideas about tropical planation have been appraised.

TROPICAL AND SUB-TROPICAL DENUDATION SYSTEMS: ASPECTS OF THEIR DYNAMICS AND DEVELOPMENT

Recent developments in the field of tropical geomorphology include the following categories (they overlap and the short list is not exclusive).

Watershed studies based on input/output analyses of mainly small catchments, little disturbed by man, seek to elucidate the nature of the denudation system. Studies in warm climates include those by Douglas in Australia (1968, 1973) and of Cleaves *et al.* in America (1970, 1974). Such work seriously questions many less rigorous studies in the tropics (F. Fournier, 1960; J. Corbel, 1959, 1964). A major contribution of such experimental work is to establish the relative importance of mechanical and chemical denudation, and the results are revealing. Rougerie many years ago (1960) estimated the ratio of dissolved to suspended load in forested catchments of the Ivory Coast at 60–80 per cent, and figures exceeding 50 per cent have been obtained by Douglas (1973) and by Cleaves *et al.* (1970, 1974) for the eastern piedmont of Maryland.

These latter studies are of particular interest here for they point to possible inconsistencies in arguments about weathering and denudation. Two small woodland catchments were studied; one over schist mantled with several metres of saprolite yielded over a 2-year period a figure over 50 per cent for the proportion of weathering effected by chemical action; the other over a serpentinite rock which is widely exposed gave a figure of over 90 per cent. A notable point about the studies is that the overall rate of denudation for the two catchments, estimated at $2{\cdot}4\,km^3/km^2/years \times 10^6$, was nearly identical. Bearing in mind the high variability of stream loads over time (Douglas, 1973), such records must be viewed with caution, but the conclusion of Cleaves *et al.* that 'the landforms of the Pond Branch watershed suggest major mechanical erosion, but our study indicates that chemical denudation is the dominant erosional process at present' (1970, p. 3030), emphasises that we still have little idea of the geomorphological implications of chemical denudation. Where loss of solutes is compensated by hydration during the synthesis of clays, Ollier (1967) has suggested that 'constant volume alteration' occurs without disturbing the rock fabric preserved in the saprolite. However, in the case of the serpentinite catchment, Cleaves *et al.* (1974) concluded that clay synthesis was inhibited by the low alumina content of the rock, while rapid wetting and drying of the rock surface produced high pH values leading to the dissolution of silica, and the loss of solutes in runoff.

These records are compatible with the widespread observation that tropical rivers are dilute in absolute terms, particularly where streams pass over deeply weathered terrain (D. A. Livingstone, 1963). Such rivers drain regoliths that have already lost most of their weatherable minerals, some of which may be continuously cycled within the ecosystem during a period of biostatic equilibrium (Douglas, 1969). However, the figures challenge a common assumption taken from tropical studies, namely, that chemical weathering and erosional lowering of bare rock surfaces are much slower than beneath a saprolite cover. Some observations have already thrown doubt on this assumption, particularly where duricrusted surfaces underlain by thick weathering profiles stand high above adjacent inselberg landscapes (J. L. D'Hoore, 1964; Thomas, 1968, 1974). In central Sierra Leone, for example, the duricrusted schists of the Sula Mountains are higher than the adjacent massive domes of the porphyroblastic granites of the Gbengbe Hills. It seems likely that the higher the proportion of the original rock that remains *in situ* as secondary or residual weathering products, the lower the component of chemical denudation is likely to be. This may help to explain the high importance given by Rapp (1960) to the chemical component of denudation in the severe environment of the Kärkevagge. None of this precludes the solutional lowering of a rock with simultaneous accumulation of residual or inherited materials, as studies of the English and French chalklands demonstrate (C. Klein, 1965; Hodgson *et al.*, 1974; Touraine, 1974).

Studies of soil geochemistry complement watershed studies for, while the latter integrate spatial differentiation above the recording station but achieve a temporal separation of events, the former are based on spatial differentiation but integrate temporal changes in soils. In that these studies record the progress of different ions through the soil and the accumulation of solid compounds at different horizons or positions in the soil or regolith, they should assist in attempts to relate relict weathering profiles or paleosols to climate and relief. In this field recent French work is of particular interest, especially that of Tardy (1971) and colleagues (1973) who relate their results to water geochemistry and to modern equilibrium concepts (Tardy, 1971). Along with work by G. Bocquier *et al.* (1973), these studies considerably advance ideas about the development of catenas, and it is through such work that questions about laterite formation by the immobilisation of iron oxides in the soil are likely to be clarified.

This approach allows a broader view of chemical denudation to embrace the geomorphological effects of the selective migration of ions and precipitation of solid compounds within the soil body. Thus where direct lowering of the land surface by loss of dissolved material is of minor importance, landform patterns may still be determined in part by geochemical differentiation within the soil or saprolite, particularly where the precipitation of iron compounds is involved. The course of denudation on the tropical shields has been profoundly influenced by these processes acting at scales varying from the individual catena or hillslope to the morphological differentiation of terrains underlain by different rocks (Thomas, 1977). Other minerals than iron may of course be involved, in what is essentially a process of duricrusting (A. S. Goudie, 1973).

Studies in palaeopedology overlap considerably with the previous category but include the evidences of soil layering which inform us about changing conditions on hillslopes over time, thus amplifying knowledge of both the spatial and temporal differentiation of landscape. Although most obviously referring to recent phases of landscape history, this developing field may be important to interpretations of remoter events (D. Yaalon, 1971; P. Buurman, 1975), including sequences of duricrust formation. Also from this work has come a more detailed understanding of sequences of ground-surface stability and instability (B. E. Butler, 1959, 1967; H. Fölster, 1969). Many results from the tropical shields record the local redistribution of regolith materials in complex patterns of sedimentation and soil development (R. V. Ruhe, 1956; Fölster, 1969; M. J. Mulcahy and E. Bettenay, 1971).

Studies of off-shore sedimentation have proliferated during the last 10 years of continental shelf research. As well as giving information about marine paleotemperatures (S. M. Savin *et al.*, 1975), there is some prospect of obtaining estimates of rates of continental denudation. A recent study by W. H. Mathews (1975), though not from the tropics, has some significance in this discussion. Mathews studied Cenozoic sedimentation along the western Atlantic continental shelf, estimating an overall rate of denudation 'not counting the contribution by chemical weathering of dissolved matter via streams to the open sea' at 1·6 km^3/km^2/over 60 million years. This figure, as with all such work, integrates the denudation over the contributing landmass, and in this case over a long timespan. It is therefore a very coarse figure to use, and is on completely different spatial and temporal scales to the work of Cleaves *et al.* (1970, 1974) in the same area. The difference in estimated rates of erosion, however, is of the order of 10^3, which might be reduced to 10^2 if chemical denudation is allowed for (Cleaves *et al.*, 1974). The figures as such may not be very useful, but if the order of magnitude is correct, then the overall rate of denudation has been very low over this period. Mathews reflects that this does not suggest massive erosion of the continent by Pleistocene ice sheets as suggested by White (1971). Comparable studies for tropical areas are not so far available.

Studies of continental sedimentation, particularly within enclosed basins, have yielded important insights into the environmental and geomorphological history of the continents. From such evidence Millot (1971) concluded that a great humid period of 'lateritic' (or ferrallitic) weathering reached its peak in the Eocene, extending geographically from the present-day Sahara to central and western Europe. Parallels between the sequences of sedimentation in the basins of north-western Africa and those peripheral to the Massif Central, for instance, are striking in this respect, recording the 'lessivage' of both sub-continents during an epoch of 'biostasie', during which the evacuation of dissolved products of weathering dominated mechanical erosion and sedimentation. The dismantling of these weathered mantles has proceeded since the mid-Eocene, resulting from crustal movements and environmental changes leading to conditions of 'rhexistasie' and the reworking of successively lower zones of the deep weathering profiles. Using the Massif Central as his model, Millot (1971) shows that basins such as Aquitaine record the reworking of the lateritic crusts to form ferruginous oolites (the 'sidérolithique'); of the pallid zone or 'lithomarge' to form the kaolinitic clays and sands, and of the transitional zones of partial weathering to provide the feldspathic sandstones. If the record of sedimentation between Europe and Africa is indeed similar, we should expect parallels in terms of both the chronology and especially the mode of landform evolution.

That fragments of a similar record can be found in Britain is clear from the reviews of Montford (1970) and Dury (1971). The Eocene London Clays and Reading Beds are probably by origin ferrallitic deposits comparable with the *in situ* 'Red Beds' preserved between the Antrim lava flows. The sedimentary sequences in basins of north and south Devon have been the subject of disagreement, but the derivation of material from kaolinised residues is explained for the Petrockstow Basin by C. M. Bristow (1968) and for the Haldon Gravels by R. J. O. Hamblin (1973). The question of hydrothermal alteration of the Dartmoor granite intrudes here, as in the interpretation of tors, but recent evidence suggests this has been much over-stressed, or at least over-simplified. Hamblin's work suggests an important solutional component in the denudation which led to the accumulation of unbraded flint in the Haldon Gravels of south Devon, and points also to the accumulation of kaolinitic material from the granite. In the Petrockstow Basin, Bristow (1968) thought the kaolinisation had been of the Culm Measures and was undoubtedly owing to meteoric weathering.

Before discussing further the application of these findings to the geomorphological interpretation of British landscapes, it is perhaps useful to relate them to the general problem of planation in warm climates, at least during the period in question. Anglophone literature on the tropics has been preoccupied with the concept of pediplanation since the work of King in the 1950s. However, Lester King's type areas in Natal occur in a zone of massive, probably Plio-Pleistocene, uplift. Rivers are incised nearly 1000 m, and the enigmatic Drakensberg forms one of the world's greatest escarpments over 1500 m high. It is not perhaps surprising that King found the influence of some 15–100 m of weathering of rather small importance to his account of landform evolution. In this area of layered sub-horizontal sedimentary and volcanic rocks, evidence for differential erosion by lateral scarp retreat is persuasive and pervasive (King, 1976).

On the crystalline shields, however, differential erosion is commonly between steeply dipping metasediments and adjacent granitoids, and where uplift and dissection have been modest, denudation has been dominantly by the stripping of the weathered mantle (Falconer, 1912; Wayland, 1933; Pallister, 1956; Ollier, 1960; J. C. Doornkamp, 1968), and the principles adduced above in this paper all come into play.

Moreover, the limited extent of rock-cut pediments in many tropical landscapes has attracted increasing attention (Büdel, 1957, 1970; Handley, 1954; R. U. Cooke, 1970; Thomas, 1965, 1966; Twidale, 1967, 1976), although it remains clear that pediment-type slopes do occur quite commonly in the tropics and sub-tropics (Pugh, 1956; Mabbutt, 1966; Swan, 1970). The persistence of landform patterns during overall denudation of the terrain is now becoming widely recognised (Thomas, 1968, 1974; Twidale, 1976) by geomorphologists, and is implicit in the geological interpretations of European relief since the Mesozoic. In this case, however, the tendency for basins to subside while intervening crystalline massifs have become elevated is a major factor (Touraine, 1974).

Most of the evidence cited above accords well with the outlines of etchplanation and with Büdel's concept of 'Doppelten Einebnungsflächen' (1957), even if the generality requires much modification for specific areas or environments.

THE TROPICAL LEGACY AND LANDFORM DEVELOPMENT IN BRITAIN

The application of these concepts to the reconstruction of past landscapes in Britain has only been tried in restricted contexts, as with the discussion of tors (Linton, 1955; Eden and Green, 1971), and our appreciation of the role of etchplanation in relation to pediplanation is still crude.

The occurrence of pediment slopes in Europe has been considered quite frequently, and Büdel (1957) considers them restricted to mainly bench-like features of *Rumpftreppen*, of limited extent. They are discussed also by J. F. Gellert (1970) in relation to inselberg forms, while J. Gjessing (1967) has called attention to possibilities of recognising such features in Norway. The important paper by Bakker and Levelt (1964) referred enigmatically in its title to 'a polyclimatic development of peneplains and pediments (etchplains) in Europe. . .', and it seems to me that the implications of this paper as well as of Millot's (1971) synthesis have still to be fully considered in relation to British landscapes. A preliminary approach to this problem is offered here.

The persistence of positive and negative elements in the relief of Europe, including Britain, throughout the Cenozoic, together with the occurrence of certain sedimentary sequences that can be determined geochemically as resulting from the progressive removal of a deep tropical saprolite from the crystalline uplands, and also from exposed sedimentary formations, both argue for some model of etch-planation as the primary mode of landform development during the last 60–100 million years at least. If pediplanation, with the retreat of steep slopes in solid rock, had been the primary mode of development, I do not see how such sedimentary sequences could occur widely, nor is it apparent how the major lineaments of relief could remain so persistent.

The retreat of escarpments within the layered sedimentaries of the English lowlands, however, remains largely undisputed, though recent interpretations of the long debated 'clay-with-flints' deposits of the Chalk are directly relevant to this discussion. More than a decade of research (J. Loveday, 1962; J. M. Hodgson *et al.*, 1970, 1974; D. M. Pepper, 1973) has led to a reappraisal of our understanding of the English Chalklands. If the conclusion of Hodgson *et al.* (1974) that these deposits were 'formed by the reorganisation of early Tertiary sediments more or less *in situ*' is taken together with the occurrence of unrolled flint, then the lowering of the Chalk by solution and the local redistribution of the Eocene clays suggests a persistence of a form of etch-planation over the Chalk, long after the removal of the deep ferrallitic residues from the crystalline uplands. This view accords with Klein's (1965) account of the *argiles à silex* of the Paris Basin in which *l'abbaisement conjugué du plancher et du toit de la nappe des argiles à silex nous paraît avoir constitué une donnée fondementale de l'évolution géomorphologique des bordures occidentales et meridionales du bassin de Paris* (p. 193). To this application of the *Doppelten Einebnungsflächen* concept of Büdel, Klein added that the processes of lowering responded to bioclimatic and tectonic rhythms, resulting in a form of 'acyclic' planation. It is not clear why Pepper (1973) finds this view unacceptable, but the presence of Tertiary *remaniés* deposits in the 'clay-with-flints' may be interpreted as a result of slope retreat in the cover materials. However, the evidence suggests rather a combination of solutional lowering of the chalk with local redistribution of overlying materials, comparable with the events recorded for the granitic basement of Western Australia (H. M. Churchward, 1969).

The stripping of an early Tertiary, or indeed Mesozoic, weathering cover from the granites of Dartmoor and from other crystalline massifs is implicit in the prior development of many Eocene deposits, and indicates the nature of the *growan* and comparable sandy weathering types known more generally as *arènes* which today are widespread in Dartmoor (Eden and Green, 1971) as well as in Europe (Millot, 1971). In the *arènes*, the formation of secondary minerals is very slight, and the clay-sized fraction (less than 2 or 4 μm) is generally only 2–7 per cent, while the loss of primary minerals varies from as low as 5 per cent to perhaps 20 per cent. A review of this phenomenon cannot be undertaken here, but it may be noted that 'arenisation' is common in a wide range of igneous rocks including quartz diorities (J. B. Borras *et al.*, 1975) and even diorites and gabbros in Aberdeenshire (I. R. Basham, 1974). It attains depths of 50 m in central Europe though most occurrences in Britain are less than 20 m, 2–5 m being common (Eden and Green, 1971; Basham, 1974).

Arguments about the environments of formation of such deposits are still inconclusive (Bakker and Levelt, 1964; Eden and Green, 1971; Basham, 1974; Clayton, 1974), but suggestions that they may represent the lower zones of early Tertiary, tropical profiles are probably unacceptable: first, because the depth of arenisation is much greater than known thicknesses of little altered rock in the tropics, and second, in the case of basic rocks, such *arènes* are almost unknown in tropical profiles. In fact, the transition from fresh rock to a highly altered saprolite rich in clays occurs quite close to the weathering front in most profiles in the tropics. On the other hand, the depths of arenisation are in places too great for Holocene or even interglacial weathering (Bakker and Levelt, 1964), and some writers opt for a pre-glacial (possibly Pliocene), warm temperate climate as the dominant environment of formation. On available evidence it therefore seems possible to suggest a period of reformation of a weathered mantle, under much cooler conditions than those of the early Tertiary, and occurring perhaps in Pliocene or early Pleistocene times. This must have been followed by more recent removal or local redistribution in periglacial or glacial environments. This account departs far from tropical conditions, but applies the principles of etch-planation developed in the tropics to a wider range of conditions.

Two major observations about British landscapes arise from these observations. First, there is a need to recognise complex land-surfaces on British uplands comparable with those suggested by Gjessing (1967) for Norway and V. Kaitanen (1969) for Lappland. These recall Mabbutt's (1965) recognition of a 'weathered land surface' in central Australia which he describes as an 'entire landscape which was subject to weathering. It is a cyclically complex surface, consisting of pre-Cretaceous uplands, mainly on resistant sandstone, and younger duricrusted plains on crystalline and softer sedimentary rocks.' (1965, p. 109). Views concerning a complex early Tertiary plain over the Chalk converge with this concept, as well as with earlier views of Ph. Pinchmel (1954), and conditions on the crystalline massifs probably reflect a succession of etch-planation/pediplanation periods in a similarly complex fashion. Increasing emphasis is being placed on 'biochemical planation' by certain French writers including Klein (1965) and Touraine (1972, 1974) who at the same time admit that a quantitative estimate of biochemical lowering of the landscape is still elusive.

The recognition of pediments in Britain (Dury, 1972) is not in itself controversial, but their origins and place in an overall scheme of landform development have yet to be elucidated. Wherever lineal dissection proceeds below the depth of weathering, slopes in fresh rock must subsequently undergo an independent evolution. The concept of the two-storey landscape was applied by Ruxton and Berry (1961) to granitoid terrains in the tropics, and it is probably necessary to invoke these ideas in any account of British landscapes.

With respect to the relationship of glacial 'interrruptions' to these ideas, there are some clear pointers. The survival of palaeoforms in the Scottish uplands, for instance, is well-known in general (Linton, 1955, 1959; A. Godard, 1965; Sugden, 1968), but as part of a picture of zones of glacial erosion (K. M. Clayton, 1974) it remains problematic. The severity of glacial erosion in north America has been the subject of renewed debate (T. Feininger, 1971; W. A. White 1972; F. Gravenor, 1975), and aspects of this situation are relevant in Britain. Feininger (1971) called attention to the similarity between the form of the weathering front in tropical areas and the character of mammillated (in Scotland, 'knoch-and-lochan') landscapes of supposed glacial scour, suggesting that glacial erosion had in fact been minimal in such areas. Several authors (Kaitanan, 1969; Feininger, 1971; Gravenor, 1975) have also called attention to the mineralogy and morphology of certain tills. Gravenor (1975) has shown how the hornblende content of tills increases from 12 to 62 per cent with successive glaciations in America, while the iron-oxide (mainly limonite) content shows an inverse relationship. This appears to signify that the first glaciations removed mainly the preglacial weathering cover; glacial scour of fresh rock became more important subsequently. It is difficult to say how pervasive this glacial erosion was in, say, Scotland, but the possibility that areas such as Rannoch Moor remain little altered from pre-glacial times must be entertained, while the origins of the more controversial 'knoch-and-lochan' landscapes of north-west Sutherland must remain doubtful, though the amount of glacial erosion below ancient levels of planation appears to be slight (see D. R. Stewart, 1972 and also Godard, 1965).

It is hoped that this review has highlighted some issues of geomorphological importance that emerge from work on the geomorphology of the tropics and from studies of the paleoclimatic history of Britain. One must not pretend that either the models that can be derived from the studies of landforms in warmer climates or the details of the past history of British landscapes are without difficulty or dispute. What is claimed here is that the study of denudation in the tropics and sub-tropics is directly relevant to the interpretation of British landscapes within which the inheritance from pre-glacial conditions may be as important as the effects of glacial and periglacial modifications. Where such palaeoforms do not survive, the inheritance may be indirect, through the influence of pre-glacial form on subsequent events (Linton, 1959; Bakker, 1965; H. Klimaszewski, 1964).

As a postscript to the discussion, it may be recalled that acceptance in Britain of ideas that have come from the study of the tropics has been restricted, and that the work of European geomorphologists in this field has, at least until recently, been neglected. The broader view has gathered momentum during the last decade, but whether British interest in, and commitment to research in, tropical environments is stronger as a result seems doubtful. The institutional framework for the encouragement of this work is weak, and where fundamental work depends on process studies, only resident research workers can make much impact. Awareness of the relevance of studies in warm climates can come ultimately only through experience of them, which can arise initially in the training of undergraduates and from the funding of research overseas for post-graduates. The peculiarly 'national' character of geomorphology in Britain, as well as in some other countries, has broadened in the last decade to embrace much north American research, and has also strengths of its own. However, a need for greater interest in the warmer environments, their peculiarities and common features, their landforms and their legacies in Britain, is, I hope, one legitimate conclusion that may be drawn from this study.

REFERENCES

Bakker, J. P. (1965), 'A forgotten factor in the development of glacial stairways', *Z. Geomorph.* N.F., **9**, 18–34.

—— (1967), 'Weathering of granites in different climates', in P. Macar, 'L'évolution des versants', *Congrès Colloques Univ. Liège* **40**, 51–68.

—— and Levelt, Th. W. M. (1964), 'An enquiry into the problems of a polyclimatic development of peneplains and pediments (etch-plains) in Europe during the Senonian and Tertiary period', *Publs Serv. Carte géol. Luxemb.* **14**, 27–75.

Basham, I. R. (1974), 'Mineralogical changes associated with deep weathering of gabbro in Aberdeenshire', *Clay Minerals* **10**, 189–202.

Beckinsale, R. P. (1976), 'The international influence of William Morris Davis', *Geogr. Rev.* **66**, 448–66.

Bocquier, G., Millot, G. and Ruellan, A. (1974), 'Différenciation pédologique et géochimique dans des paysages Africain, Tropicaux et Mediterranéans, la pedogenèse lateral remontante', *Trans. 10th int. Congr. Soil Science* (Moscow 1974), 226–32.

Borras, J. B., Chevalier, Y. and Dejou, J. (1975), 'Évolution géochimique superficielle des diorites quartzitiques dans les régions mediterranéenes humides', *C.r. hebd. Séanc. Acad. Sci. Paris D.* **280**, 387–90.

Bristow, C. M. (1968), 'The derivation of the Tertiary sediments in the Petrockstow Basin, North Devon', *Proc. Ussher Soc.* **2**, 29–35.

Bryan, K. (1922), 'Erosion and sedimentation in the Papago Country, Arizona', *Bull. U.S. geol. Surv.* **730**, 19–90.

Büdel, J. (1948), 'Das System der klimatischen Geomorphologie', *Verh. dt. GeogrTags* **27**, 65–100.

—— (1957), 'Die "Doppelten Einebnungsflächen" in den feuchten Tropen', *Z. Geomorph.* N.F. **1**, 201–88.

—— (1963), 'Klimagenetische Geomorphologie', *Geogr. Rdsch.* **15**, 269–85.

—— (1965), 'Die relieftypen der Flächenspülzone: Sud-Indiens am Ostabfall Dekans gegen Madras', *Colloquium geogr.* (Bonn) **8**, 100 pp.

—— (1970), 'Pedimente, Rumpflächen und Rückland-Steilhange', *Z. Geomorph.* N.F. **14**, 1–57.

Buurman, P. (1975), 'Possibilities of paleopedology', *Sedimentology* **22**, 289–98.

Butler, B. E. (1959), 'Periodic phenomena in landscapes as a basis for soil studies', *C.S.I.R.O., Australia, Soil Publs* **14**, 430–44.

—— (1967), 'Soil periodicity in relation to landform development in south east Australia', in Jennings, J.N. and Mabbutt, J.A. (eds.), *Landform studies from Australia and New Guinea* (Cambridge), 231–55.

Campbell, J. M. (1917), 'Laterite', *Mineralog. Mag.* **17**, 67–77; 120–8; 171–9; 220–9.

Churchward, H. M. (1969), 'Erosional modification of a lateritised landscape over sedimentary rocks, its effects on soil distribution', *Austr. J. Soil Res.* **8**, 1–19.

Clayton, K. M. (1974), 'Zones of glacial erosion', *Inst. Br. Geogr. Spec. Publ.* **7**, 163–76.

Clayton, R. W. (1956), 'Linear depressions (Bergfüssneiderungen) in savanna landscapes', *Geogr. Stud.* **3**, 102–26.

Cleaves, E. T. and Godfrey, A. E. (1970), 'Geochemical balance of a small watershed and its geomorphic implications', *Bull. geol. Soc. Am.* **81**, 3013–32.

——, Fisher, D. W. and Bricker, O. P. (1974), 'Chemical weathering of serpentinite in the Eastern Piedmont of Maryland', *Bull. geol. Soc. Am.* **85** (1), 437–44.

Cooke, R. U. (1970), 'Morphometric analysis of pediments and associated landforms in the Western Mojave Desert, California', *Am. J. Sci.* **269**, 26–38.

Corbel, J. (1959), 'Vitesse de l'érosion', *Z. Geomorph. N.F.* **3**, 1–28.

—— (1964), 'L'érosion terrestre, étude quantitative (méthodes–téchniques–résultats)', *Annls Géogr.* **73**, 385–412.

Cotton, C. A. (1947), *Climatic accidents in landscape making* (Whitcomb and Tombs, New Zealand; reprinted 1969, Hafner, New York), 354 pp.

—— (1961), 'The theory of savanna planation', *Geography* **46**, 89–96.

D'Hoore, J. L. (1964), *Soil map of Africa, scale 1:5 million. Explanatory Monograph, C.C.T.A. Publ.* **93** (Lagos).

Demangeot, J. (1975), 'Recherches géomorphologiques en Indie du Sud', *Z. Geomorph. N.F.* **19**, 229–72.

De Swardt, A. M. J. (1946), 'Recent erosional history of the Kaduna Valley, near Kaduna township', *Rep. geol. Surv. Nigeria* 1946, 39–45.

—— (1964), 'Lateritisation and landscape development in equatorial Africa', *Z. Geomorph. N.F.* **8**, 313–32.

—— and Trendall, A. F. (1964), 'The physiographic development of Uganda', *Overseas Geol. Miner. Resour.* **10**, 241–88.

Dixey, F. (1922a), 'Notes on lateritisation in Sierra Leone', *Geol. Mag.* **57**, 211–20.

—— (1922b), 'The physiography of Sierra Leone', *Geogr. J.* **60**, 41–65.

Doornkamp, J. C. (1968), 'The role of inselbergs in the geomorphology of northern Uganda', *Trans. Inst. Br. Geogr.* **44**, 151–62.

Douglas, I. (1967a), 'Erosion of granite terrains under tropical rain forest in Australia, Malaysia and Singapore', *Publs int. Ass. Scient. Hydrol., Symp. River Morphology, Gen. Assembly* (Bern), **75**, 31–9.

—— (1967b), 'Natural and man-made erosion in the humid tropics of Australia, Malaysia and Singapore', *Publ. int. Ass. scient. Hydrol.* **75**, 17–29.

—— (1967), 'The efficiency of humid tropical denudation systems', *Trans. Inst. Br. Geogr.* **40**, 1–16.

—— (1973), 'Rates of denudation in selected small catchments in Eastern Australia', *Univ. Hull Occas. Pap. Geogr.* **21**, 127 pp.

Dury, G. H. (1971), 'Relict deep weathering and duricrusting in relation to the paleoenvironments of middle latitudes', *Geogr. J.* **137**, 511–22.

—— (1972), 'A partial definition of the term "pediment" with field tests in the humid climate areas of southern England', *Trans. Inst. Br. Geogr.* **57**, 139–52.

—— and Langford-Smith, T. (1964), 'The use of the term peneplain in descriptions of Australian landscapes', *Austr. J. Sci.* **27**, 171–5.

Eden, M. J. (1971), 'Some aspects of weathering and landforms in Guyana (formerly British Guiana)', *Z. Geomorph.* **15**, 181–98.

—— and Green, C. P. (1971), 'Some aspects of granite weathering and tor formation on Dartmoor, England', *Geogr. Annlr* **53A**, 92–9.

Erhart, H. (1955), 'Biostasie et rhexistasie: esquisse d'un théorie sur le rôle de la pedogenèse en tant que phénomène géologique', *C.r. hebd. Séanc. Acad. Sci. Paris* **241**, 1218–20.

Falconer, J. D. (1911), *The geology and geography of Northern Nigeria* (Macmillan, London), 295 pp.

—— (1912), 'The origins of kopjes and inselbergs', *Rep. Br. Ass. Advmt Sci., Proc. Sect. C.*, 476.

Feininger, T. (1971), 'Chemical weathering and glacial erosion of crystalline rocks and the origins of till', in 'Geological survey research', *U.S. geol. Surv. Prof. Pap.* **750-C**, C65–81.

Fölster, H. (1969), 'Slope development in south western Nigeria during late Pleistocene and Holocene', *Ber. Göttinger Bodenkundl.* **10**, 3–56.

Fournier, F. (1960), *Climat et érosion: la relation entre l'érosion du sol par l'eau et les précipitations atmospheriques* (Paris, P.U.F.), 201 pp.

Gellert, J. F. (1970), 'Climatomorphology and paleoclimates of the central European Tertiary', in Pecsi, M. (ed.) *Problems of relief planation* (Akademiai Kaido, Budapest), 107–11.
Gjessing, J. (1967), 'Norway's Paleic surface', *Norsk. geogr. Tidschr.* **21**, 69–132.
Godard, A. (1965), *Recherches de géomorphologie en Écosse du Nordouest* (Les Belles Lettres, C.N.R.S., Paris), 701 pp.
Goudie, A. S. (1973), *Duricrusts in tropical and sub-tropical landscapes* (Clarendon Press, Oxford), 174 pp.
Gravenor, C. P. (1975), 'Erosion by continental ice sheets', *Am. J. Sci.* **275**, 594–604.
Grove, A. T. (1958), 'The ancient erg of Hausaland and similar formations on the southern side of the Sahara', *Geogr. J.* **124**, 528–33.
—— and Pullan, R. A. (1963), 'Some aspects of the Pleistocene paleogeography of the Chad Basin', in Howell, F.C. and Boulière, F., *African ecology and human evolution* (Methuen, London), 230–45.
Hamblin, R. J. O. (1973), 'The Halden gravels of south Devon', *Proc. Geol. Ass.* **84**, 454–76.
Handley, J. R. F. (1954), 'The geomorphology of the Nzega area of Tanganyika with special reference to the formation of granite tors', *19th int. geol. Congr.* (Algiers), 201–10.
Harrison, J. B. (1910), 'The residual earth of British Guiana commonly termed "laterite",' *Geol. Mag.* **7**, 439–52, 488–95, 553–62.
—— (1933), 'The katamorphism of igneous rocks under humid tropical conditions', in Hardy, F. (ed.) *Imperial Bureau of Soil Sci.* (Harpenden), 79 pp.
Hodgson, J. M., Catt, J. A. and Weir, A. H. (1967), 'The origin and development of clay-with-flints and associated soil horizons on the South Downs', *J. Soil Sci.* **18**, 85–102.
—— and Catt, J. A. (1974), 'The geomorphological significance of clay-with-flints on the South Downs', *Trans. Inst. Br. Geogr.* **61**, 119–29.
Jennings, J. N. and Bik, M. J. (1962), 'Karst morphology in Australia and New Guinea', *Nature, Lond.* **194**, (4833), 1036–8.
Jutson, R. T. (1934), 'The physiography (geomorphology) of Western Australia', *Bull. geol. Soc. Austr.* (Perth), **95**, 366 pp.
Kaitanen, V. (1969), 'A geographical study of morphogenesis in northern Lappland', Fennia **99** (5).
Kesel, R. (1973), 'Inselberg landform elements: definition and synthesis', *Revue Géomorph. dyn.* **22**, (3), 97–108.
King, L. C. (1953), 'Canons of landscape evolution', *Bull. geol. Soc. Am.* **64**, 721–52.
—— (1957), 'The uniformitarian nature of hillslopes', *Trans. Edinb. geol. Soc.* **17**, 81–102.
—— (1962), *The morphology of the Earth* (Oliver and Boyd, Edinburgh), 699 pp.
—— (1976), 'Planation remnants upon high lands', *Z. Geomorph.* N.F. **20**, 133–48.
Klein, Cl. (1965), 'La significance géomorphologique des argiles à silex', *C.r. hebd. Séanc. Acad. Sci., Paris,* **261**, 191–4.
Klimaszewski, M. (1964), 'On the effect of preglacial relief on the course and the magnitude of glacial erosion in the Tatra Mountains', *Geogr. Polonica* **2**, 11–21.
Linton, D. L. (1955), 'The problem of tors', *Geogr. J.* **121**, 470–87.
—— (1959), 'Morphological contrasts between eastern and western Scotland',

in Miller, R. and Watson, J. W. (eds.) *Geographical essays in memory of Alan G. Ogilvie* (Edinburgh), 16–45.
Livingstone, D. A. (1963), 'Chemical composition of rivers and lakes', *U.S. geol. Surv. Prof. Pap.* **440-G**, 64 pp.
Loveday, J. (1962), 'Plateau deposits of the southern Chiltern Hills', *Proc. Geol. Ass.* **73**, 83–99.
Mabbutt, J. A. (1961), 'A stripped landsurface in Western Australia', *Trans. Inst. Br. Geogr.* **29**, 101–14.
—— (1965), 'The weathered landsurface of central Australia', *Z. Geomorph.* N.F. **9**, 82–114.
—— (1966), 'The mantle-controlled planation of pediments', *Am. J. Sci.* **264**, 78–91.
McConnell, R. B. (1968), 'Planation surfaces in Guyana', *Geogr. J.* **134**, 506–20.
Maignien, R. (1966), 'Review of research on laterites', *Nat. Resour. Res.* (UNESCO, Paris), **4**, 148 pp.
Mathews, W. H. (1975), 'Cenozoic erosion and erosion surfaces of eastern North America', *Am. J. Sci.* **275**, 818–24.
Michel, P. (1967), 'Les bassins des fleuves Sénégal et Gambie', *Étud. géomorph. O.R.S.T.O.M. Mém.* **63** (3 Vols), 752 pp.
Millot, G. (1964), *Géologie des argiles, alternations, sedimentologie, géochemie* (Masson et Cie, Paris; Engl. Transl. 1970, Chapman and Hall), 429 pp.
Mohr, E. C. and Van Baren, F. A. (1954), *Tropical soils* (La Haye), 498 pp.
Montford, H. M. (1970), 'The terrestrial environment during Upper Cretaceous and Tertiary times', *Proc. Geol. Ass.* **81**, 181–203.
Mulcahy, M. J. and Bettenay, E. (1971), 'The nature of old landscapes', *Search* **2**, 433–4.
Ollier, C. D. (1960), 'The inselbergs of Uganda', *Z. Geomorph.* N.F. **4**, 43–52.
—— (1965), 'Spheroidal weathering, exfoliation and constant volume alteration', *Z. Geomorph.* N.F. **9**, 285–304.
Pallister, J. W. (1956), 'Slope development in Buganda', *Geogr. J.* **122**, 80–7.
Palmer, J. and Neilson, R. A. (1962), 'Origin of granite tors on Dartmoor, Devonshire', *Proc. Yorks. geol. Soc.* **33**, 315–40.
Penck, W. (1924), *Die morphologische Analyse* (Geogr. Abhandlung 2, R.H.2, Stuttgart, Engl. trans.: Czech, H. and Boswell, K. C. (1954), *Morphological analysis of landforms,* Macmillan, London), 429 pp.
Pepper, D. M. (1973), 'Comparisons of the "argile à silex" of northern France with the "clay-with-flints" of southern England', *Proc. Geol. Ass.* **84**, 331–52.
Pinchmel, Ph. (1954), *Les plaines de craie du nord-ouest du bassin Parisien et du sud-est du bassin de Londres et leur bordeurs* (Thèse lettres, A. Colin, Paris), 502 pp.
Playford, P. E. (1954), 'Observations on laterite in Western Australia', *Austr. J. Sci.* **17**, 11–14.
Prescott, J. A. and Pendleton, R. L. (1952), 'Laterites and lateritic soils', *Tech. Commun. Commonw. Bur. Soil Sci.* (Harpenden) **47**, 51 pp.
Pugh, J. C. (1956), 'Fringing pediments and marginal depressions in the inselberg landscape of Nigeria', *Trans. Inst. Br. Geogr.* **22**, 15–31.
Rapp, A. (1960), 'Development of mountain slopes in Kärkevagge and surroundings, northern Scandinavia', *Geogr. Annlr* **42**, 65–287.
——, Berry, L. and Temple, P. H. (eds.) (1972), 'Studies of soil erosion and sedimentation in Tanzania', *Geogr. Annlr* **54A**, 379 pp.
Rhodenburg, H. (1971), *Einführung in die klimategenetische Geomorphologie* (Lenz-Kerlag, Giessen), 350 pp.

Rougerie, G. (1960), 'Le façonnement actuel des modelés en Côte d'Ivoire forestière', *Mém. Inst. fr. Afr. noire* **58**, 542 pp.
Ruhe, R. V. (1956), 'Landscape evolution in the High Ituri, Belgian Congo', *Publs. Inst. natn. Etude agron. Congo belge, Sér. Sci.* **66**.
Ruxton, B. P. (1958), 'Weathering of granite and sub-surface erosion in granite at the piedmont angle, Balos, Sudan', *Geol. Mag.* **95**, 353–77.
—— and Berry, L. (1957), 'Weathering of granite and associated erosional features in Hong Kong', *Bull. geol. Soc. Am.* **68**, 1263–92.
—— and —— (1961), 'Weathering profiles and geomorphic position on granite in two tropical regions', *Revue Géomorph. dyn.* **12**, 16–31.
Savigear, R. A. G. (1960), 'Slopes and hills in West Africa', *Z. Geomorph. Suppl. Bd* (*Morphologie des versants*), 156–71.
Savin, S. M., Douglas, R. C. and Stehli, F. G. (1975), 'Tertiary marine paleotemperatures', *Bull. geol. Soc. Am.* **86**, 1499–510.
Scrivenor, J. B. (1931), *The geology of Malaya* (London), 217 pp.
Stewart, A. D. (1972), 'Precambrian landscapes in north-west Scotland', *Geol. J.* **8**, 111–24.
Stoddart, D. R. (1969), 'Climatic geomorphology: review and reassessment', *Progr. Geogr.* **1**, 160–222.
Sugden, D. E. (1968), 'Selectivity of glacial erosion in the Cairngorm Mountains, Scotland', *Trans. Inst. Br. Geogr.* **45**, 79–92.
Swan, S. B. StC. (1970), 'Piedmont slope studies in a humid tropical region, Johor, southern Malaya', *Z. Geomorph. Suppl. Bd* **10**, 30–9.
—— (1972), 'Landsurface evolution and related problems with reference to a humid tropical region, Johor, West Malaysia', *Z. Geomorph.* N.F. **16**, 160–81.
Sweeting, M. M. (1958), 'The karstlands of Jamaica', *Geogr. J.* **124**, 184–99.
Tardy, Y. (1971), 'Characterisation of the principal weathering types by the geochemistry of waters from some European and African crystalline massifs', *Chemical Geol.* **7**, 253–71.
——, Bocquier, G., Paquet, H. and Millot, G. (1973), 'Formation of clay from granite and its distribution in relation to climate and topography', *Geoderma* **10**, 271–84.
Tator, B. A. (1952/3), 'Pediment characteristics and terminology', *Ann. Ass. Am. Geogr.* **42**, 295–317 and **43**, 47–53.
Thomas, M. F. (1965a), 'Some aspects of the geomorphology of domes and tors in Nigeria', *Z. Geomorph.* N.F. **9**, 63–81.
—— (1965b), 'An approach to some problems of landform analysis in tropical environments', in Wood, P. D. and Whittow, J. B. (eds.) *Essays in geography for Austin Miller* (Reading), 118–44.
—— (1966), 'Some geomorphological implications of deep weathering patterns in crystalline rocks in Nigeria', *Trans. Inst. Br. Geogr.* **40**, 173–95.
—— (1968), 'Some outstanding problems in the interpretation of the geomorphology of tropical shields', *Br. geomorph. Res. Grp. Publ.* **5**, 41–9 (mimeo.)
—— (1974), *Tropical geomorphology* (Macmillan, London), 332 pp.
—— (1977), 'Chemical denudation, lateritisation and landform development in Sierra Leone', in Alexandre, J. (ed.) *Environmental geomorphology in the Tropics* (Univ. Zaire), 00–00.
Touraine, F. (1972), 'Erosion et planation', *Revue Géogr. alp.* **60**, 101–21.
—— (1974), 'Qu'est ce donc qu'une argilite ferrugineuse?' *Revue Géogr. alp.* **62**, 433–54.
Trendall, A. F. (1962), 'The formation of apparent peneplains by a process of combined lateritisation and surface wash', *Z. Geomorph.* N.F. **6**, 183–97.

Tricart, J. and Cailleux, A. (1965), *Introduction à la géomorphologie climatique* (Paris, Transl. 1972, De Junge, C. J. K., Longmans, London, 1972), 295 pp.
Tuan, Y. F. (1959), 'Pediments in southeastern Arizona', *Univ. Calif. Publs Geogr.* **13**, 20 pp.
Twidale, C. R. (1964), 'A contribution to the general theory of domed inselbergs', *Trans. Inst. Br. Geogr.* **34**, 91–113.
—— (1967), 'Hillslopes and pediments in the Flinders Ranges, South Australia', in Jennings, J. N. and Mabbutt, J. A. (eds.) *Landform studies from Australia and New Guinea* (Cambridge), 95–117.
—— (1976), 'On the survival of paleoforms', *Am. J. Sci.* **276**, 77–95.
—— and Bourne, J. A. (1975), 'Episodic exposure of inselbergs', *Bull. geol. Soc. Am.* **86**, 1473–81.
U.N.E.S.C.O. (1969), *Soils and tropical weathering* (*Proc. Bandung Symp. U.N.E.S.C.O. 1969, Paris*), 149 pp.
Van Dijk, D. C., Riddler, A. M. H. and Rowe, R. K. (1968), 'Criteria and problems in ground surface correlations with reference to a regional correlation in south east Australia', *9th int. Congr. Soil Sci.* (Adelaide), **4**, 131–7.
Wayland, E. J. (1933), 'Peneplains and some other erosional platforms', *Bull. geol. Surv. Uganda, Ann. Rep.,* Notes 1 and 74, p. 366.
White, W. A. (1972), 'Deep erosion by continental ice sheets', *Bull. geol. Soc. Am.* **83**, 1037–56.
Wilhelmy, H. (1958), *Klimamorphologie der Massengesteine* (Westermann, Braunschweig), 238 pp.
Williams, G. E. (1973), 'Precambrian landscapes in north-west Scotland—a discussion', *Geol. J.* **8**, 397–8.
Woolnough, W. G. (1927), 'The duricrust of Australia', *J. Proc. R. Soc. N.S.W.* **61**, 24–53.
—— (1930), 'Influence of climate and topography in the formation and distribution of products of weathering', *Geol. Mag.* **67**, 123–32.
Yaalon, D. H. (ed.) (1971), *Paleopedology* (Israel Univ. Press., Jerusalem), 350 pp.
Young, A. (1972), *Slopes* (Oliver and Boyd, Edinburgh), 288 pp.

13

RESEARCH IN COASTAL GEOMORPHOLOGY: BASIC AND APPLIED

H. J. WALKER
(*Louisiana State University*)

Coasts have been a major interest of man throughout most, if not all, of his long history and will certainly continue to be so in the future. Although the nature of this concern has changed through time, it has probably always included elements of aesthetics, economics and philosophy. C. O. Sauer (1962) went so far as to write: 'When all the lands will be filled with people and machines, perhaps the last need and observance of man still will be, as it was at his beginning, to come down to experience the sea.'

Probably as much as any other type of environment the shoreline has caused man to pause and admire, to wonder at and analyse the workings of nature. In his attempts to gain an understanding of nature he has turned to art, philosophy, religion and science. Although all of these disciplines (and others) have the same objective–to render nature intelligible–their practitioners utilise different approaches, tools and techniques.

This paper attempts to analyse how the practitioners of one of these groups, those devoted to science, have proceeded, what they have accomplished and where they are heading in their studies of coastal morphology. The stock-in-trade of the scientist is research, the type, location, duration and intensity of which is affected by many factors including, in addition to the physical nature of the coast itself, those of finance, equipment, politics, Gross National Product, perseverance and so forth. There is no indication that such controls will disappear in the future. Some of these factors are especially important in determining the relative number of individuals engaged in, and the relative amount of support allocated to, either basic or applied research.

These two types of research are closely related: indeed, in many instances it is difficult to distinguish one from the other. However, as applied research is aimed at solving specific problems, usually those with pressing economic significance, it is easily and often emphasised at the expense of basic research. Such emphasis will almost certainly lead to fewer fundamental discoveries, discoveries that actually make applied research meaningful. The relative proportions of basic and applied research appear to be closely tied to the state of the economy and in fact are probably a good economic indicator in their own right. Although the above comments are generally applicable to all scientific research, they are especially pertinent to present-day coastal research.

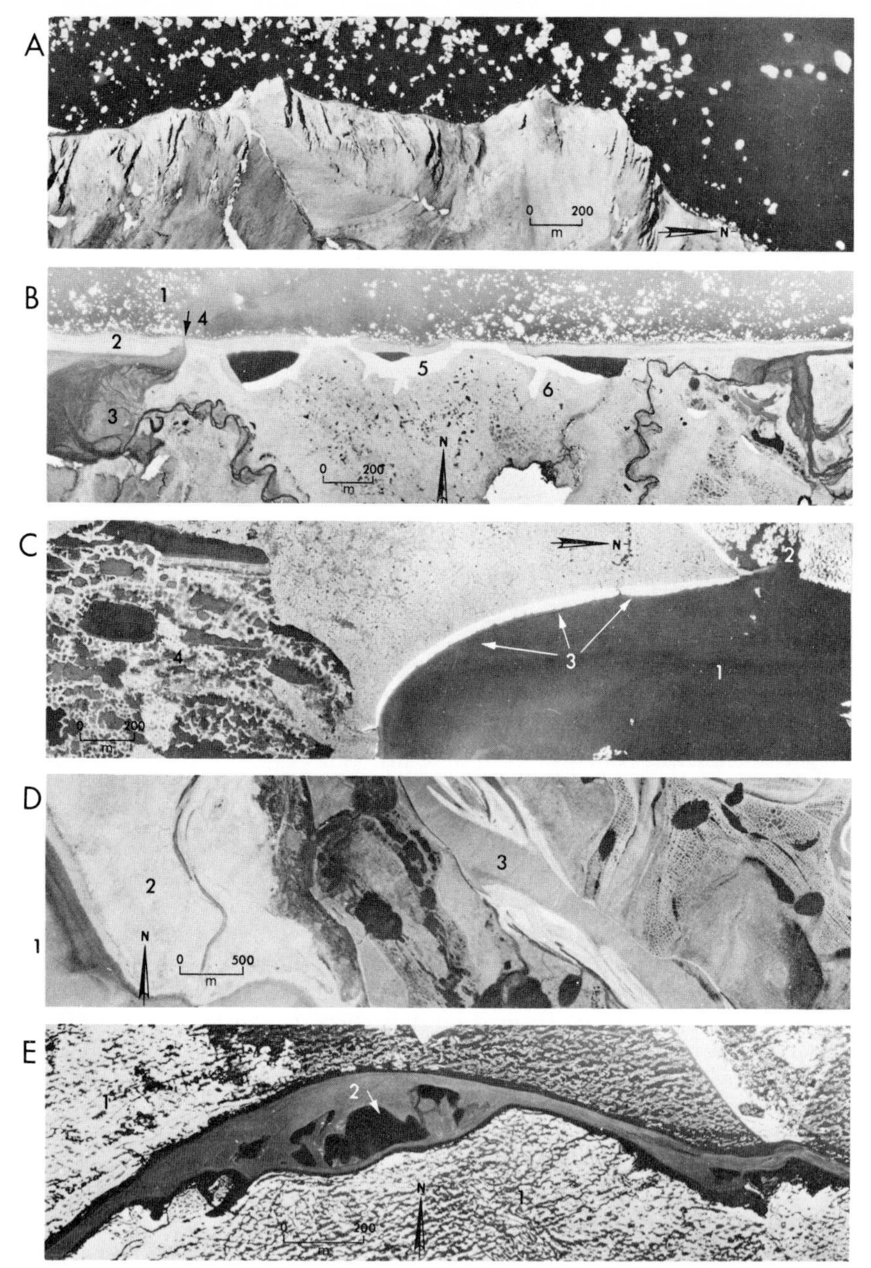

FIG. 13.1. Variation of form along the Arctic coast of Alaska. A. Coastal cliffs at Cape Lisburne. B. Barrier bars near Cape Beaufort: 1. Sea ice. 2. Bar. 3. Lagoonal delta. 4. Tidal inlet. 5. Snow ramp. 6. Gully with remnant snow. C. Oriented lake shoreline: 1. Arctic Ocean. 2. Sea ice. 3. Snow ramp at base of former oriented lake bluff. 4. Drained oriented lake. D. Deltaic coast: 1. Arctic Ocean. 2. Mudflat. 3. Delta distributary. E. Offshore barrier bar: 1. Sea ice. 2 Bar pond (from Walker, 1973).

Despite the fact that the coast is essentially linear and therefore finite in areal extent, it is extremely varied and complex (Fig. 13.1). Variability arises because of the many contrasting characteristics associated with land, sea and air. Among these are such variables as geological structure, climate, biotic composition and numbers, water chemistry, waves, currents and tides. The shoreline, some 450 000 km long, is a place of concentrated interactions, a place where many of the oceanographic, atmospheric and terrestrial processes are accelerated and accentuated.

It is not surprising then, to find that the accumulated literature about coasts is monumental and that it is expanding at an ever-increasing rate. Most studies in coastal morphology emphasise one or more of the following: material, process, time and human modification. These four topics form the framework of this essay within which an attempt is made to:

1. characterise the major coastal forms and illustrate presently-held beliefs about their evolution,[1]
2. provide a sample of the most recent morphologically-related research on coasts,[2] and
3. indicate those areas of research which, in the author's opinion, will be emphasised in the near-future.

LITHOLOGY, STRUCTURE AND CONFIGURATION

The character of any coastal form, whether it be relatively permanent or ephemeral, or whether it be large or small, is dependent on the material of which it is composed. Indeed, from one viewpoint, form is nothing more than a container, a container presupposing a content (S. R. Aiken, 1976). It is believed by some (R. J. Russell, 1967) that when it is devised the most valuable classification will have its primary distinctions based on lithological and structural contrasts. Yet so far relatively few research projects have used material as their major point of focus. It is the geologist (mineralogist, sedimentologist) who has contributed most in this field (P. D. Komar, 1976).

The variety of materials of which coastal forms are composed is great, as are their size, shape and arrangement. These conditions are important whether we are dealing with individual grains or large masses of consolidated material with or without distinctive joints, beds, bands, cleavages or brecciation.

A recent preliminary report (E. Pruitt *et al.*, 1975) states: 'If we consider the two extremes of the spectrum of geomorphic characteristics that can be exhibited by coastal localities, we would find, at one extreme an area totally deficient in sediment, a cliffed coast; and at the other end, an area that is continuously accreting, or building out, a series of beach ridges'.

Such a spectrum presents a wide range in texture, structure and attitude as well as composition, and in addition may reflect availability and source region.

[1] Emphasis is placed on those specific items that are likely to be central to most future research.

[2] It is believed that most future research (at least that to be undertaken in the near-future) will be mainly a modification of that presently under way. Although 'breakthroughs' do and will occur they are seldom predictable. Indeed, for the most part, they are based on current research activities. It is hoped that discussion of current research will suggest problems, potentials and therefore future research. Further, most of the references contain excellent bibliographies and can be used as a starting point for further study.

Plunging cliffs of durable rock such as Cape St Vincent, Portugal, remain relatively unchanged with time despite severe wave action. Differential erosion, resulting from contrasting composition, is illustrated well by stacks along present-day shorelines and also of past shorelines as evidenced by the jasper remnants that rise above Quaternary terraces along the central coast of California. Overhangs and benches often reflect bedding planes; volcanic craters and columnar basalts have their own distinctive appearance in coastal settings as do glacially scoured granites.

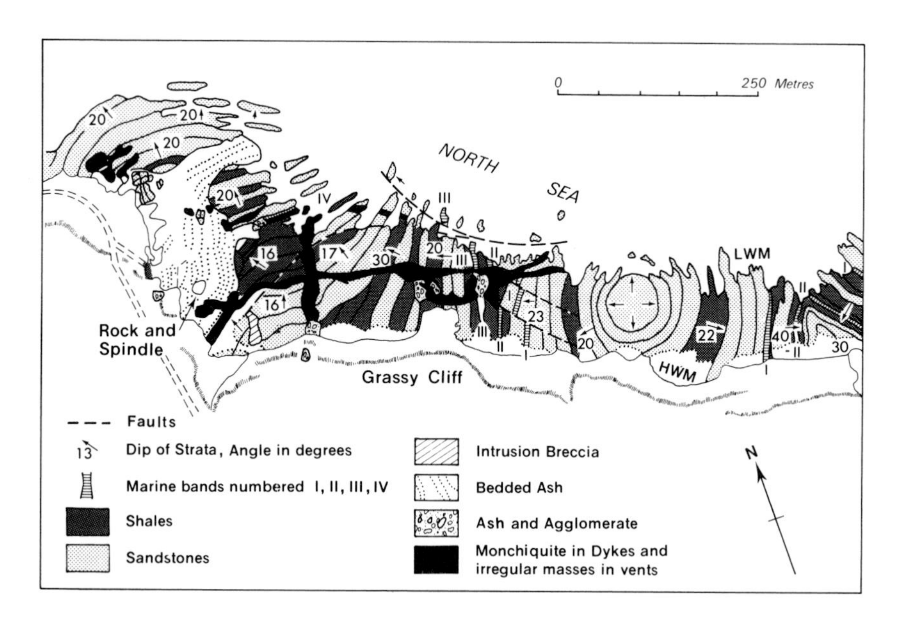

FIG. 13.2. The rock and spindle coast, Scotland (adapted from MacGregor, 1968).

Topography, lithology and structure also may account for the appearance of peninsulas and islands that occur along many shores (G. Wilson, 1952). Possibly as good an illustration of such relationships as one can find is the well-known rock and spindle coast just south of St Andrews, Scotland (Fig. 13.2). The contorted nature of this shoreline makes sense when mapped geologically.

Some of the material along coasts is chemically and biologically formed, much of it coming from local sources such as that moving from headlands to coves, and some is brought to the coast by rivers from distant locations. The sediment reaching the shore depends mainly on the size, climate and lithology of drainage basins.

An examination of a world drainage map (Fig. 13.3) shows that the bulk of the land drains into the Atlantic Ocean. The correlation that exists between drainage-basin area on the one hand, and discharge and sediment supply on the other, is frequently poor. Arid and semi-arid streams usually have low discharges but relatively high sediment loads. For example, the Colorado River ranks 52nd

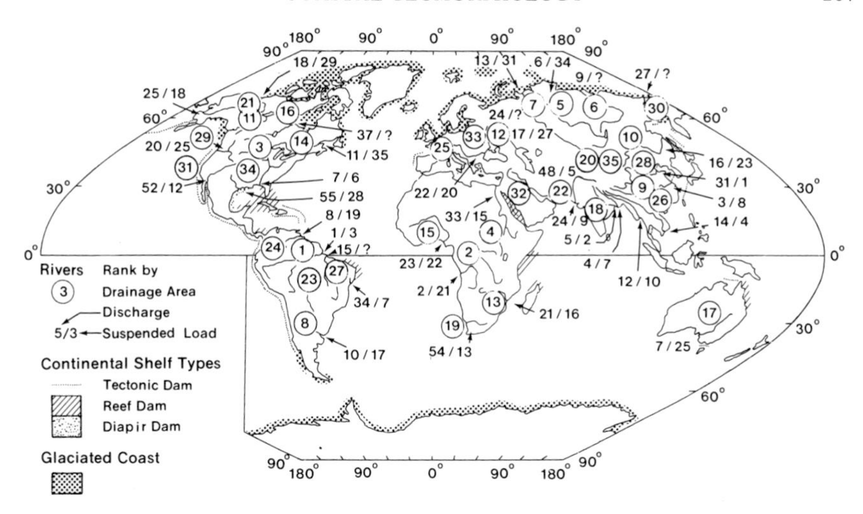

FIG. 13.3. Worldwide pattern of drainage systems and continental shelf types. Riverine data calculated from Inman and Nordstrom (1971). Shelf types after Emery (1969).

in terms of discharge but 12th in terms of sediment load. Of course, in the case of the Colorado River this relationship is essentially academic because of the presence of Boulder Dam which has virtually eliminated the possibility of sediment reaching the Colorado Delta. The transport of material to the coast may be in solution, suspension or as bedload. The relative proportions vary greatly.

As Komar notes (1976), geological research on beach material has been 'concentrated mainly on the physical properties of the sediments that compose the beach–size distributions of sediment particles, particle shapes (roundness and sphericity)–as well as sedimentary structures typical of beach sediments. . .' Most of this work has been on beaches in the temperate regions where quartz and feldspar grains predominate. However, in recent years research on biologically and chemically formed beach materials has received increased emphasis (R. F. McLean, 1974). Possibly the most exciting recent results in research on beach material have come from geochemists, especially from those who have been concentrating on the nature of the cementing agents found in recently consolidated materials common along many of the world's beaches (C. Moore, 1976).

Whereas sandy beaches, deltas and coral reefs (see below) have been relatively intensively studied, muddy coasts and to a lesser extent cliffed coasts, both of which can be found worldwide, are awaiting detailed research. Research along muddy coasts involves detailed consideration of the electrochemical state of the muds, flocculation and high suspended loads (J. Coleman, personal communication).

PROCESS

Most coastal research today emphasises process, an emphasis that does not appear destined to change in the near future. The actual appearance of a coast depends not only on material, as discussed above, but also on the degree of natural modification that has occurred to the original form. Physical, chemical and biotic agents all operate although their relative importance varies with several factors. Many of these studies are along the lines of what is now referred to as process-form research (M. Hernandez-Avila and H. Roberts, 1974).

The most important processes are those associated with water movement which in the nearshore environment is complex. Although unidirectional flow occurs, from the standpoint of coastal morphology it is the undulatory, repetitive ocean wave that causes most coastal change (Walker, 1975). Since World War II theoretical and experimental studies of the nature of the turbulent boundary layer have been undertaken by several institutes. Boundary roughness, change in flow from smooth to rough boundaries, and forces needed to initiate sediment movement have been considered. Generally the results have not been tested in the field where conditions are much more complex (J.R. Hails, 1974).

In 1972, J. R. Weggel classified oceanic flow according to type of generating force: astronomical, impulsive and meteorological. Astronomical forces produce the tides which range in level from near zero to over 15 m and determine the size of band upon which other oceanic processes operate. Type of tide, whether diurnal, semi-diurnal or mixed, determines the frequency of inundation and length of exposure of intertidal areas.

Although the importance of tides has long been recognised and tide tables for many parts of the world are available, surprisingly little research has actually been completed on the role of tides in coastal morphology. The location that has the highest tidal range in the world is claimed to be Cobequid Bay, an arm of the Bay of Fundy. It is of interest to note that the factors controlling such high tides depend on the dimensions of the Bay, dimensions achieved only within the past 4000 years. Recent research in the Bay of Fundy by V. Middleton (1972), W. Dalrymple (1973) and others concentrated on determining the paths of sediment transport in the tidally dominated setting and on interpreting the textural, mineral and form distributions. Two tentative conclusions are that tidal-current time-velocity asymmetry is a pervasive factor in tidal environments, and that mega-ripple amplitude and wavelengths are strongly influenced by current velocity and water depth.

The relevance of tides in such areas as the Bay of Fundy is conspicuous. Such is not the case in low tide areas, such as the Arctic coast of Alaska or the Gulf Coast of the United States where tidal effects are much more subtle. L. Wright and Coleman (1974) have examined the importance of tidal range and tidal currents on river-mouth flow and depositional patterns in the subaqueous portions of the Mississippi Delta. They found that salt-wedge behaviour and associated circulation are affected by tidal phase even though the astronomical tide averages only 33 cm. They also showed that patterns resulting from riverine flow into the Gulf of Mexico vary with tide and that the boundary between the effluent and ambient water masses is farther seaward during ebb than flood.

The second generating force, that of an impulsive nature, is caused by earthquakes, landslides and volcanic eruptions, and is responsible for the production of tsunamis which may result in spectacular morphological modifications along the coast.

In terms of coastal modification, meteorological forces are more important than either the astronomical or impulsive type. Wind-produced waves, upon approaching shore, redistribute their energy in ways that may result in erosion or deposition. Waves are also transporting agents, moving sediment inshore, offshore and alongshore. Not all wave-induced modification is easily observed or measured. A recent field study of ripple marks has shown a correlation between type of ripple and relative velocity of offshore bottom flow under the waves that generated them (T. Machida *et al.*, 1974). Relatively accurate wave-measurement techniques, only now being developed (V. Klemas *et al.*, 1976), hold promise for the future researcher of surf-zone problems.

As pointed out by Hails (1974), 'there has been considerable progress in our knowledge of nearshore sediment dynamics in recent years but many important questions remain unanswered about the mechanics of wave motion and how wave energy is transformed and dissipated in and near the surf zone'. Also relatively unanswered is the nature of the transfer of atmospheric energy to the ocean in the first place. Few studies, especially in the field, have been made of the surface shear-stress of various wind systems. One such is that of S. Hsu (1970) who has measured temperature and wind profiles under sea breezes along the Gulf Coast of the United States. The study of sea breezes and their effect on beaches has only begun (C. Sonu *et al.*, 1973) and is a fertile field for the future.

Most of the progress in longshore current and sediment transport research has been achieved along mid-latitude sandy shores. Thus it is likely that much of the future research in the role of long-shore currents in coastal morphology will be conducted along muddy coasts, rocky coasts and Arctic coasts, all of which provide unique conditions.

The climatic zonation of the earth combined with the shape of the ocean basins and the arrangement of the continents results in a pattern of wave environments on the coasts of the world. J. L. Davies (1973) published a map of wave environments, a modification of which is shown as Figure 13.4. Such a compilation is leading toward a more meaningful correlation between wave type

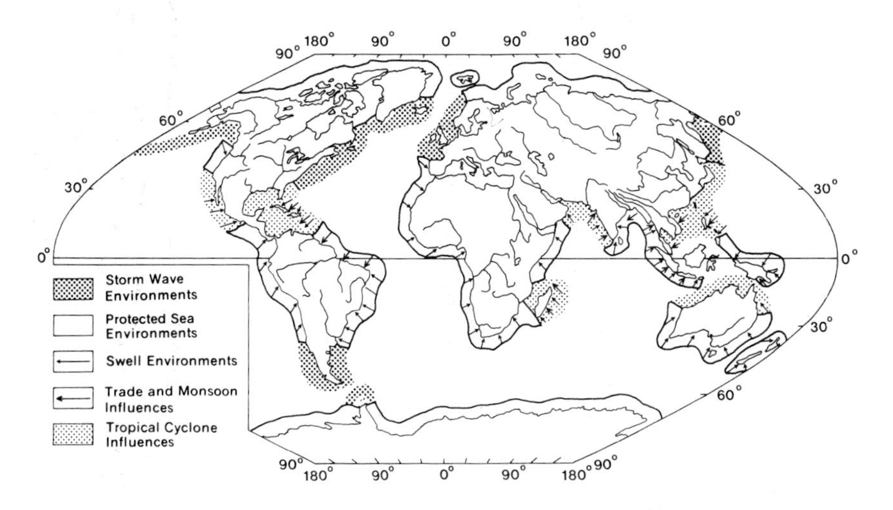

FIG. 13.4. Worldwide wave environments. Adapted from Davies (1973).

and coastal form. However, the questions raised by it serve to emphasise the fact that the study of wave climate is only beginning (B. Hayden and R. Dolan, 1976).

Although shear stresses of wind systems have received little attention, the effect of large hurricane-generated waves on coasts has been the subject of several reports, especially for the coasts of the southern and eastern United States but also for many island coasts in tropical cyclone belts. The major importance of severe storms to the coastal morphologist is that coastal systems experience amounts of erosion and deposition within a few hours that often exceed those that would occur in years of normal activity.

Because of increasing population, industry and other developments along many storm-prone coasts, the hurricane is receiving increased attention. The Bureau of Economic Geology of the University of Texas (L. Brown *et al.*, 1974) has developed a schematic model of hurricane effects on the Texas coastline. This model accommodates the changes occurring during a storm's approach, passage and decay.

The storm surge and associated waves that accompany the approach of a hurricane erode the normal beach and foredunes and create breach channels across the barrier islands. Sediment transported through these channels is deposited as fans on the bay margin and spread throughout the lagoon. In turn, bay-bottom sediments are moved to the mainland.

The landfall of hurricanes is accomplished by the highest wind velocities and by a change in the direction of current movement and wave approach. On the left-hand side of the storm, water and sediment are moved from bays back into the Gulf. As waves shift their angle of approach with storm passage, they bring about a change in the direction of long-shore currents.

Although hurricanes usually dissipate rapidly after moving inland, their morphological work is by no means complete. The water that has accumulated in the bays drains seaward through storm breach channels and deposits sediment in the channels and the Gulf. Subsequently, and usually within only a few days, long-shore currents construct bars and close hurricane channel mouths although the abandoned channels themselves may retain water for months. Gradually, also, the beach profile is restored and the hurricane deposits in lagoons and bays are reworked.

Although this model was developed for a specific coast, it appears that with appropriate modifications it might fit most storm-frequented low-lying coasts. Undoubtedly, it will be tested frequently in years to come.

From the standpoint of coastal morphology, wind is important mainly through the waves it generates. Nonetheless, it may also be a coastal modifier directly. Wind, a moving fluid like water, is able to erode, transport and deposit. Because it is the most important climatic element in any study of sand dunes, knowledge of sand transport by wind is essential. The Dutch in recent years have studied wind transport both in wind tunnels and in the field. Recent field measurements on a natural beach (J. Svasek and J. Terwindt, 1974) show that R. A. Bagnold's (1954) theory can be used to calculate the amount of sand transport, but only on clean dry beaches. At low wind velocities and over wet beaches the great deviations observed are prompting further research.

Most coasts of the world are subjected to waves of one intensity or another throughout the year. However, in those latitudes where temperatures are low enough to freeze sea water, waves are eliminated as an active agent for part of the year. Possibly the most important effect of the formation of shorefast ice is the protection it provides the shore. Even long-lasting snow drifts assist in reducing bank erosion. In summer, the presence of ice off-shore results in a reduction of fetch and therefore of wave size. Further, floating ice along some Arctic coasts may be blown on-shore at any time.

Much of the research on this subject has been conducted at Point Barrow in northern Alaska although within the past few years similar work has been initiated in the Canadian Arctic (S. B. McCann, 1973). At Point Barrow, J. Hume and M. Schalk (1967) found that ice is so effective in reducing waves that the resulting longshore transport tends to be small. It appears that it is of the order of only one-thirtieth of that occurring along coasts with similar topography in ice-free environments. Although wave action is normally not heavy, occasional storms do occur and sizeable waves accelerate erosion. Ice moving on to the beach may glide over it (even up to heights of 6 m) or it may plough into the beach. The end-result of this and other ice-related processes is a highly irregular surface.

Waves are also important during freeze-up. The sequence of beach freeze-up has recently been analysed (McCann and R. Taylor, 1975). The first stage begins when salt water in the form of slush and spray freezes on the beach. This process continues until the beach is protected by a covering of ice. However, while this covering is forming, blocks of pack-ice are driven toward the shore and become grounded. These floes protect the beach from what wave action remains and also serve to anchor further accumulations. The last stage of freeze-up ends with beach immobility. Because of the recently rejuvenated interest in Arctic areas, research along its shores is likely to increase in priority in the near-future.

Although it is generally considered that wave action is the most important process in coastal modification, other processes occur along all coasts and may dominate some. Two of the most significant are the result of chemical and biological agents.

The vast number of chemicals contained in sea water includes gases, compounds and elements in solution and particles in suspension, some of which are picked up by the wind and carried inland. The interstitial waters of the bottom sediments also contain similar materials. Diagenesis, the changes in sediments occurring during and after deposition, may result from organic action, solution and redisposition, or chemical replacement. In one sense, studies of these biogeochemical alterations are in the province of soil science. Soil scientists have not tended to direct much attention to soils under water (save in such agricultural situations as rice paddies). However, in 1975 an entire issue of *Soil Science* (Vol. 119, Number 1, January 1975) was devoted to this subject. It carried articles with such titles as 'Geochemical facies of sediments', 'Humic matter in natural waters and sediments' and 'Submerged soils in the northwestern Mediterranean sea and the process of humification'. The last article by F. Gadel and associates from Marseilles is excellent and provides a good orientation for future important research. The authors analyse 'the role of sedimentary lithology, physico-chemical characteristics of the mineral matrix, and the role of biological activity on the distribution of organic matter' in several near-shore environments

(Gadel *et al.*, 1975).

From the standpoint of coastal modifications, the chemical combinations which are relatively soluble and precipitable in the oceanic environment are very important. Included is $CaCO_3$, which being chemically weak, is easily etched, resulting in a rough surface. The relative importance of such a chemical solution especially where biotic action is also present is debated.

The reverse process, that is, precipitation, is significant in the cementation process. The products can be very different morphologically as in beach rock (Russell, 1967). Intertidal cements, as the cementing precipitates in the coastal zone have been called, are highly variable even at one general locality. Studies within the past couple of years are becoming so refined that cement chemistry can be used as an indicator of the climatic, hydrological, geochemical, sedimentary and biological conditions under which cementation occurred (Moore, in press).

As in the case of the physical and chemical processes, the biological can be both erosional and depositional or, if it is preferred, constructional. Organisms, plant and animal, are found in both the subaerial and subaqueous zones along all coasts and their effectiveness as geomorphological agents is complex. Biologists over the past 100 years have developed many schemes on the intertidal zonation of marine biota (J. W. Hedgpeth, 1976). These schemes illustrate the complex inter-relationships among plants, animals, tidal levels and bottom type. Although there is a degree of orderliness in the banding of life on seashores (Fig. 13.5), the actual nature of the bands is quite different between, say, the rocky shores in the state of Washington and the coral reefs in the tropics. For that matter,

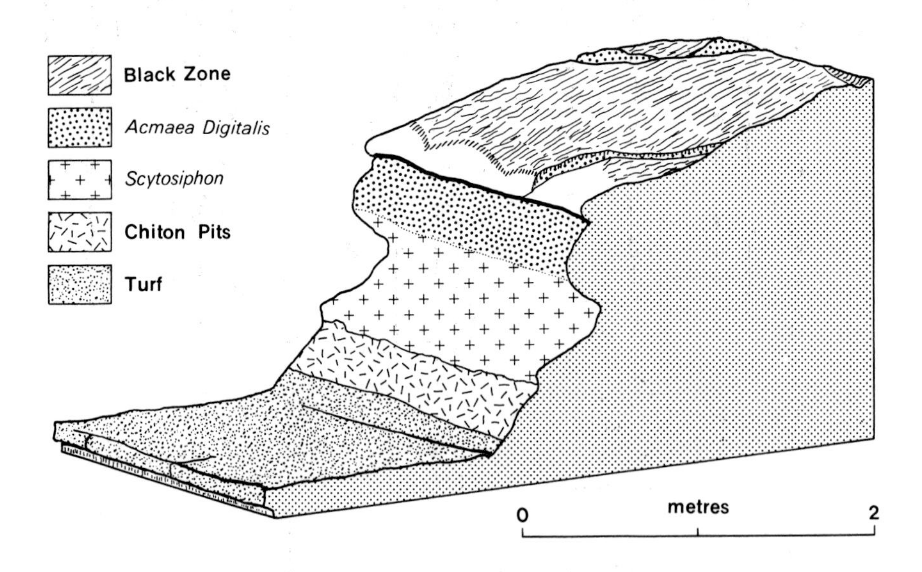

FIG. 13.5. Block diagram of a strip of soft sandstone, La Jolla, California (after Stephenson and Stephenson, 1972).

the zonation differs greatly between these same Washington rocky shores and adjacent sandy shores (R. McGreevy, no date).

Bio-erosion, a much neglected study, is one destined to attract increasing numbers of students (McLean, 1974). As Russell (1971) wrote, 'Snails, sea urchins, and other animals are capable of rasping even the hardest rock while clams, sponges, worms, algae, and other organisms weaken it by boring or secreting chemically active solvents'. Distinguishing between it and erosion resulting from chemical or physical change is a difficult task and one that has practical as well as academic interest. In most situations all three are somewhat involved although the relative proportions vary with climate, wave energy, and biotic abundance and type. Bio-erosion is not limited to the hard substrate; unconsolidated zones also undergo biotic change.

The availability of calcium carbonate and silicon dioxide in sea water make possible the construction of many coastal forms. Some rarely examined types include those constructed by bryozoa, serpulids and oysters. Oyster reefs, for example, are common along the Gulf Coast of the United States and, although commercially important, are practically unstudied.

The best researched type of organic coastal form is the coral reef, a form that received attention, especially from the biologist, as much as 100 years ago. Nonetheless, most coral reef studies to date have been confined generally to the horizontal. In recent years, with the development of scuba gear, vertical factors have come under study. J. C. Lang (1974) reported on the changes in growth in the vertical zonation of a Jamacian Reef and found that, at a depth of 70 m, the normal hermatypic corals were replaced by the sclerosponges.

In contrast to coral reefs, algae reefs have been little studied. Algae reefs or flats frequently consist of a series of steps in which ponded water remains at low tide. These flats provide a rather unique niche along the shore, the importance of which is biotically little understood.

Coastal plants of a higher order than algae, and especially those which are subaerial for the most part, are often effective constructional agents. Lagoons and low-energy coasts are favourable environments for this type of plant growth. Mangroves in low latitudes and salt marshes generally in higher latitudes are but two examples. There is little doubt that both types promote build-up and help to stabilise the shoreline by trapping sediment and contributing organic matter. However, the actual importance of mangroves as a land-builder is debated. B. G. Thom (1967) found in Tabasco, Mexico, 'that the geomorphic competence of mangrove is directly related to the type of habitat in which the trees grow'.

Although we have concentrated on the living form, biotic remains may also be significant, as in the case of guano on the coast of Peru and drift wood along many coasts including those of the Arctic. Recent mining of driftwood from the beaches of the coast of Oregon, U.S.A., is apparently leading to increased cliff erosion, a subject worthy of applied research (Stembridge, personal communication).

The role of the atmosphere and the sea in the growth and maintenance of coastal biota is both direct and indirect. Hurricanes are usually considered destructive agents, but recent studies show that the disordering energy resulting from their passage is beneficial to mangrove growth and it is suggested that the same may be true in coastal marshes (Day and Smith, in preparation). Although wind shear as an atmospheric phenomenon is conspicuous, more subtle is the role played by salt spray as a contributor of nutrients to vegetal growth (H. Art *et al.*, 1974).

TIME

Coasts, often referred to as the most changeable of environments, have been subjected to the primary coastal processes ever since our rotating earth acquired its threefold combination of a lithosphere and hydrosphere surrounded by an atmosphere. Destruction, transport and construction by waves, tides, currents and rivers operated long before man, or any type of life for that matter, began to inherit the earth. Such purely physical processes existed until biota began to

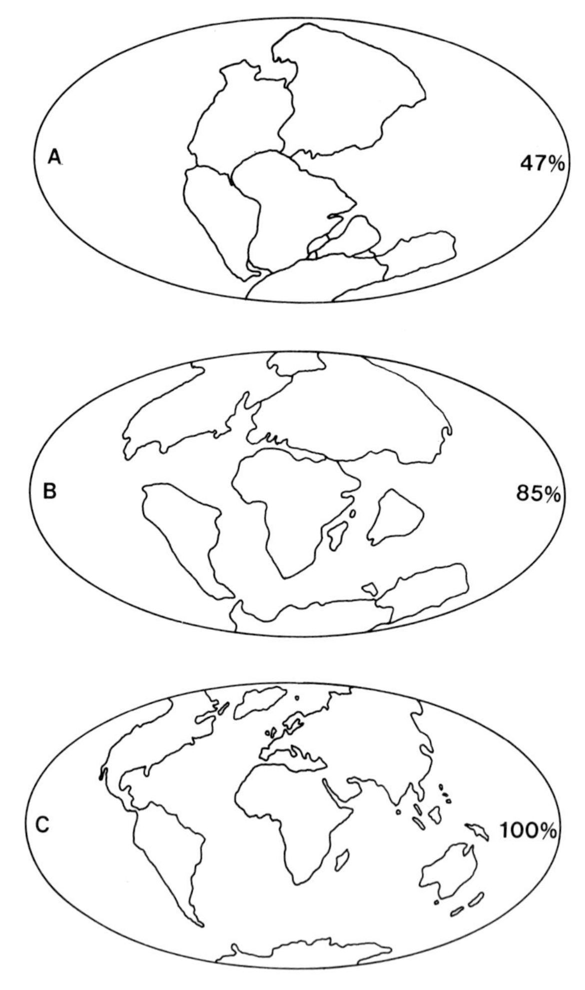

FIG. 13.6. Increasing length of shoreline with continental separation. A–200 million years ago, B–65 million years ago, C–present time. The percentages represent the lengths of shorelines of each period relative to that of the present.

evolve. Man as a geomorphological agent must be considered only the most recent, even if the most versatile, in a long evolutionary line.

The intensity and effectiveness of coastal processes have varied through time with continental configuration and climate. Accepting the concept of sea-floor spreading and plate tectonics, we see that the coastline of today is more than twice as long as it was at the time rifting was initiated (Fig. 13.6). Continued rifting, as along the Great African Rift and the Gulf of California is likely to increase the total length of the world's coast only slightly more.

Even more important, revived thinking about continental drift places emphasis on both continents and oceans, and therefore on the junction between the two, the junction which is the realm of the coastal morphologist. Plates that are being created at one edge and consumed at the other vary in size, shape and stability. Of the three basic types of boundary existing between plates, the most important to coastal morphology is the one where convergence is occurring. Coastal configuration along such boundaries depends mainly on the density and thickness of the colliding plates and on the rate of movement. For example, where one plate is thin and the other one thick, a mountainous coast with bordering trench will develop as along the western part of the Americas. Not all coastlines occur along the leading edges of plates. Some are embedded and, like the eastern coast of the Americas, act as trailing edges.

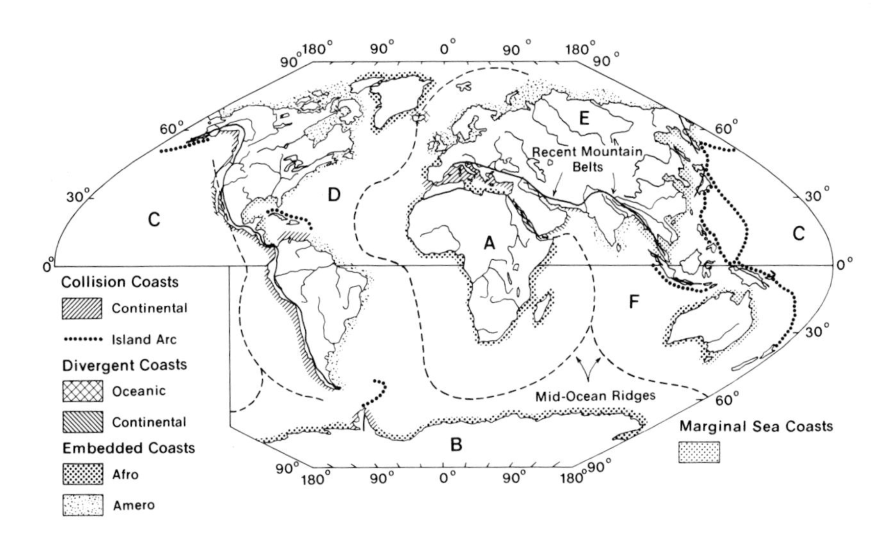

FIG. 13.7. Worldwide distribution of coastal types (modified from Inman and Nordstrom, 1971 and Davies, 1973) and the major plates and their boundaries.

Based on plate characteristics and the tectonics of plate movement, D. Inman and C. Nordstrom (1971) classified the coasts of the world into four categories (Fig. 13.7). As an initial approach to a world-wide categorisation of coasts it has exciting and promising possibilities. It is likely that this framework will be used as a basis for theoretical research at a global scale for some time to come.

Possibly during all but 3 per cent of earth history has the Earth been ice-free, or, if it is preferred, during 97 per cent of all earth history there has been a full ocean. However, during that 3 per cent of time, ice has been an integral physical constituent of the earth's surface and the sea has been subjected to extreme fluctuations in level. Each change in level was accompanied by a changing relationship between sea and land and therefore between coastal process and form. Inundation causes the length of the shoreline to increase. The length of the shoreline may be as much as three times as long at the end of sea-level rise as at its beginning when sea level approximated to the position of the edge of the continental shelf, between 150 and 200 m below the present level (Fig. 13.8).

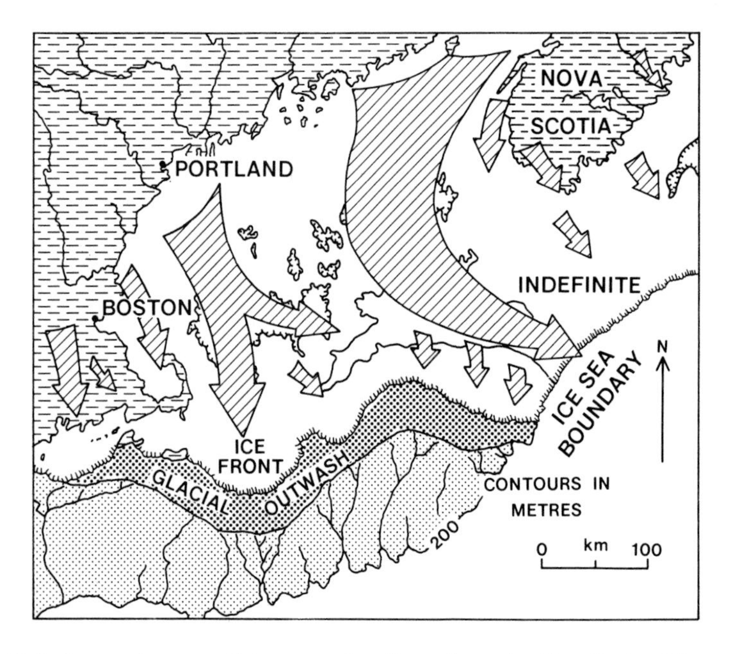

FIG. 13.8. Present-day and Pleistocene shorelines along the New England-Nova Scotia coasts (after Healey, 1972).

The virtual elimination of the continental shelf at low stands combined with changing atmospheric conditions not only altered oceanic circulation patterns but also affected biotic composition and distribution. These changes affected the type and intensity of most physical, chemical and biotic processes operating along coasts.

With a falling sea-level, rivers incised their courses as they extended their routes across newly exposed shelves. Increased erosion provided additional sediment for deposition in shallow coastal waters. However, as the level approached a minimum, greater and greater amounts of riverine sediment were lost to the continental slope.

Once sea-level began to rise, coastal valleys that had been cut to a low base-level began to fill and vast quantities of sand and gravel formerly deposited on the wide coastal plain were reworked. A rapid rise (of the order of 1m/100 years) allowed littoral belts to be subjected to shore-forming processes for short periods only. Relatively little change was possible. Nonetheless, as sea level rose over old coastal plains, new areas of sand were attacked by waves; the size of beaches increased and surplus sand was the base for extensive dune systems.

From the standpoint of present-day coasts, possibly the most important event since the Pleistocene glaciation began was the establishment of a virtual still-stand of sea-level some 4–5000 years ago. The coastline must have been at about its maximum length—estuaries were numerous and many possessed lengthy shorelines.

Since that time these estuaries have been filling rapidly. Mobile Bay, an estuary of considerable proportions 4000 years ago, has been more than three-quarters filled. The marshes and lagoons of the world have also rapidly changed. As the rate of rise in sea-level slackened and still-stand became established, sand supplies along the shore became more and more restricted to those brought by rivers. Sand along the shore dwindled and dune formation lessened. In addition, old dunes became stabilised as inundation ceased. There may have been a slackening of wave impact as well because of the increasing width of the near-shore area.

The discussion thus far has been mainly about eustasy—change in levels which has been more or less continuous during the Quaternary. However, such changes in level are often accompanied by isostatic and tectonic variations. Within the past 5000 years, reduced rates of eustatic change have provided rebound and tectonic uplift with relatively greater significance. For example, rebound in Canada and Scandinavia is steadily exposing formerly glaciated and submerged terrain to coastal action.

One of the most debated of all topics of significance to coastal morphologists is the nature of change of sea-level, especially in the last 6–7000 years. One group advocates a stand 2–3 m higher than present sea-level, the other that since the Holocene transgression began, sea-level has not risen above present-day level (see Komar (1976) for a recent discussion on this debate).

Although much of the research on sea-level fluctuations and coastal processes has been academic, applied research utilising such data has been increasing in amount. For example, R. Moberly *et al.*, (1975) recently assessed the sand resources of the island of Oahu. Their research included the mapping of sand deposits that cover submarine terraces and beaches and old drainage channels cut through the reef while the sea was at a lower level (Fig. 13.9).

Between 1972 and 1976, data on changes of the world's coasts during the past century were compiled by the Working Group on the Dynamics of Shoreline Erosion of the International Geographical Union. As E. C. F. Bird (1976) writes, 'It was hoped that the project would stimulate research in countries where coastal information was previously scanty, and encourage closer investigation of shoreline dynamics on coasts that are comparatively well known, and on both counts a measure of success can be claimed'. The data collected show that widespread erosion has prevailed during the past century. Along many coasts the resulting retreat follows a long period of progradation. Many hypotheses have been advocated including those related to a steady rise in sea-level, man-made structures and increased storminess. Both the collection of data

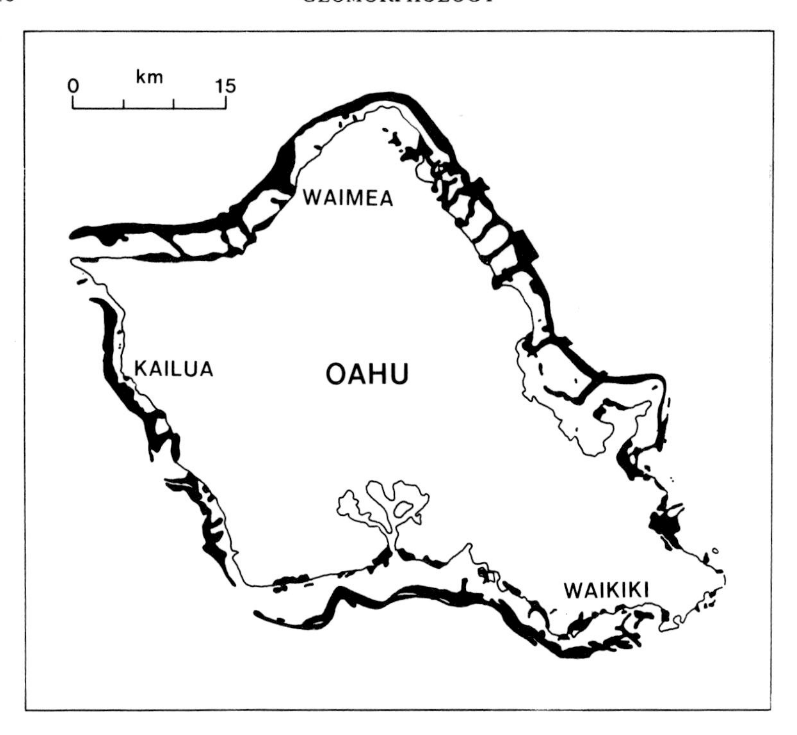

FIG. 13.9. Sand resources around the island of Oahu, Hawaii (adapted from Moberly *et al.*, 1975).

about, and the development of explanations for, shoreline changes are in their infancy and will certainly be amplified in future research.

In the section above, tectonics and plate movement are discussed in terms of millions of years. However, tectonic activity may be very dramatic and virtually instantaneous. Such was the case in the 1964 Good Friday earthquake in southern Alaska when a portion of one island was raised by 11 m and an expanse of former ocean bottom 400m wide was exposed. This same earthquake also caused subsidence of 2m in Turnagain Arm at the head of Cook Inlet in Alaska. Characterised by shifting sand bars and channels, Turnagain Arm has a high tidal range and around its edge are swamps, bogs, grasslands and forests. Although the inhabitants of Portage, a small town at the head of the embayment, realised that the earthquake had been bad, they remained in their houses. Several days later, a 10 m spring tide broke through the ice which still covered the tidal flats and inundated Portage causing its abandonment, killing vegetation as much as 3 km inland and depositing sediment to depths of 2 m.

The U.S. Geological Survey began a study of the recovery of the area in 1973 in which they mapped the post-earthquake silt deposits. The rate of sedimentation has been so rapid that by 1974 some revegetation occurred and it appears that the Portage area is returning to its pre-earthquake landscape.

Cores taken in the area show two other soil horizons below the 1964 level.

These levels indicate that the sequence of events occurring since 1964 may have been repeated throughout the Holocene—that is, subsidence caused by earthquakes, inundation with its devastation of plants, the deposition and eventual rejuvenation of the area and revegetation. S. Bartsch-Winkler and A. Ovenshine (1976) of the U.S. Geological Survey suggest that the buried zones can act as a 'sedimentary seismograph' recording tectonic episodes and resulting sedimentation. The big question is this: can such evidence be used to predict an earthquake interval for the Anchorage area?

Since the increasing attention being given to the 'catastrophic' event (hurricanes, volcanic eruptions, earthquakes, etc.) is likely to continue, it appears that coastal research modelled after that of the U.S. Geological Survey should prove a fruitful endeavour for coastal morphologists in the future.

MAN AS A COASTAL AGENT

The attention being given to man as a coastal agent is increasing more rapidly than that being devoted to any other coastal modifier. Man's actions along the coast are extremely important and as in the case of other biota may be either destructive or constructive. In addition, these changes may be either intentional or unintentional.

As long as 4000 years ago the Minoans were constructing breakwaters and the Chinese salt pans. But even before that, over-grazing in the Middle East may well have caused increased rates of silting in estuaries and deltas. Many of the early modifications (as is true of those of today) had unanticipated results. Strabo (*c.* 7 B.C.) wrote that the harbour of Ephesus was 'made narrow by engineers, but they, along with the king who ordered it, were deceived as to the result, . . .silt, made the whole of the harbour, . . ., more shallow'. Some even question whether our success today is much better. Examples are legion and appear almost daily in the newspapers.

Man lengthens and shortens the shore line; he constructs beaches where none existed before, he fills in lagoons for housing projects and he constructs islands. Such modifications, whether appropriate or not, are usually conspicuous. Whereas many structures such as groynes tend to lengthen the coastline, others such as dykes serve just the opposite function. The Dutch, in constructing polders, have decreased the effective oceanic shoreline of The Netherlands to less than one-quarter of its former extent.

When considering coastal aggradation at the hands of man, we often think of the bulldozer, the bucket dredge and the suction dredge. The Dutch have developed suction dredges of such capacity that they are being effectively used in closing tidal channels, extending harbour moles, and nourishing eroding beaches (J. Terwindt and A. Walther, 1976). Other methods, such as the use of polypropylene seaweed to initiate deposition, are being developed as well.

One of the most fascinating areas for research that has yet to be investigated in detail deals with unintentional coastal modifications, many of which result from the activities of non-coastal peoples. An example is overgrazing, to which could be added hydraulic mining. Gold mining in California was responsible for contributing over 10^9 m^3 of sediment to San Francisco Bay, evidence of which can be seen at the north end of the Bay on ERTS imagery (Hedgpeth, 1975).

An example of a different kind results from the construction of dams. Between Los Angeles and San Diego there are fourteen drainage basins leading to the Pacific Ocean. All have been dammed. Today the area of unobstructed drainage to the Pacific is less than one-third of its former size. Peak discharges, and therefore flood transport of silt, have been reduced or eliminated. Sediment is now trapped behind the dams and the contribution of water itself to the sea has been reduced because of withdrawal from the streams. The net result is that all beaches along this coast are being starved (L. Berkman, 1975).

An additional example of man as an unintentional geomorphological agent along the coast has to deal with his role in subsidence, the results of which in Italy have even become a tourist attraction (J.-C. Fontes, 1975). Subsidence, as a natural process, has received considerable attention, especially as it occurs in deltaic areas (P. Morgan, 1972), but only recently has subsidence caused by man come into focus. Fluid withdrawal from below the subsurface, at least in large quantities, is a rather recent human accomplishment, and one that is certainly not abating. The primary fluids withdrawn are water, oil and gas, withdrawal of which may lead to the compaction of fine-grained sediments and subsequent subsidence. When such subsidence occurs along the coast, land may be submerged by the sea and other low-lying areas become more subject to storm-induced flooding. In Baytown, Texas, subsidence resulting from oil withdrawal was recognised in the 1920s but attracted relatively little attention. However, today the problem is more extensive because vast quantities of water are also being pumped from aquifers beneath the area. As low-lying areas are subject to hurricane-induced flooding, concern has resulted in applied research. One of the objectives of the Texas study has been to gather data in support of the contention that pumping should be discontinued. Some researchers have suggested that recharging the aquifer might reverse the present trend and in some aquifers this procedure might actually prove workable. However, water in the subsurface clays will percolate into the sand aquifers when drained. As the clay is inelastic, most compaction is permanent. It has been calculated that no more than 10 per cent rebound can be expected from total recovery of artesian pressure (C. W. Kreitler, 1976).

SUMMARY AND CONCLUSIONS

During most of history, applied research along coasts has been concerned with ocean transport (especially harbours) and with those wars which involved coastal activities. Such research was mostly conducted by governmental (including military) and industrial organisations. In recent years, however, an almost endless list of conditions and circumstances has given impetus to unprecedented amounts of coastal research. Responsible stimuli include population pressures, conflict between user groups, recent environmental crusades, oil spills, governmental regulations, recreation, the creation of the Sea Grant Program (in the U.S.A.) and an increasing dependence on coastal resources (Walker, 1975).

As a result, workshops, conferences and symposia designed to identify, analyse and solve coastal problems are being convened in increasing numbers. Most of these sessions have coastal morphology as an exclusive or major component. Some are very specialised and nearly all have resulted in pertinent published volumes. A few of the most recent are: *Waves on beaches and resulting sediment transport* (R. E. Meyer, 1972), *Coastal geomorphology* (D. R. Coates, (ed.) 1973), *Contributions to coastal geomorphology* (R. W. Fairbridge, 1975), *Nearshore sediment dynamics and sedimentation* (Hails and A. P. Carr, 1975), the aforementioned issue of *Soil Science* (Tedrow, 1975), *Coastal research* (Walker, 1976) and *Research techniques in coastal environments* (Walker, in press).

The contents of such volumes emphasise some trends in coastal morphology research that are destined to continue. A variety of sensors is being continuously developed and improved; the capabilities of computers for data analysis and model production are improving; an increasing emphasis is being put on quantitative description, dynamic analysis and adoption of the process-response approach; and laboratory and field facilities that will continue the present-day emphasis on interdisciplinary cooperation are expanding.

Such a variety of approaches and techniques, combined with an increasing interest in coasts by the layman and politician as well as the scientist, suggest a bright future for coastal research.

REFERENCES

Aiken, S. R. (1976), 'Towards landscape sensibility', *Landscape* **20**, (3) 20–29.

Art, H., Bormann, F., Voigt, G. and Woodwell, G. (1974), 'Barrier island forest ecosystem: role of meteorologic nutrient inputs', *Science* **184**, 60–2.

Bagnold, R. A. (1954), *The physics of blown sands and desert dunes* (Methuen, London), 265 pp.

Bartsch-Winkler, S. and Ovenshine, A. (1976), 'Coastal modifications at Portage, Alaska, resulting from the Alaska earthquake of March 27, 1964' in Walker, H. (ed.) *Research techniques in coastal environments* (*Geoscience and Man*, **XVIII**, Louisiana State Univ., Baton Rouge).

Berkman, L. (1975), 'Erosion puts southland beaches on endangered list', *Los Angeles Times*, 10 August 1975.

Bird, E. C. F. (1976), 'Shoreline changes during the past century: a preliminary review', *23rd int. geogr. Congr. Moscow*, 1976 (Melbourne, Australia), 54 pp.

Brown, L., Morton, R., McGowen, J., Kreitler, C., and Fisher, W. (1974), 'Natural hazards of the Texas coastal zone', *Bureau of Economic Geology, Univ. of Texas at Austin*, 13 pp.

Bullard, E. (1969), 'The origin of the oceans', *Scient. Am.* **221**, 66–75.

Coates, D. R. (ed.) (1973), *Coastal geomorphology* (State Univ. of New York, Binghamton), 404 pp.

Coleman, J. and Wright, L. (1971), 'Analysis of major river systems and their deltas: procedures and rationale, with two examples', *Tech. Rep. Coastal Stud. Inst.* (Louisiana State Univ.) **95**, 125 pp.

Dalrymple, W. (1973), 'Preliminary investigations of an intertidal sand body, Cobequid Bay, Bay of Fundy', *Maritime Sedim.* **9 (1)**, 21–8.

Davies, J. L. (1973), *Geographical variation in coastal development* (Hafner, New York), 204 pp.

Emery, K. O. (1969), 'The continental shelves', *Scient. Am.* **221**, 106–22.

Fairbridge, R. W. (ed.) (1975), 'Contributions to coastal geomorphology', *Z. Geomorph., Suppl.* 22, 170 pp.

Fontes, J.-C. (1975), 'Why venice is sinking', *C.N.R.S. Res.* 2, 13–19.

Gadel, F., Cahet, G. and Bianchi, A. (1975), 'Submerged soils in the northwestern Mediterranean Sea and the process of humification', *Soil Sci.* **119**, 106–12.

Hails, J. R. (1974), 'A review of some current trends in nearshore research', *Earth Sci. Rev.* **10**, 171–202.

—— and Carr, A. P. (1975) (eds.), *Nearshore sediment dynamics and sedimentation* (Wiley, London), 316 pp.

Hayden, B. and Dolan, R. (1976), 'Seasonal changes in the planetary wind system and the most severe coastal storms', in Walker, H. (ed.) *Research techniques in coastal environments* (*Geoscience and Man,* **XVIII**, Louisiana State Univ., Baton Rouge),

Healey, W. R. (ed.) (1972), *Proceedings of the second New England Coastal Zone Management Conference, Sept. 1971* (New England Council, Inc., Boston, Mass.), 98 pp.

Hedgpeth, P. W. (1975), 'San Francisco Bay', in Walker, H. J. (ed.) *Coastal resources* (*Geoscience and Man,* **XII**, Louisiana State Univ. Baton Rouge), 23–30.

Hedgpeth, J. W. (1976), 'The living edge' in Walker, H. J. (ed.) *Coastal research* (*Geoscience and Man,* Louisiana State Univ., Baton Rouge), **XIV**, 17–52.

Hernandez-Avila, M. and Roberts, H. (1974), 'Form-process relationships on island coasts', *Tech. Rep. Coastal Stud. Inst.* (Louisiana State Univ.) **166**, 76 pp.

Hsu, S. (1970), 'Coastal air-circulation system: observations and empirical model', *Mon. Weather Rev.* **98(7)**, 487–509.

Hume, J. and Schalk, M. (1967), 'Shoreline processes near Barrow, Alaska: a comparison of the normal and the catastrophic', *Arctic* **20(2)**, 86–103.

Inman, D. and Nordstrom, C. (1971), 'On the tectonic and morphologic classification of coasts', *J. Geol.* **79**, 1–21.

Klemas, V., Davis, G. and Wang, H. (1976), 'Combining satellites, aircraft and drogues to study the dynamic characteristics of coastal waters and pollutants', in Walker, H. (ed.) *Research techniques in coastal environments* (*Geoscience and Man,* **XVIII**, Louisiana State Univ., Baton Rouge).

Komar, P. D. (1976), *Beach processes and sedimentation* (Prentice-Hall, New Jersey), 429 pp.

Kreitler, C. W. (1976), 'Subsidence in the Texas coastal zone', in Walker, H. (ed.) *Research techniques in coastal environments* (*Geoscience and Man,* **XVIII**, Louisiana State Univ., Baton Rouge).

Lang, J. C. (1974), 'Biological zonation at the base of a reef', *Am. Scient.* **62**, 272–81.

MacGregor, A. R. (1968), *Fife and Angus geology* (Blackwood, London), 266 pp.

Machida, T., Inokuchi, M., Matsumoto, E., Ishii, T. and Ikeda, H. (1974), 'Sand ripple patterns and their arrangement on the sea bottom of the Tatado Beach, Izu Peninsula, Central Japan', *Sci. Rep. Tokyo Kyoiku Daigaku,* Sect. C. (Geo., Geol. and Mineralogy), **12**, 113, 1–16.

McCann, S. B. (1973), 'Beach processes in an arctic environment', in Coates, D. R. (ed.) *Coastal geomorphology* (State Univ. of New York, Binghampton), 141–55.

—— and Taylor, R. (1975), 'Beach freeze-up sequence at Radstock Bay, Devon Island, Arctic Canada', *Arct. alp. Res.* **7 (4)**, 379–86.

McGreevy, R. (no date), 'Seattle shoreline environment', (City of Seattle, Department of Community Development and Washington Sea Grant Program), 41 pp.

McLean, R. F. (1974), 'Geologic significance of bio-erosion on beachrock', *Proc. 2nd int. Coral Reef Symp.* (2. Great Barrier Reef Comm., Brisbane), 401–8.

Meyer, R. E. (ed.) (1972), *Waves on beaches and resulting sediment transport* (Academic Press, New York), 462 pp.

Middleton, V. (1972), 'Brief field guide to intertidal sediments, Minas Basin, Nova Scotia', *Maritime Sedim.* **8 (3)**, 114–22.

Moberly, R., Campbell, J. and Coulbourn, W. (1975), 'Offshore and other sand resources for Oahu, Hawaii', Univ. of Hawaii-Seagrant-TR-75-03, 36 pp.

Moore, C. (1976), 'Beach rock origin: some geochemical, mineralogical and petrographic considerations', in Walker, H. (ed.) *Research techniques in coastal environments* (*Geoscience and Man,* **XVIII**, Louisiana State Univ., Baton Rouge).

Morgan, P. (1972), 'Impact of subsidence and erosion on Louisiana coastal marshes and estuaries', *Proc. Coastal Marsh Estuary Management Symp.* (Louisiana St Univ.), 217–33.

Pruitt, E., Alexander, L., Mitchell, J., Psuty, N. and Zinn, J. (1975), *Coastal zone* (Preprint; Marine Geography Committee, Ass. Am. Geogr.)

Russell, R. J. (1967), *River plains and sea coasts* (Univ. of California Press, Berkeley), 173 pp.

—— (1974), 'Coastal features', *Encyclopaedia Britannica* (15th ed.) **4**, 795–802.

Sauer, C. O. (1962), 'Seashore—primitive home of man?' *Proc. Am. Phil. Soc.* **106** (1) 41–7.

Sonu, C., Murray, S., Hsu, S., Suhayda, J. and Waddell, E. (1973), 'Sea breeze and coastal processes', *Trans. Am. geophys. Un.* **54, (9)**, 820–33.

Stephenson, T. and Stephenson, A. (1972), *Life between tidemarks on rocky shores* (Freeman, San Francisco), 425 pp.

Strabo (*c.* 7 B.C.) *Geography* **14.1.24.**

Svasek, J. and Terwindt, J. (1974), 'Measurements of sand transport by wind on a natural beach', *Sedimentology* **21**, 311–22.

Terwindt, J. and Walther, A. (1976), 'Current trends of coastal research in the Netherlands', in Walker, H. (ed.) *Coastal research* (*Geoscience and Man,* Louisiana State Univ., Baton Rouge), **XIV**, 73–87.

Thom, B. G. (1967), 'Mangrove ecology and deltaic geomorphology: Tabasco, Mexico', *J. Ecol.* **55**, 301–43.

—— (1974), 'Coastal erosion in Eastern Australia', *Search* **5 (5)**, 198–209.

Walker, H. J. (1973), 'Morphology of the North Slope' in Britton, M. (ed.) *Alaskan Arctic Tundra* (*Arctic Inst. N. Am. Tech. Pap.*) **25**, 49–92.

—— (1975), 'Coastal morphology', *Soil Sci.* **119 (1)**, 3–19.

—— (ed.) (1975), *Coastal resources* (*Geoscience and Man*, Louisiana State Univ., Baton Rouge), **XII**, 115 pp.

—— (ed.) (1976), *Coastal research* (*Geoscience and Man,* Louisiana State Univ., Baton Rouge), **XIV**, 153 pp.

—— (ed.) (in press), *Research techniques in coastal environments* (*Geoscience and Man,* **XVIII**, Louisiana State Univ., Baton Rouge).

Weggel, J. R. (1972), 'Water motion and process of sediment entrainment', in D. J. P. Swift, D. B. Duane and O. H. Pilkey (eds.) *Shelf sediment transport* (Dowden, Hutchinson & Ross, Stroudsburg, Pa.), 1–20, 656 pp.

Wilson, G. (1952), 'The influence of rock structures on coastline and cliff development around Tintagel, North Cornwall', *Proc. Geol. Ass.* **63**, 20–48.

Wright, L. and Coleman, J. (1974), 'Mississippi River mouth processes: effluent dynamics and morphologic development', *J. Geol.* **82**, 751–78.

14

COASTAL GEOMORPHOLOGY IN THE UNITED KINGDOM

CUCHLAINE A. M. KING
(*University of Nottingham*)

Coastal geomorphological studies, like those in other branches of geomorphology, are becoming more sophisticated both in the equipment used and in the analysis of the quantitative data obtained with it. The part played by satellites, radar, various forms of photography, electronics and computers has increased, but the complexities of the coastal system still defy exact description and full understanding.

This article seeks to present some of the fields in which progress is being made in coastal studies in the United Kingdom. No reference is made to work carried out in other areas by workers from the United Kingdom, nor is any mention made of the many sophisticated and valuable studies being carried out elsewhere in the world, especially in the United States and the Soviet Union.

Five different aspects of coastal work will be referred to briefly. In the first section, basic process studies such as the monitoring of waves and wind will be mentioned. In the open continental shelf area and the North Sea waves have acquired a special significance in view of the oil extraction carried on around Britain: rigs have been designed to withstand the effect of waves 30 m high in some of the stormiest seas of the world. The fact that Britain is in the storm-wave environment is of great significance in coastal studies. Waves must also be studied in the near-shore surf zone and swash zone, where they exert an important effect on coastal processes. Particularly significant is the understanding and assessment of long-shore movement of material under wave action. For this purpose tracers have been extensively used. The tide must not be ignored, and its vital importance in some areas is being increasingly recognised and studied; both tidal streams and the features to which they give rise are being investigated. No less important is the part the tide plays in foreshore development in many areas, including the formation of salt marshes (J. R. Hails and A. P. Carr, 1975).

Another currently favoured aspect of geomorphology is its practical application and this is of considerable importance in coastal studies (J. K. Mitchell, 1968; D. H. Willis and W. A. Price, 1975). The second section is therefore devoted to some examples of the application of coastal studies to specific problems. The extraction of sand and gravel from off-shore and from the foreshore has given rise to much controversy, especially in areas such as Chesil Beach in Dorset. Coastal defence problems, such as those of erosion in Swansea Bay and Start Bay, and the protection of the Dungeness power station and the low coasts of eastern England from storm surges such as that of 1953, can also be assisted by geomorphological studies. Specific projects, such as water storage in the Wash, a tidal barrier across the Lower Thames, reclamation and water supply in various estuaries, as well as possible tidal power stations, all provide opportunities for coastal research, much of which is assisted by or carried out by Government agencies.

Studies of coastal processes often provide new insight into the development of particular areas. Some studies are deliberately designed to elucidate the geomorphological development of particular stretches of coast, with the production of general models being a secondary consideration. Studies of this type have been carried out for different purposes and with various techniques in a wide range of areas, only a few of which can be mentioned in the third section. These studies include investigation of such features as Chesil Beach, Orfordness and Spurn Point, as well as specific stretches of coast, for example East Anglia, Start Bay and Swansea Bay. Many illustrate the value of historical material, as well as other approaches, some of which link the human aspect with the physical.

The fourth section is devoted to topical studies, for example the development of salt marshes and sand dunes, which introduces the botanical and ecological aspects of coastal study, as well as other coastal processes. Coastal cliffs and coastal platforms have also received some attention lately, although they still remain little studied compared with accretional forms.

A study of coastal platforms leads into the fifth topic, that of sea-level change as deduced from a study of the coastal geomorphology. In this very complex field Britain occupies an interesting situation on the edge of a 'trailing plate' continental land-mass. This aspect of coastal geomorphology is more relevant to the longer time-scale studies that combine both the eustatic and tectonic aspects of sea-level change. Much effort has also been expended in studying Cenozoic, particularly Quaternary, changes of sea-level. In long-term studies many problems are involved, as both eustatic and isostatic changes occur over most of the country, and the evidence is usually difficult to read and often ambiguous. Nevertheless, progress has been made, and Holocene sea-level changes, including the Flandrian transgression, have been studied in great detail in several areas, notably Scotland, south-west Britain and Ireland (Fig. 14.1).

PROCESSES

Waves

It is generally agreed that waves are of fundamental importance in effecting changes on the beach. They also represent one of the most complex of natural processes, being continually variable, and consisting nearly always of a three-dimensional spectrum rather than a single wave-train. Waves are generated by the wind, which itself exerts an important influence in the coastal zone, acting with or against the waves to reverse, slow down or reinforce their movement of material both normal to and along the shore.

Off-shore zone. One problem that requires elucidation is the depth to which waves can affect the bottom. Some interesting early work has been done on this topic in North America by direct observation, while in Britain, where conditions are less favourable for direct underwater observation, some useful studies have also been made, mainly in very shallow water close to the shore.

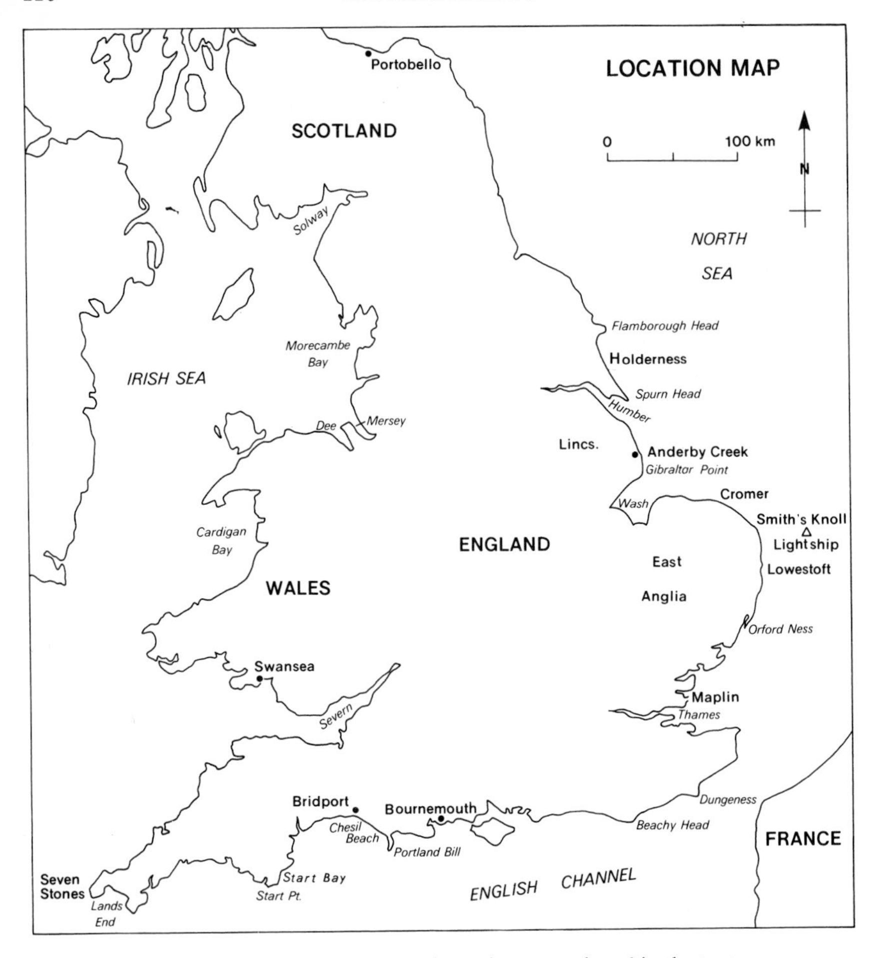

FIG. 14.1. Location map to show places mentioned in the text.

In the deeper waters of the continental shelf and the North Sea basin, oil companies have found it necessary to obtain information on waves both on the surface and on the sea floor. Direct observations are difficult to make in these stormy seas, and a theoretical approach to the problem of bottom movement has yielded interesting results (P. Holmes, 1975b; P. H. Kemp, 1975). Surface wave observations in the open Atlantic at the weather ships, Seven Stones lighthouse off Lands End, in Morecambe Bay in the Irish Sea and Smith's Knoll Lightship in the North Sea have been used. These surface wave data have made it possible, through theoretical studies of wave generation, to assess the maximum water-particle speeds likely to occur on the sea bed in various depths. It has been shown that the longer waves characteristic of the outer, western part of the continental shelf off Britain produce water movements at depths of 180 m that would be expected to exceed 43 cm/s for one day each year, and to reach values of twice this amount on that day once every 3 hours (L. Draper, 1967). Speeds of this magnitude will probably occur only in depths of about 30 m in the enclosed Irish Sea and North Sea, where wave periods are much shorter. These water-particle speeds do not necessarily move substantial amounts of sediment, and it is generally agreed that waves cannot achieve major changes of sediment pattern and movement in water deeper than about 15 to 20 m, and that erosion under these depths is unlikely to be effective through wave action alone, a matter that is relevant to the formation of wave-cut platforms. Other processes, such as internal waves and tides, must also be considered in relation to the movement of material in the off-shore zone.

Wave-measuring methods and equipment have been actively developed in Britain, and techniques for use in ships, wave-recording buoys and other methods have been successfully introduced. The wave-recording buoys provide information on the three-dimensional spectrum, which can be analysed by Fourier methods to produce wave spectra indicating the energy in different frequency bands coming from different directions. Laboratory experiments have also been carried out to complement the field observations and theoretical analyses.

Near-shore and surf zone. As waves approach the shore they become refracted, as well as undergoing other modifications. The effects of these changes upon the shore are important. Considerable effort has been devoted to the problem of computerising the process of refraction in specific areas to produce refraction patterns. The recent work in Start Bay exemplifies this type of study. Another interesting aspect of wave activity which was investigated in Start Bay was the development of edge-waves. Until recently they had never been recorded in the field. The observation of horizontal particle velocities at positions up to 25 m from Slapton Ley shoreline in Start Bay, which is a long, straight, steep beach, revealed the existence of edge-waves. They are modes of wave motion trapped against the shoreline by refraction, and they bear a direct relationship to the period of the incoming waves. At times the edge-waves are associated with cusp development and other rhythmic shoreline features, whose wave lengths are directly related to those of the long-shore edge-waves and those of the incoming wind waves. The amplitude of the edge-waves diminishes exponentially offshore with mode zero, and varies sinusoidally along-shore, thus producing the cuspate features under suitable conditions (D. A. Huntley and A. J. Bowen, 1975). These field tests in Start Bay have confirmed the results of model and theoretical analyses of edge-waves (P. Holmes, 1975a).

The importance of waves is indicated by the considerable amount of work devoted to a wide range of different types of wave analysis on the part of government agencies. This work includes studies of refraction and other effects of shoaling on waves, and studies of shear stress due to waves (Draper and P. J. Dobson, 1965; M. S. Longuet-Higgins, 1972). Model studies have also been carried out on a variety of problems relating to waves and their effects on structures.

Swash zone. The swash zone is important in coastal geomorphology because it is here that a large amount of sediment is transported. The transport of sediment influences and is influenced by the shape of the beach both in plan and profile. Furthermore the swash zone is highly variable and therefore difficult to monitor successfully. The measurement of wave particle velocity is important, and has already been mentioned in connection with the generation of edge-waves. Wave poles to record the height and direction of wave approach in the swash zone are being developed in the coastal work at Gibraltar Point in Lincolnshire. These instruments should provide a digital output that can be directly analysed by a computer.

Long-shore wave activity and sediment movement. Perhaps the field in which most effort has been expended recently is in the study and measurement of long-shore wave activity and the sediment movement to which it gives rise. This work involves theoretical and experimental approaches, together with field observation (I. P. Jolliffe, 1964). In the latter, the use of tracers has been of great significance, while the theoretical analysis of the long-shore movement of sediment has recently made considerable progress. The theoretical work must be based on an understanding of the processes of wave refraction in shallow water, as well as the distribution of energy in the waves at and within the break-point. An assessment of the local wind-generated currents is also necessary (R. S. Newton, 1968; W. R. Parker, 1973, 1975).

The situation is complicated by the interaction of the waves with the sea bed, including both the form of the bed in profile and the long-shore plan. Feedback relationships must be considered in assessing the long-shore movement of water and sediment. These small-scale, short-term interactions are important in understanding the larger-scale and longer-term changes in the coastal zone. The day-to-day variations in long-shore wave currents and changing wave characteristics determine the variability of the beach, while the longer-term coastal configuration is determined by the mean net direction of long-shore transfer of material over a longer period. This type of movement accounts for zones of coastal accretion and erosion, and the varying landforms such as spits, barriers, eroding cliffs and other features. In assessing the mean net long-shore transfer of material, a useful approach is through sediment budget calculations, such as those being attempted for the East Anglian coast (S. J. Craig-Smith and G. Cambers, 1972; M. J. Crickmore and G. H. Lean, 1962a,b; Crickmore *et al.*, 1972; A. W. Phillipps, 1963, 1971; W. A. Price, 1968).

In any study of the short-term variations in long-shore wave activity, it is usual to relate the changing wave conditions to variations in long-shore current velocities and associated beach changes. Work of this kind can be carried out by short-term tracer experiments with marked sand or shingle (D. B. Smith, 1973) or by means of current observations. The results can be analysed by time-series analysis to relate wave characteristics to long-shore transport and changing beach configuration. This type of approach is being used in work on the south Lincolnshire coast. Because it is necessary to consider the longer-term net conditions, rather than the short-term variability, in the longer-term studies of beach erosion and accretion, a study of the wave climate provides useful data.

Tides

In many areas the tide plays an important part in coastal dynamics (Bowden, 1972). The distinctive tidal morphology that has been intensively studied in the offshore zone of the continental shelf around Britain in decades since suitable geophysical instruments were developed, has also been recorded in considerable detail in the near-shore zone. The morphology itself can provide information of value concerning the processes responsible for its generation.

Tidal currents. Methods of measuring tidal streams have been developed for use in waters of varying depth. The early work on jelly bottles for bottom current measurements provided some useful results, while more recently other methods have been developed to record tidal streams. The techniques apply either at one point over a continuous time-span; alternatively, floats can be used to trace the course of a particular water body in motion. The development of the Woodhead sea-bed drifter, originally for fisheries research, has provided information that can be integrated over a longer time-span to assess the movement of water under the influence of tidal streams (A. H. W. Robinson, 1966, 1968). The disadvantage of this method is that only the time and location of the release point and the final recovery point are known. The route covered and the speed of movement cannot be resolved by this method. Nevertheless such studies have helped to elucidate the movement of water under the influence of the tides. A combination of both methods provides a clearer picture of the tidal stream pattern, although the network of observations must be fairly dense in complicated areas to provide a complete pattern. It is in such areas that a knowledge of the relationship between the tidal processes and the morphological response is particularly helpful.

Tidal morphology. Where tidal streams flow in a rectilinear pattern, and where there is plenty of mobile material to be shaped into banks, the streams tend to develop opposing channels for the ebb and the flood. Between these channels tidal banks develop with an orientation parallel to the major tidal currents. This pattern was recognised many years ago in the North Sea and occurs widely around the British coasts. The subtler details of the tidal morphology also assist in establishing the pattern of tidal streams. The sand waves and ebb and flood shields that develop around the leading edges of tidal banks provide more detailed evidence of the predominant tidal stream at any point. The sand waves have asymmetrical profiles with the steeper side facing down-current, allowing the predominant tide to be distinguished. In shallow water some of the smaller waves may reverse with the tide, and care is necessary in interpretation (M. S. Yalin and Price, 1974). A more detailed study of the sedimentological characteristics of the sand in the banks can help in this, and such methods have been used on the tidal banks at the north-west corner of the Wash. Tidal morphology can be studied by offshore survey using reflecting techniques while in areas where the tidal banks emerge at low water, aerial photographs provide evidence of the pattern of tidal activity, enabling changes during the intervals between photographic coverage to be assessed. Admiralty charts also provide useful historical evidence of changes in the banks (F. J. T. Kestner, 1970, 1975; J. R. Hardy, 1966; H. R. Wilkinson *et al.*, 1973).

A study is currently being made of a single linear tidal bank, the Sizewell Bank, south of Lowestoft off the East Anglian coast. This work involves both field observation and a mathematical model. The observations help in assessing the stability of the bank over time, while the model is developed for the pattern of tidal streams, relating these if possible to the movement of sediment. The sediment movement in turn can then be related to the changes in morphology (Robinson, 1975).

APPLIED COASTAL GEOMORPHOLOGY

Coastal geomorphology was one of the fields that was early found to be applicable to a wide range of coastal problems, such as sea defence, coastal conservation and coastal exploitation. Three aspects of applied coastal geomorphology will be briefly discussed, namely, material extraction, coastal defence and some major engineering works.

Removal of beach material. With increased building activity and the gradual depletion of inland supplies of sand and gravel, industry has turned increasingly to the foreshore and offshore zones as sources of materials. As early as the beginning of the century, sand and gravel were being taken from the beaches around Britain and the practice has continued. Many examples are cited in a government report of 1904: Portobello in Scotland, Start Bay in Devon and many areas around south-east England. In numerous cases severe coastal erosion followed the removal of beach material, and subsequently many of the licences for the removal of beach material were terminated. But even in the 1950s and 1960s the practice continued, and is still carried on in some areas (Jolliffe, 1974).

One particularly controversial areas, where pea-sized gravel is still being taken from the foreshore, is Chesil Beach in Dorset. Its extraction has raised a number of questions that can only be answered by a full and detailed study of this unique barrier beach or tombolo of shingle that links Portland Island to the mainland. This feature will be referred to again in the next section, as it is a phenomenon of great geomorphological interest, illustrating many facets of coastal geomorphology. Since the best protection for any coast is a wide, high beach, it is generally recognised that the removal of material from the foreshore is undesirable in the context of coast defence. Indeed the most satisfactory modern method of coast defence is that of artificial beach nourishment, the reverse of the extraction process (D. E. Newman, 1974).

Although the recovery of sand and gravel from the offshore zone raises further problems, it is not necessarily a harmful process; in some instances the offshore zone could prove to be a relatively safe source of essential material. The offshore zone around Britain, particularly along the east and south-east coast, is well-endowed with large amounts of loose sand and gravel, the legacy of the ice-sheets that once occupied the North Sea. The Irish Sea also provides glacial sands and gravels since the ice sheet in this area, which once reached the Scilly Islands, deposited large volumes of suitable material.

Much of the sand and gravel has been shaped by the tidal streams into the banks and channels mentioned in the last section. These banks and channels are in some areas intimately linked with the foreshore conditions and beach supply, so that their disturbance for aggregate extraction requires careful consideration. Even though the movement of shingle under wave action is probably very slow at depths exceeding 12 m, it would not be safe to assume that removal of material from greater depths would necessarily cause no change on the foreshore. Material can probably be moved at much greater depths by tidal streams, while the removal of material can change the bottom configuration. Work is being done, for example, on the effect of artificial holes on the wave-refraction pattern, and changes in this pattern could in turn effect changes on the foreshore (J. M. Motyka and D. H. Willis, 1974).

Coastal geomorphology is by no means the only aspect of the environment that can be affected by the offshore removal of sand and gravel. The effects on the living ecosystem can be even more severe, affecting fishing and many other aspects of the ecosystem.

Coastal defence. Coastal defence ideally involves the creation or maintenance of a wide, high beach. Lack of such a beach was clearly shown to be a fundamental cause of severe coastal damage during the 1953 east-coast storm surge. Various methods of maintaining an adequate beach defence have been used, including the placing of groynes across the beach, the reinforcement and maintenance of coastal sand-dunes, and the actual creation of new beaches by the dumping of artificial fill on the foreshore. Other experimental methods have been tried in some areas. The fixing of artificial seaweed in the nearshore zone off the beach at Bournemouth is an interesting example, although the effect was not entirely as predicted (Price *et al.*, 1969). The beach did not build up as much as the area offshore where the seaweed was fixed, while changes occurred farther along the coast, thus again emphasising the importance of longshore sediment movement.

There has been considerable study of the effect of groynes on specific stretches of coast, supplemented by experimental work using wave tanks (Price and K. W. Tomlinson, 1969). The problem of the terminal groyne has long been recognised, for although the groynes may help the beaches where they are installed and in the up-drift direction, they usually result in intensified erosion in the down-drift zone. An example of this has long been known from the seriously eroding coast of Holderness, where erosion is intensified to the south of the termination of solid sea defences and groynes. Indeed it is recognised that sea-walls and groynes are usually necessary evils as far as coast defence is concerned. This applies to the Lincolnshire coast where sea-walls and groynes were rebuilt after the 1953 surge damage; their strength, however, has been greatly increased, enabling them successfully to withstand the very high tides of early January, 1976.

The more flexible and natural methods of sea defence, such as beach replenishment by artificial dumping of suitable material and the improvement of coastal dunes, are normally preferable. The environment is usually improved by these methods, while hard, impermeable sea defences are not necessarily desirable for any purpose other than their own essential *raison d'être.* Thus coastal conservation and coastal defence can often be combined to produce not only a more desirable coast but also a safer one. The dual aspect of coastal preservation should be encouraged as far as possible, and can be achieved by working with the natural processes rather than trying to work against them. This requires a clear and complete knowledge of how these processes work—hence the vital part that can be played by the coastal geomorphologist, working with the coastal planner, coastal engineer and all coastal users.

Major engineering works. In the last decade, several feasibility studies have been undertaken for a variety of major engineering works in coastal locations. Major barrages across estuaries have been suggested for a variety of purposes: water supply, tidal power generation, amenity and access requirements, and for defence against the sea and excessively high tides. Barrages have been proposed across the Wash, Morecambe Bay and the Dee estuary, mainly for water-supply purposes. The Severn Estuary, with its macrotidal system, has been suggested as the most suitable site for tidal power generation. The Thames Estuary barrage has as its main purpose the protection of London from excessively high tides generated by surges in the North Sea, which are becoming increasingly menacing as sea-level rises and London subsides.

The creation of a third London airport at Maplin, now temporarily shelved, would also have involved considerable coastal modification and reclamation (Sir Frank Marshall, 1974). Other reclamation schemes have been considered in the Humber Estuary associated with the improvement of navigation. All these projects involve great expenditure of money and resources so that feasibility studies to assess engineering requirements, economic returns and the environmental impact of such major schemes are essential. Preliminary work on these schemes usually involves the building of a scale model, as well as making relevant field observations of present conditions and the processes currently operating. Once the model can be made to reproduce known changes, then the effect of the barrage or other engineering works can be determined more accurately.

Increasing use is also being made of mathematical models for major schemes of various types. A mathematical model of the Wash water-storage scheme has been developed to assess the effect of the barrages on the tidal streams. Another type of mathematical model, designed to study the generation and propagation of tidal surges in the North Sea has been developed with a practical application in view. This will enable a more effective warning system based on relevant meteorological observations to be instituted, and will also improve theoretical knowledge of surges.

Studies for these large-scale engineering projects have provided a great deal of useful information on many aspects of coastal geomorphology. The collection of wave data, for example, has greatly increased, while interesting phenomena, such as the cyclic movement of the banks in Morecambe Bay, have been revealed (Kestner, 1970). As yet none of the major schemes has been implemented, and it is doubtful whether some of them will ever be, apart possibly from the Thames barrier and the small bunded reservoirs in the Wash. They have nevertheless stimulated important coastal research: the Maplin study is a good example.

Investigation for the Maplin project was started in December 1971 and a considerable amount of preliminary work was undertaken before the project was temporarily abandoned. Hydraulic investigations included the following: (1) wave recording, (2) wave refraction exercises, (3) a new outer Thames estuary model, (4) accommodation for the new model, (5) development of a mathematical model for the outer Thames estuary, (6) fieldwork for physical and mathematical models, (7) field studies for the River Crouch, Havengore Creek and elsewhere, (8) armouring and wave over-topping exercise for sea-wall design, (9) use of radioactive tracers to study material movement on the sea bed, (10) investigations of the infills of navigational channels for dredging requirements, and (11) an historical survey to assess long-term changes in estuary régime. Other engineering investigations included: (12) a prospecting survey for sources of reclamation material, (13) a study of sea-wall foundations, (14) a geological survey of the total reclamation area, and (15) an economic survey for sources of sea-wall materials. The other major item was the construction in January, 1973, of a trial bank on Maplin Sands to assess, at the site and on a full scale, the effects of waves on the proposed structure. This bank has been monitored since its construction.

The physical model was made on a scale of 1:1000 horizontally and 1:100 vertically, covering the whole outer Thames estuary from Felixstowe in the north to Margate in the south. The maximum dimensions of the model are 140 × 90 m. The purpose of the physical model was to test the various shapes proposed for the reclamation scheme. It was the largest model in Europe and was filled with water in January, 1973. The full-scale trial bank as originally constructed was 3500 m long and 50 m wide. After 10 months it had been reduced to a length of 3000 m and a width of about 30 m, and had migrated slightly eastwards.

The site consists of a foundation of London Clay 20 to 30 m below chart datum, overlain by sand and gravel, layers of silty clay, and a thick sand deposit mostly exceeding 10 m in thickness. A study of these sediments required an extensive boring programme of 98 bores, as well as geophysical surveys. Prospecting for material for reclamation was also carried out and an ample supply of suitable material was located near-by. Stage 1 would need 160 million m^3 to reclaim 3200 ha and build 30 km of sea-defence works. The sea-defence works were to be in the form of gravel beaches, with a gradient of 1:8, built up to 7–8 m above O.D. The work would resemble a natural shingle beach and there would be the great advantage of flexibility. Flume tests were run to ascertain the wave run-up levels for different wave sizes, while the trial bank illustrated the natural reaction of such a beach to the local waves, reaching a stable profile without seriously altering the original bank.

The extensive and comprehensive preliminary work that has been carried out in connection with reclamation at Maplin has provided useful data on a wide range of topics, such as the wave régime of the outer Thames estuary, the availability of sand and gravel in the offshore zone, and geological and sedimentological data. The project also illustrates the wide range of disciplines involved in a major undertaking of this type, and the important role that coastal geomorphology can play in applied studies (Figs. 14.2, 14.3 and 14.4).

LOCAL STUDIES

Much of the work in coastal geomorphology of a more strictly academic character is based on the study of local stretches of coastline. Some of the studies are concerned with particular features, others with a particular stretch of coast, and yet others with a coastal system or a particular problem (C. Kidson, 1964). A few examples are selected.

Start Bay, Devon. The series of papers published in 1975 concerning the coast of Start Bay indicate the wide range of data needed to understand a coastal system, which must include the relevant offshore area. The work is concerned with the submarine geology, the sediment distribution and the Quaternary history of the area, and illustrates the value of an interdisciplinary approach to coastal problems. The bay forms a coastal system in that it receives little sediment from beyond its confines so that the beach depends on material circulated within the bay by both tides and waves. It is possible to separate areas within the bay that are mainly dependent for their material and morphology on wave processes

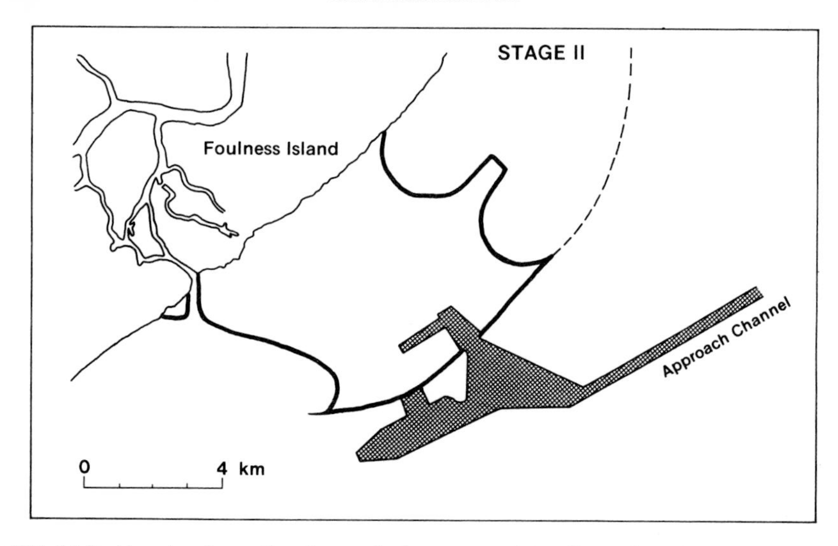

FIG. 14.2. Map to show the shape of the proposed Maplin reclamation area and the approach deep-water channel and port.

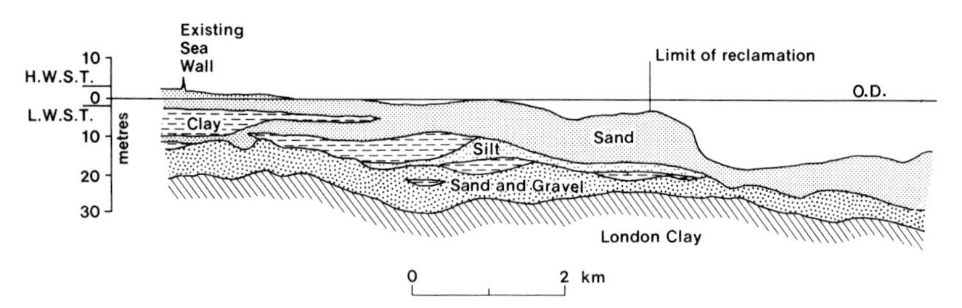

FIG. 14.3. Geological section to show the nature of the superficial deposits overlying the London Clay under the proposed Maplin reclamation area.

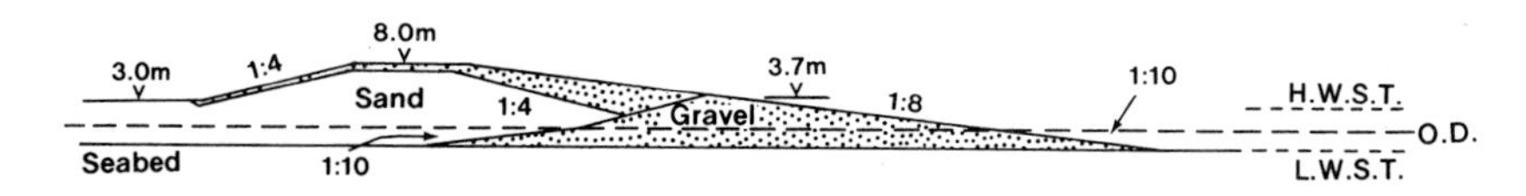

FIG. 14.4. Profile of the proposed sea defence banks around the Maplin reclamation area. The plan is similar to that of the trial bank already constructed.

from those that depend largely on tidal currents.

Serious beach erosion can be attributed both to natural and man-made phenomena. The removal of beach material during the period 1897–1902 lowered and weakened the beach at Hallsands to the point where it was unable to withstand the natural storm conditions of high tides, strong winds and storm waves in combination that created major damage to the coast at this point in 1917.

Study of the distribution and characteristics of the material, both at the coast and within the bay, enabled the processes operating within the area to be distinguished. The deposits could be categorised according to their genesis, allowing barrier, bay and bank material to be recognised and mapped. The evidence of the foraminifera and sediments has also provided valuable insights into the historical development of the bay and the important part played by the Flandrian transgression in the migration of features within the bay. Present-day processes have also been studied; the waves within the bay have been analysed in detail, while the morphology of the bay is suitable for the study of the generation of edge-waves at the shoreline. Such studies have a great academic interest in themselves, but also provide much useful information concerning, for example, the availability of sand and gravel from offshore, coast erosion and protection (Fig. 14.5) (Hails, 1975; Holmes, 1975; J. McManus, 1975).

Swansea Bay, South Wales. Swansea Bay is being investigated in a similar manner in order to study the causes of erosion along its eastern side—which could again be the result of extracting material both from the beach and from the offshore zone. Natural changes in the offshore zone could be associated with changes in the distribution of energy at the coastline through the effects of wave refraction.

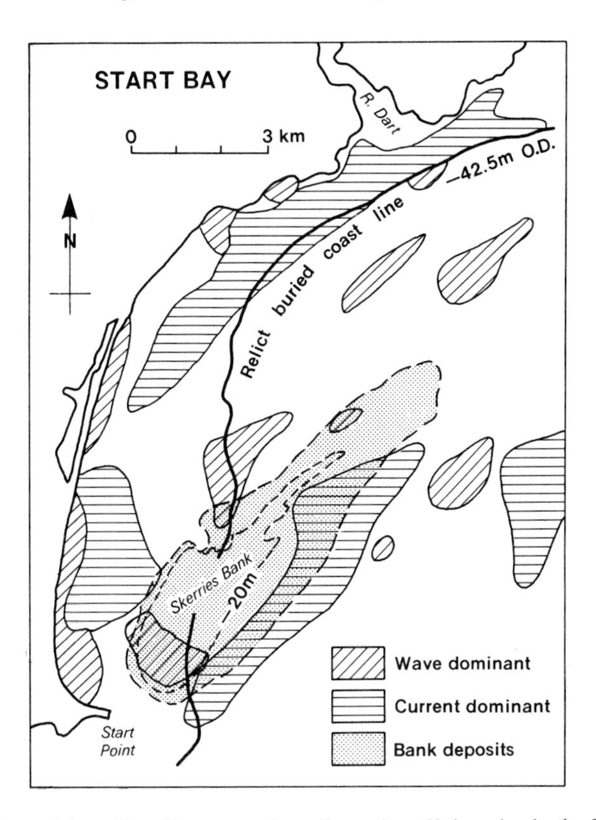

FIG. 14.5. Map of Start Bay, Devon, to show the major off shore bank, the Skerries Bank, and the areas covered by wave-related sediments and current-related sediments.

A tidal harbour has recently been developed in the area, and the effect of this structure on the coastal régime requires elucidation. An attempt is being made to quantify the sediment budget in the area.

Chesil Beach, Dorset. Chesil Beach (References: *see* Carr, Carr *et al.*, P. R. Corbyn, D. J. M. Neate) is one of the three major shingle structures in southern Britain, and is a feature that has long intrigued geomorphologists, especially on account of the very regular grading of shingle along its length from potato-sized material at the south-eastern end near Portland, to pea-sized material at the north-western end near Bridport. The beach has already been mentioned in connection with the extraction of shingle and the disquiet which this engenders.

In the last decade much work has been carried out on the beach, including a detailed study of a large number of pebbles in order quantitatively to assess the distribution of particle size, both above and below water-level, along the beach and at depth within the beach. The latter required the sinking of boreholes through the beach. The quantitative data on pebble dimensions have been statistically analysed.

The beach well illustrates how relict features that are not at present being actively supplied with new material can be brought into equilibrium with the processes operating, both in terms of orientation and morphology, and in terms of the distribution of material along them. These aspects are intimately interrelated and illustrate the importance of studying morphology in terms of both material and process. In fact, these three elements of the geomorphological situation provide the most useful results and fullest understanding when studied together. Because of the relict nature of the material that forms Chesil Beach, it is dangerous to remove even small quantities from it. In any case it must be a wasting asset on account of attrition of its material.

South-east England. The coast of south-east England (References: *see* K. R. Dyer, R. W. Hey, M. P. Kendrick, V. J. May) is under heavy pressure of population for many purposes, and suffers also from serious erosion problems. The increasing compartmentalisation of the coast, through the building of sea defences to arrest the erosion, has itself led to increasing difficulties in some areas. The development of this coast and its stabilisation provides a useful laboratory in which the significance of the natural coastal unit can be fully appreciated.

The building of the atomic power station at Dungeness, a shingle structure comparable in size to Chesil, was a subject of great controversy that has not been without its scientific repercussions. Borehole investigations for the foundations of the power station have provided valuable evidence on the original accumulation of the vast amount of shingle involved. The study of longshore movement of material along the eroding southern shore of the cuspate foreland showed the need to allow shingle to by-pass the structure, and shingle is now artificially moved westwards to balance the natural eastern drift. The influence of particle size on pebble movement was also established by observations in this area. The evolution of Dungeness has been established to a considerable extent both by studying its present morphology and by reference to historical material. In this instance the records go back to the Roman invasion by Caesar and include such events as the destruction of Winchelsea in the stormy thirteenth century.

Eastern England. The east coast of England possesses many interesting coastal phenomena, including features of both erosion and deposition. A study of the coast of East Anglia has involved the measurement of a series of profiles and an assessment of the sediment budget. The latter is based on calculations of longshore transport of material, derived from regular observations of wave characteristics by local observers, whose reliability had been checked. The human response to problems of coastal erosion is also being studied by questionnaire survey, thus providing an integrated physico-human study of an extremely vulnerable coast in the context of erosion problems. The establishment of coastal cells related to the net direction of longshore transport is an interesting application of the concept of coastal units to an actively changing coast which has a generally smooth outline diversified by five shallow 'nesses' of accumulation that are mobile. Extensive coastal defences are another significant characteristic, for, apart from the nesses, erosion occurs along most of the coast (Fig. 14.6) (References: *see* G. de Boer, Carr, Carr *et al.*, R. S. Clymo, J. T. Greensmith *et al.*, King, R. E. Randall).

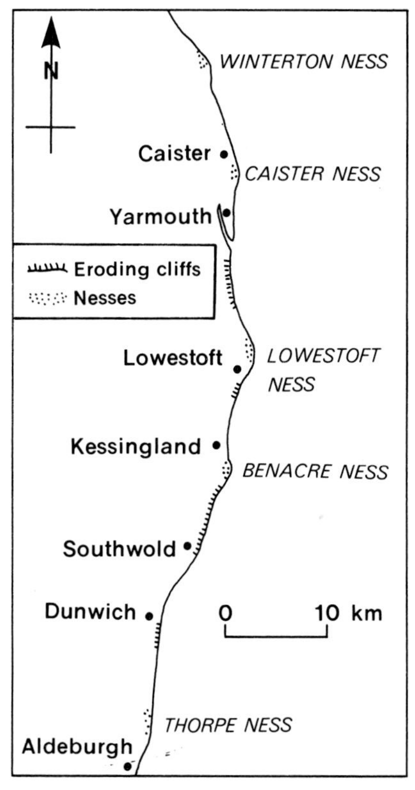

FIG. 14.6. Map to show the position of the East Anglian nesses and the eroding cliffs between them.

The results of recent work do not altogether agree with earlier interpretations of the migration of the nesses. They have been related to the tidal streams and their associated morphology, and are thought to occur where material comes ashore along net tidal residual paths. In some instances these are flood-dominated and engender a southerly movement; elsewhere, particularly in the south, the ebb residual results in a northerly movement. Some recent work, however, suggests an offshore sediment movement at the nesses. Nevertheless, it is generally agreed that both tides and waves together influence coastal processes in these areas, and only more detailed work will establish their precise interaction.

The significance of the tide is also admirably seen in the Gibraltar Point area of south Lincolnshire where tidal morphology is extremely well developed. Tidal banks and channels form in response to the rectilinear tidal streams and these features exert an all-important influence on the development of the foreshore. The presence of ridges and runnels on the foreshore can be directly attributed to the amount of material on the beach, which in turn is derived from the offshore zone primarily through the action of tidal streams, while the waves build the ridges on the foreshore. The ridges and runnels in turn provide the foundations on which sand-dunes and salt marshes develop, giving the typical pattern of arcuate dune ridges and intervening strip saltings. This area of marked accretion lies adjacent to an area of long-continued and often severe erosion, in which artificial sea-walls are essential to protect the low-lying coastlands of reclaimed marsh behind them.

In order to study the interaction of tides and waves, monitoring systems are being developed to study in detail the way in which material is transferred both onshore and alongshore in this area. The contrasts in the character of the sea bed in the zones of erosion and accretion in the nearshore area provide a clue to the importance of the tidal streams in this area. The sea bed is flat and generally lacks loose sediment in the zone of erosion, while tidal banks and channels are well developed off the zone of accretion. The south Lincolnshire coast also emphasises the need to study both the foreshore and nearshore zones, which together constitute the coastal cell or system. The sand-dunes and salt marshes of the backshore zone are also part of this complex coastal system.

The erosion of Holderness clearly illustrates the great significance of the longshore movement of material in coastal dynamics. It also illustrates the importance and often deleterious effect of coastal defences on the area down-drift of their position. The material lost from the drift cliffs of Holderness travels southwards along the coast, and provides material that helps to build the constructional features of the east coast. These include Spurn Head, which has been shown to exhibit a cycle of development. The elucidation of the stages in the cycle clearly demonstrates the value of a careful study of the relevant historical documents and maps relating to coastal events. It also provides an excellent example of cyclic coastal development, a geomorphological phenomenon that deserves further investigation in many fields beyond the well-known cycle of erosion.

TOPICAL STUDIES

Some studies have been concerned with special topics rather than particular areas. These include work on sand-dunes and salt-marshes, coastal cliffs and platforms.

Sand-dunes and salt-marshes. A common feature of these two different environments is the important role played by vegetation in their development. Sand-dune studies have been concerned with the pattern and orientation of dunes in relation to the formative wind and sand sources. Dunes can be divided into (a) *frontshore dunes,* comprising (i) foreshore dunes, (ii) dune islands or barriers, (iii) dunes on spits or nesses, and (b) *hindshore systems,* consisting of (i) bay dunes, (ii) climbing dunes, (iii) hindshore dunes, and (iv) machair-type flattened dune systems, extending some way inland. The machair dunes are distinguished by their highly calcareous soils and are composed mainly of shell sand, and thus are specially valuable for grazing.

Dunes are also important for their protective capacity. They provide a reservoir of mobile sand that can be drawn down to the beach by waves under storm or surge conditions, thus helping to preserve the coast. An instance of this protective role was demonstrated during the 1953 storm surge on the Lincolnshire coast, where at Anderby Creek, although cliffed to a height of 6 m, the dunes prevented a breach occurring. Methods of artificially encouraging and conserving dunes therefore have an important practical value. They include the planting of suitable plants, and the protection of the sand before the vegetation can gain a hold, as well as other methods of dune preservation. The role of dunes must not, however, be taken for granted, as they are not always a suitable means of coastal defence, although around the British coasts they normally exert a beneficial role (R. S. Colquhoun, 1969; W. Ritchie, 1972; Ritchie and K. Walton, 1972).

Salt-marsh plants also have a practical importance in that they help to trap silt and thus to raise the level of the marsh, facilitating and accelerating natural processes of accretion in reclamation projects, such as those now being undertaken in the Wash. Studies of the processes operating, particularly under the influence of the tide in salt marshes are, therefore, of considerable interest (M. E. Marker, 1967). One particular salt-marsh plant, *Spartina townsendii* and its hybrids, has been studied in great detail. This plant, which first appeared on the south coast in about 1870, has subsequently spread widely and played an important part in marsh development all over the country.

A study of the processes operating in the marshes of the Wash in connection with the construction of barrages for water storage has provided valuable information concerning the natural processes of sedimentation. The role played by tidal creeks in facilitating the deposition of sediment is analysed, because it is through the creeks that the silt-laden waters of the flood-tide gain access to the salt-marsh. In its natural state the system is one of dynamic balance, in which deposition can only take place in areas where the tidal waters have low velocities. These areas are of limited extent. Engineering works provide shelter and greatly speed-up accretion in the sheltered zones. The process takes place exponentially, with rapid accretion at first, slowing down markedly with time. Because the water overflowing the creeks supplies the sediment, the vegetated marsh outline develops a cuspate plan, with more rapid growth along the creek banks. Reclamation is still proceeding actively in the Wash, so that studies of this type have a practical application (Clymo, 1967; G. Evans, 1965; Evans and M. B. Collins, 1975).

Coastal cliffs and platforms. Studies of cliffs and wave-cut platforms are closely related. Cliffs can be studied from several points of view according to the speed with which they develop; this in turn depends on their setting and character. Some cliffs erode extremely rapidly, so that measurements of cliff retreat and the study of processes of retreat over relatively short periods can give valuable results. The detailed work on the cliffs of Holderness and East Anglia, which are mainly formed of glacial drift, illustrates the value of comparative map studies in relation to cliff character and setting. Relationships can be established between rates of erosion and relevant variables, including changes in sea-level, cliff exposure, cliff height and character of material, processes of mass movement, and especially longshore movement of beach material under the action of waves and tides. The relevance of mass movement to studies of cliff erosion is well exemplified by the work done on the cliff slumping and related processes in southern England and northern Ireland. The combination of subaerial and marine processes must be appreciated in understanding cliff processes in areas of soft rock and drift cliffs, which predominate in the south-east and east coast of England (May, 1971).

On the hard-rock cliffs of south-west England and much of the western coasts of the British Isles, the problems of cliff and platform study are very different. The cliffs show signs of great antiquity, often including relics of interglacial age. Their rate of change is so slow that it cannot be appreciated during a single lifetime, nor indeed over several centuries or even millenia. Studies of these ancient cliffs usually aim to describe their profiles and to relate these to their stages of development. The character and meaning of the coastal bevel, the upper, gently sloping section of many cliff profiles, have given rise to controversy. It is generally held to represent a period or periods of down-wasting by subaerial processes during low sea levels, often under glacial or periglacial conditions.

Another line of study has been to relate the level of the cliff foot to the relevant controls, such as wave exposure, tidal range and rock type. The considerable range found in the heights of currently active features is a warning when elevations of the raised, older platform-cliff intersections are being used to establish former sea levels. The morphology of, and processes acting upon, the shore platforms along the south and east coasts of England and Wales have also been the subject of recent research. A statistically significant relationship between platform gradient and tidal range illustrates the significance of the tide in erosional as well as depositional features. A relationship between cliff height and fetch on the one hand, and platform gradient on the other can also be established. The attempt to classify and explain coastal platforms has value beyond the immediate purpose of obtaining a deeper understanding of their character and formation. Such work is an essential preliminary to the use of raised shore platforms in establishing former sea levels (C. L. So, 1965 *et seq.*; A. S. Trenhaile, 1971 *et seq.*; A. Wood, 1968; L. W. Wright, 1967, 1970).

SEA-LEVEL CHANGES

One aspect of coastal geomorphology that has received considerable attention is the evidence for, and elucidation of, sea-level changes. The problem is extremely complex, as both eustatic, glacio-isostatic and tectonic influences are involved. These influences vary with the time-span involved in the investigation. Evidence for Tertiary changes of sea-level has been sought in planation surfaces, involving studies of denudation chronology. This aspect will not be considered, although it does raise important points concerning the differentiation of subaerial and marine-cut surfaces, and the ability of waves and other marine processes to form surfaces. Detailed work being carried out on currently developing platforms may eventually help to sort out some of these problems, although the time and tempo of change impose restrictions on such approaches.

The studies of Quaternary and Recent changes of sea-level are more directly relevant to coastal geomorphology. The evidence is found in the coastal zone in the form of raised shorelines, while other features,including buried channels, underwater shorelines and drowned subaerial deposits, such as submerged forests, indicate lower sea-levels. The evidence available for such studies varies with the location. In the northern part of the country where glacio-isostatic rebound has been effective, there are well-marked raised shorelines that have been studied in great detail and considerable accuracy, for example in eastern and western Scotland and Northern Ireland. Farther south in England and Wales the rising sea of the Flandrian transgression has dominated many coastal areas, and thus much of the evidence is drowned or buried. Sub-surface studies of peat and other deposits, together with their pollen content, have proved of great value in the investigation of such features as buried channels. The raised features on the whole appear to date from the older periods of the Pleistocene, when interglacial, high sea-levels were effective, although they may subsequently have been reworked and modified (References: *see* C. E. Everard *et al.*, G. D. Gaunt and M. J. Tooley, J. M. Gray, C. Kidson *et al.*, D. Linton, J. Rose, J. B. Sissons *et al.*, D. E. Smith *et al.*, N. Stephens *et al.*, F. Synge *et al.*, Tooley).

Another interesting aspect of the study of changing sea levels is that of the contemporary changes now in progress. The tide-gauge data related to various precise levelling programmes have provided some fairly reliable evidence of the present trend of sea-level change. The tilt of Britain down to the south-east and up to the north-west, where isostatic recovery is still not complete, is probably still proceeding (J. E. Prentice, 1972).

These movements continue ancient trends of warping and subsidence. The subsidence of the North Sea basin has been proceeding since at least the Carboniferous, and is still actively continuing and accelerating. There are several other subsiding basins around the coast of Britain, in the Irish Sea, off western Scotland and elsewhere. These basins are also of considerable age, mostly dating back to the Mesozoic at least. They may well be related to the position of Britain on the continental margin of the trailing edge of a plate, which is splitting along the mid-Atlantic ridge. It has been suggested that the distortion of the convection current carrying the plate as it meets the continental block, where the crust thickens, may be the cause of the basin subsidence. The raising of the British Isles between the subsiding basins is another manifestation of the process. It is, therefore, possible to relate current sea-level trends, and past evidence of sea-level fluctuations caused by tectonic activity, to the global position of Britain. If this connection is real, then the outline of Britain, the presence of the Irish Sea and North Sea, and the contrast between subsiding basins and uprising blocks, can be associated with the world-wide forces that cause the crustal plates to migrate, split and move over the face of the earth. The outline of the British coast is, therefore, probably in essentials very old. By contrast there is ample evidence that most of the details of the coastal features are very young. The beaches, spits, marshes, barriers, cliffs and most of the other coastal features owe their characteristics to the waves, tides and winds of the recent period, and particularly to the Flandrian transgression of the last 15 000 years.

CONCLUSION

Research in coastal geomorphology in Britain, as elsewhere, is becoming more sophisticated. Increasing use is being made of modern equipment, including computers to analyse the large amounts of data that modern recording instruments provide. Remote-sensing techniques are being rapidly developed, from satellite photographs and other sensing devices to remote-control of underwater monitoring systems. The need to study both offshore and onshore problems in conjunction is realised, and the resulting analyses are more complete and satisfactory.

Increasing pressure on the coastal zone from many points of view is adding urgency to the problems of coastal development, and the part played by the coastal geomorphologist in guiding the development along the most satisfactory lines is a vital one. The geomorphologist has the ability to appreciate the problems from a wide range of angles, and knows that it is essential to treat the coast as a whole system, within which separate units interact. The size of the coastal system and its constituent units varies with the character of the coast, from small enclosed bays on a crenulate coast to large stretches of beach and the offshore zone on a more open shoreline. The units include the offshore zone as far as it influences the incoming waves, which are the major process operating on the coast, while in many areas the tide also plays an essential role.

The study of waves and tides is basic to coastal geomorphology. Advances in this field are limited by the complexity of waves, which still defy exact mathematical analysis. Spectral analysis of the three-dimensional wave spectra can provide some order out of the apparent chaos of the natural wave field. Tidal streams are rather simpler and more easily predicted, owing to their greater regularity, but much work remains to be done in recording them accurately and in assessing their effect on the movement of material and its supply to the foreshore, especially in areas of complex, tide-induced morphology.

In coastal studies, as in other fields of geomorphology, a useful approach is to link the study of material, process and morphology in order to obtain a fuller understanding of the latter in terms of the two former elements. The study of material includes its character, such as the particle-size distribution of sand, the shapes of pebbles, and the source of material, while much effort has been expended in studying the movement of material on the beach, both on and offshore and especially alongshore. The nature of the solid rock or drift of the cliffs and platforms must not be ignored in the longer-term studies of coastal development, as opposed to the day-to-day variability of the beaches. In fact coastal studies can be divided into two broad groups; one is related to the coast and is a long-term study, the other to the beach and is a short-term study of currently-acting processes.

In long-term coastal studies, changes of sea-level must be considered. The rapid Flandrian transgression, which rapidly raised sea-level in all but those areas undergoing rapid isostatic uplift, has had a profound effect on all coastlines. It is only during the last 4000 years or so that eustatic sea-level has been relatively static, and the coastline has been able to adjust to the present level in stable areas. Thus nearly all coasts are young, and are actively undergoing change. Both erosion and deposition are active, as material becomes redistributed along the coast and equilibrium is being sought to conform to the present pattern of waves and tides. In achieving this equilibrium, longshore transfer of material is fundamental, especially now that the supply from offshore is reduced in many areas, as the offshore bottom becomes adjusted to the present sea-level. Apart from the longshore source of material and erosion of soft, drift cliffs, the only supply of material of significance is that from inland via the rivers. The contribution that these make along the coast of Britain is probably small as far as gravel and sand are concerned, owing to the low gradients of many of the lower parts of British rivers. Thus longshore movement is of increasing importance, and hence there is a need to consider the pattern of coastal systems within which material can migrate alongshore. Within a coastal system, the material is probably largely now relict to the extent that it cannot be rapidly replenished by natural means. This applies, for example, to features such as Chesil Beach, so that the removal of material from the foreshore and the interruption of the natural longshore flow of material must be viewed with great concern. Man's activities on the coast are thus of great significance, and they must be planned on a large scale and with a sound understanding of the processes involved.

Coastal research is one of the fields attracting a high proportion of geomorphologists according to the 1975 register of current research in British geomorphology (Table I). Sixty-two entries out of 420 listed express an interest in

TABLE I

Number of studies in progress on particular aspects of coastal geomorphology (Source: Register of Current Geomorphological Research, B.G.R.G. (1975))

Sedimentology	9
Sea level and coastal platforms	7
Salt marshes	7
Beaches	6
Water flow, tides	5
Cliffs	4
Dunes	3
General	3
Spits	2
Total	46*

*Entries listing several interests have been counted as a half each, and work outside Britain has been excluded.

coasts and coastal processes. The only other topics to exceed this number are Applied Geomorphology (65) and Glacial and Quaternary studies. Of those working on the coastline of Britain, the sedimentological aspects seem to attract the greatest number of workers. Coastal research in Britain thus seems to have an adequate work force to tackle the many problems and topics that require further elucidation and understanding (Fig. 14.7).

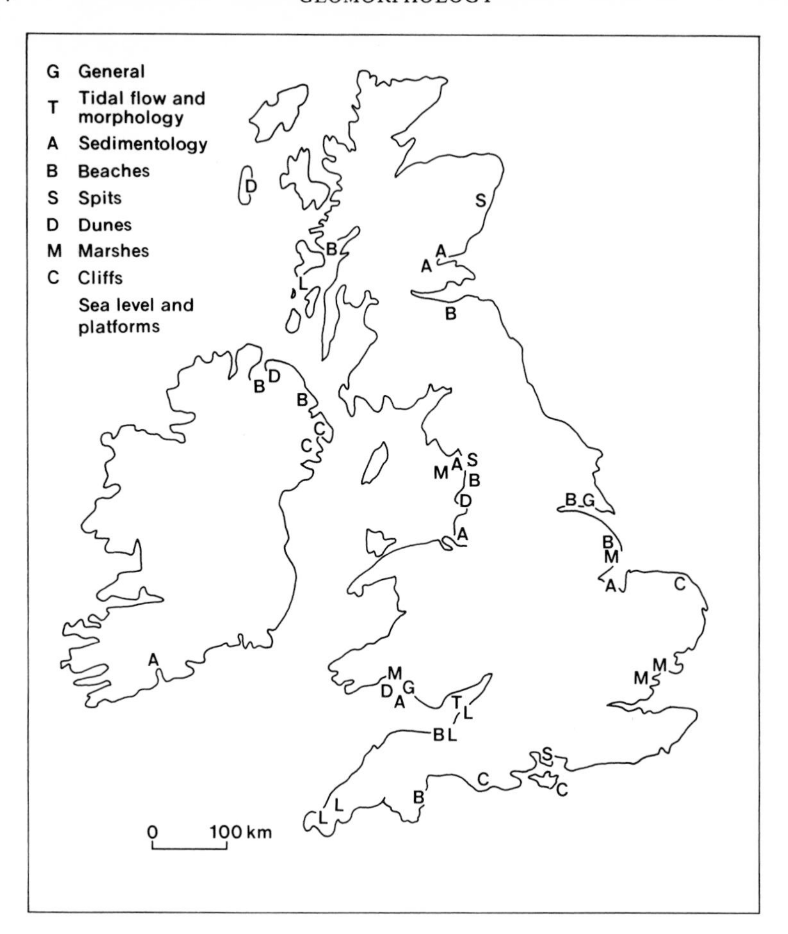

FIG. 14.7. Map to indicate areas where coastal research is now in progress around the coastline of Britain, and the category of work being undertaken. Data derived from B.G.R.G. *Register of Current Geomorphological Research* (1975).

REFERENCES

Boer, G. de (1967), 'Cycles of change at Spurn Head, Yorkshire, England', *Shore Beach* **35**, 13–20.

Bowen, A. J. (1972), 'Tidal regimes' in 'A discussion on problems associated with the subsidence of southeast England', *Phil. Trans. R. Soc. Lond.* **A 272**, 187–99.

Bowen, D. Q. (1973), 'Time and place on the British Coast', *Geography* **58**, 207–16.

Buller, A. T., Gaunt, C. D. and McManus, J. (1975), 'Dynamics and sedimentation in the Tay in comparison with other estuaries', in J. Hails and A. Carr (eds.), *Nearshore sediment dynamics and sedimentation* (Wiley, London), 201–50.

Carr, A. P. (1965), 'Coastal changes at Bridgwater Bay, 1957–1964,' *Proc. Bristol Nat. Soc.* **31**, 91–100.

—— (1965), 'Shingle spit and river mouth short-term dynamics', *Trans. Inst. Br. Geogr.* **36**, 117–29.

—— (1969), 'Size grading along a pebble beach, Chesil Beach, England', *J. sedim. Petrol.* **39**, 297–311.

—— (1970), 'The evolution of Orfordness, Suffolk before 1600 A.D.: geomorphological evidence', *Z. Geomorph.* **14**, 289–300.

—— (1971), 'Experiments on longshore transport and sorting of pebbles: Chesil Beach, England', *J. sedim. Petrol.* **41**, 1084–104.

—— (1972), 'Aspects of spit development and decay: the estuary of the River Ore, Suffolk', *Fld Stud.* **3**, 633–53.

—— and Baker, R. E. (1968), 'Orford Suffolk: evidence for the evolution of the area during the Quaternary', *Trans. Inst. Br. Geogr.* **45**, 107–23.

—— and Blackley, M. W. L. (1969 for 1968), 'Geological composition of the pebbles of Chesil Beach', *Proc. Dorset nat. Hist. archaeol. Soc.* **90**, 133–40.

—— and Blackley, M. W. L. (1973), 'Investigations bearing on the age and development of Chesil Beach, Dorset and the associated area', *Trans. Inst. Br. Geogr.* **58**, 99–112.

—— and Gleason, R. (1971), 'Chesil Beach, Dorset and the cartographic evidence of Sir John Coode', *Proc. Dorset nat. Hist. archaeol. Soc.* **93**, 125–31.

——, Gleason, R. and Hardcastle, P. J. (1973), 'The significance of wave parameters in the sorting of beach pebbles', *Estuar. coastal mar. Sci.* **1**, 11–18.

Clymo, R. S. (1967), 'Movement of the main shingle beach at Blakeney Point, Norfolk', *Trans. Norfolk Norwich Nat. Soc.* **21**, 3–6.

—— (1967), 'Accretion rate in two of the salt marshes at Blakeney Point, Norfolk', *Trans. Norfolk Norwich Nat. Soc.* **21**, 17–18.

Colquhoun, R. S. (1969), 'Dune erosion and protective works at Pendine, Carmarthenshire, 1961–8', *Proc. 11th Conf. Coastal Engng 1968,* **1**, 708–18.

Corbyn, P. R. (1967), 'The size and shape of pebbles on Chesil Beach', *Geogr. J.* **133**, 54–5.

Craig-Smith, S. J. and Cambers, G. (1972), 'Report of a field meeting to the Norfolk coast at Yarmouth and Overstrand', *Bull. geol. Soc. Norfolk* **21**, 29–40.

Crickmore, M. J. and Lean, G. H. (1962a), 'The measurement of sand transport by means of radioactive tracers', *Proc. R. Soc.* A, **266**, 402–21.

—— and Lean, G. H. (1962b), 'The measurement of sand transport by the time integration method and radioactive tracers', *Proc. R. Soc.* A, **207**, 27–47.

Crickmore, M. J., Waters, C. B. and Price, W. A. (1972), 'The measurement of offshore shingle movement', *Proc. 13th Conf. Coastal Engng* **2**, 1005–25.

Crofts, R. (1972), 'Coastal sediments and processes around St. Cyrus', in C. M. Clapperton (ed.), *North East Scotland Geographical Essays* (Univ. of Aberdeen), 15–19.

—— and Mather, A. S. (1972), *The beaches of Wester Ross* (Dept. of Geography, Univ. of Aberdeen), 108 pp.

—— and Ritchie, W. (1973), *The beaches of Mainland Argyll* (Dept. of Geography, Univ. of Aberdeen), 160 pp.

Dobbie, C. H. (1969), 'Case histories of two estuaries: River Spey in Moray Firth and Dawlish Warren', *Proc. 11th Conf. Coastal Engng 1968,* **2**, 1295–1303.

Donovan, D. T. and Stride, A. H. (1973), 'Three drowned coast lines of probable late Tertiary age around Devon and Cornwall', *Mar. Geol.* **19**, M35–M40.

Draper, L. (1967), 'Wave activity at the sea bed around northwestern Europe', *Mar. Geol.* **5**, 133–40.

—— and Dobson, P. J. (1965), 'Rip currents on a Cornish beach', *Nature, Lond.* **206**, 1249.

Durrance, E. M. (1969), 'The structure of Dawlish Warren', *Proc. Ussher Soc.* **2**, 91–101.

Dyer, K. R. (1971), 'The distribution and movement of sediment in the Solent, south England', *Mar. Geol.* **11**, 175–87.

Evans, G. (1965), 'Intertidal flat sediments and their environments of deposition in the Wash', *Q. J. geol. Soc. Lond.* **121**, 209–41.

—— and Collins, M. B. (1975), 'The transportation and deposition of suspended sediment over the intertidal flats of the Wash', in J. Hails and A. Carr (eds.) *Nearshore sediment dynamics and sedimentation* (Wiley, London), 273–306.

Everard, C. E. *et al.* (1965 for 1964) 'Raised beaches and marine geomorphology', in *Some aspects of the geology of Cornwall* (ed. K. F. G. Hosking & G. J. Shrimpton; Royal Geological Soc. of Cornwall), 283–310.

Gaunt, G. D. and Tooley, M. J. (1974), 'Evidence for Flandrian sea level changes in the Humber estuary and adjacent areas', *Bull. geol. Surv. Gt Br.* **48**, 25–41.

Gray, J. M. (1974), 'Late-glacial and post-glacial shorelines in western Scotland', *Boreas* **3**, 129–38.

—— (1974), 'The main rock platform of the Firth of Lorn, west Scotland', *Trans. Inst. Br. Geogr.* **61**, 81–100.

—— (1975), 'The Loch Lomond Readvance and contemporaneous sea levels in Loch Etive and neighbouring areas of western Scotland', *Proc. Geol. Ass.* **86**, 227–38.

—— (1975), 'Measurement and analysis of Scottish raised shoreline altitudes', (Dept. of Geography, Queen Mary College, London), *Occas. Pap.* **2**, 40 pp.

Greensmith, J. T. and Tucker, E. V. (1966), 'Morphology and evolution of inshore shell ridges and mud-mounds on modern intertidal flats near Bradwell, Essex', *Proc. Geol. Ass.* **77**, 329–46.

—— and Tucker, E. V. (1973), 'Holocene transgression and regressions on the Essex coast, outer Thames estuary', *Geologie Mijnb.* **52**, 193–202.

—— and Tucker, E. V. (1975), 'Dynamic structures in the Holocene Chenier Plain setting of Essex, England', in J. Hails and A. Carr (eds.), *Nearshore sediment dynamics and sedimentation* (Wiley, London), 251–72.

Greenwood, B. (1972), 'Modern analogues and the evaluation of a Pleistocene sedimentary sequence', *Trans. Inst. Br. Geogr.* **56**, 145–70.

Hails, J. R. (1975), 'Submarine geology, sediment distribution and Quaternary history of Start Bay, Devon', *J. geol. Soc. Lond.* **131**, 1–5 and 19–35.

—— and Carr, A. P. (eds.) (1975), *Nearshore sediment dynamics and sedimentation* (Wiley, London).

Halliwell, A. R. and O'Connor, B. A. (1969), 'Shear velocity in a tidal estuary', *Proc. 11th Conf. Coastal Engng 1968*, 2, 1377–96.

Hardy, J. R. (1966), 'An ebb-flood channel system and coastal changes at Winterton, Norfolk', *E. Midld Geogr.* **4**, 24–30.

Hey, R. W. (1967), 'Section in the beach plain deposits of Dungeness, Kent', *Geol. Mag.* **104**, 361–70.

Hodson, F. and West, I. M. (1972), 'Holocene deposits of Fawley, Hants., and the development of Southampton Water', *Proc. Geol. Ass.* **83**, 421–42.

Holmes, P. (1975a), 'Wave conditions in Start Bay, Devon', *J. geol. Soc. Lond.* **131**, 57–62.

—— (1975b), 'Wave conditions in coastal areas', in J. Hails and A. Carr (eds.) *Nearshore sediment dynamics and sedimentation* (Wiley, London), 1–16.

Huntley, D. A. and Bowen, A. J. (1975), 'Comparison of the hydrodynamics of steep and shallow beaches', in J. Hails and A. Carr (eds.) *Nearshore sediment dynamics and sedimentation* (Wiley, London), 69–100.

Jardine, W. G. (1973), 'Observed beach modification, Kiloran Bay, Isle of Colonsay, Argyll', *Scott. J. Geol.* **9**, 213–18.

Jolliffe, I. P. (1964), 'An experiment designed to compare the relative rates of movement of different sizes of beach pebbles', *Proc. Geol. Ass.* **75**, 67–86.

—— (1974), 'Beach-offshore dredgings: some environmental consequences', *Offshore Tech. Conf. Pap.* OTC 2056.

Kemp, P. H. (1975), 'Wave asymmetry in the nearshore zone and breaker area', in J. Hails and A. Carr (eds.), *Nearshore sediment dynamics and sedimentation* (Wiley, London), 47–68.

Kendrick, M. P. (1972), 'Siltation problems' in 'A discussion on problems associated with the subsidence of southeast England'. *Phil. Trans. R. Soc.* **A 272**, 223–43.

Kestner, F. J. T. (1970), 'Cyclic changes in Morecambe Bay', *Geogr. J.* **136**, 85–97.

—— (1975), 'The loose-boundary regime of the Wash', *Geogr. J.* **141**, 388–414.

Kidson, C. (1964), 'Dawlish Warren, Devon: late stages in sand spit evolution', *Proc. Geol. Ass.* **75**, 167–184.

—— and Heyworth, A. (1973), 'The Flandrian sea level rise in the Bristol Channel', *Proc. Ussher Soc.* **2**, 565–84.

—— and Manton, M. M. (1973), 'Assessment of coastal change with the aid of photogrammetric and computer-aided techniques', *Estuar. coastal mar. Sci.* **1**, 271–84.

—— and Wood, R. (1974), 'The Pleistocene stratigraphy of Barnstaple Bay', *Proc. Geol. Ass.* **85**, 223–38.

King, C. A. M. (1973), 'Dynamics of beach accretion in south Lincolnshire, Enlgand', in D. R. Coates (ed.) *Coastal geomorphology* (Proc. 3rd Annual Geomorphology Symposium, St Univ. New York, Binghampton, Publs. in Geomorph.), 73–98.

Linton, D. L. (1971), 'The low raised beach of southern Ireland, south Wales, Cornwall, Devon and Brittany and its relation to earlier weathering and later gelifluxion', *Quaternaria* **15**, 91–8.

Longuet-Higgins, M. S. (1972), 'Recent progress in the study of longshore currents', in R. E. Meyer (ed.), *Waves on beaches and resulting sediment transport* (Acad. Press, New York), 203–48.

Marker, M. E. (1967), 'The Dee Estuary: its progressive silting and salt marsh development', *Trans. Inst. Br. Geog.* **41**, 65–71.

Marshall, Sir Frank, Chairman (1974), *Maplin Development Authority Rep. 1973/4: Engineering design*, 23–62.

Mather, A. S. and Crofts, R. S. (1972), *Beaches of west Inverness and north Argyll* (Dept. of Geography, Univ. of Aberdeen), 201 pp.

May, V. J. (1966), 'A preliminary study of recent coastal changes and sea defences in southeast England', (Univ. of Southampton, Research Series in geography), **3**, 3–24.

—— (1969 for 1968), 'Reclamation and shoreline change in Poole Harbour, Dorset', *Proc. Dorset nat. Hist. archaeol. Soc.* **90**, 141–54.

—— (1971), 'The retreat of chalk cliffs', *Geogr. J.* **137**, 203–6.

McCann, S. B. (1966), 'The main post-glacial raised shoreline of western Scotland from the Firth of Lorne to Loch Broom', *Trans. Inst. Br. Geog.* **39**, 87–100.

McManus, J. (1975), 'Quartile deviation–median diameter analysis of surface and core sediments from Start Bay', *J. geol. Soc. Lond.* **131**, 51–6.

Mitchell, J. K. (1968), 'A selected bibliography of coastal erosion, protection and related human activity in North America and the British Isles', *Natural Hazard Res. Working Pap.* (Univ. of Toronto, Canada) **4**, 66 pp.

Motyka, J. M. and Willis, D. H. (1974), 'The effect of wave refraction over dredged holes', *Proc. 14th Conf. Coastal Engng* (Copenhagen).

Neate, D. J. M. (1967), 'Underwater pebble grading of Chesil Beach', *Proc. Geol. Ass.* **78**, 419–26.

Newman, D. E. (1974), 'A beach restored by artificial nourishment', *Proc. 14th Conf. Coastal Engng* (Copenhagen).

Newton, R. S. (1968), 'Internal structure of wave-formed ripple marks in the nearshore zone', *Sedimentology* **11**,275–92.

Oldfield, F., Carter, R., Kitchener, K. and Wilcock, F. (1972), 'Report of an investigation into coastal erosion and accretion along the coastlines of County Antrim and County Londonderry', *Coastal Res. Proj. 1968–72* (New Univ. of Ulster), 47 pp.

Parker, W. R. (1973), 'Folding and intertidal sediments on the west Lancashire coast, England', *Sedimentology* **20**, 615–24.

—— (1975), 'Sediment mobility and erosion on a multibarred foreshore, southwest Lancashire, U.K.' in J. Hails and A. Carr (eds.) *Nearshore sediment dynamics and sedimentation* (Wiley, London), 151–80.

Phillips, A. W. (1963), 'Tracer experiments at Spurn Head, Yorkshire, England', *Shore Beach* **31**, 30–5.

—— (1971), 'Present coastal changes on Walney Island', (Dept. of Geography, Univ. of Lancaster) *Res. Pap.* **8**, 13–36.

Prentice, J. E. (1972), 'Sedimentation' in 'A discussion on problems associated with the subsidence of southeast England', *Phil. Trans. R. Soc.* **A**, **272**, 115–19.

Price, W. A. (1968), 'Variable dispersion and its effects on the movements of tracers on beaches', *Proc. 11th Conf. Coastal Engng* **1**, 329–34.

—— and Tomlinson, K. W. (1969), 'The effect of groynes on stable beaches', *Proc. 11th Conf. Coastal Engng* **1**, 518–25.

——, Tomlinson, K. W. and Hunt, J. N. (1969), 'The effect of artificial seaweed in promoting the build-up of beaches', *Proc. 11th Conf. Coastal Engng* **1**, 570–8.

——, Tomlinson K. W. and Willis, D. H. (1972), 'Predicting changes in the plan shape of beaches', *Proc. 13th Conf. Coastal Engng* **2**, 1321–9.

Randall, R. E. (1973), 'Shingle Street, Suffolk: an analysis of a geomorphic cycle', *Bull. geol. Soc. Norfolk* **24**, 15–35.

Report of a Working Party on Marine Sedimentation, U.K. (N.E.R.C. Research in the Physical marine sciences: Papers Ser. B), 8 (1973), 50–9.

Richards, A. (1969), 'Some aspects of the evolution of the coastline of northeast Skye', *Scott. geogr. Mag.* **85**, 122–31.

Ridgway, K. and Scotton, J. B. (1973), 'Whistling sand beaches in the British Isles', *Sedimentology* **20**, 263–80.

Ritchie, W. (1972), 'The evolution of coastal sand dunes', *Scott. geogr. Mag.* **88**, 19–35.

—— and Walton, K. (1972), 'The evolution of the sands of Forvie and the Ythan Estuary' in C. M. Clapperton (ed.) *Northeast Scotland, Geographical Essays* (Univ. of Aberdeen) 12–14.

Robinson, A. H. W. (1966), 'Residual currents in relation to shoreline evolution of the East Anglian coast', *Mar. Geol.* **4**, 57–84.

—— (1968), 'The use of sea-bed drifters in coastal studies with particular reference to the Humber', *Z. Geomorph., Suppl. Bd* **7**, 1–23.

—— (1975), 'Cyclical changes in shoreline development at the entrance to Teignmouth Harbour, Devon, England' in J. Hails and A. Carr (eds.), *Nearshore sediment dynamics and sedimentation* (Wiley, London), 181–200.
Rollinson, W. (1971), 'Coastal changes on Walney Island: an historical appraisal' (Dept. of Geography, University of Liverpool), *Res. Pap.* **8**, 1–12.
Rose, J. (1975), 'Raised beach gravels and ice-wedge casts at Old Kilpatrick, near Glasgow', *Scott J. Geol.* **11**, 15–21.
Sissons, J. B. (1974), 'Late-glacial marine erosion in Scotland', *Boreas* **3**, 41–8.
——, Smith, D. E. and Cullingford, R. A. (1966), 'Late-glacial and post-glacial shorelines in southeast Scotland', *Trans. Inst. Br. Geogr.* **39**, 9–18.
Smith, D. B. (1973), 'The use of artificial radioactive tracers in the United Kingdom' in *Tracer Techniques in Sediment Transport* (*Int. Atomic Energy Agency,* Vienna), 97–102.
Smith D. E., Sissons, J. B. and Cullingford, R. A. (1969), 'Isobases for the main Perth raised shoreline in southeast Scotland, as determined by trend surface analysis', *Trans. Inst. Br. Geogr.* **46**, 45–52.
Smith, J. S. and Mather, A. S. (1974), 'Beaches of east Sutherland and Easter Ross', (Dept. of Geography, Univ. of Aberdeen), 97 pp.
So, C. L. (1965), 'Coastal platforms of the Isle of Thanet', *Trans. Inst. Br. Geogr.* **37**, 147–56.
—— (1966), 'Some coastal changes between Whitstable and Reculver, Kent', *Proc. Geol. Ass.* **77**, 475–90.
—— (1971, 1972), 'Early coast recession around Reculver, Kent', *Archaeol. cantiana* **86**, 93–7.
Steers, J. A. (1973), *The coastline of Scotland* (Cambridge Univ. Press), 335 pp.
Stephens, N. and Synge, F. M. (1966), 'Pleistocene shorelines', in G. H. Dury (ed.) *Essays in Geomorphology* (Heinemann, London), 1–51.
—— (1970), 'The coastline of Ireland', in N. Stephens and R. E. Glassock (eds.), *Irish Geographical Studies* (Queen's University, Belfast), 125–45.
Synge, F. M. and Stephens, N. (1966), 'Late and post-glacial shorelines and ice limits in Argyll and Northeast Ulster', *Trans. Inst. Br. Geogr.* **39**, 101–25.
Thomas J. M. (1966), 'Sedimentation on Instow Beach, North Devon', *Proc. Ussher Soc.* **1**, 257–8.
Tooley, M. J. (1974), 'Sea-level changes during the last 9000 years in northwest England', *Geogr. J.* **140**, 18–42.
Trenhaile, A. S. (1971), 'Lithological control of high-water rock ledges in the Vale of Glamorgan', *Geogr. Annlr* **53 A**, 59–69.
—— (1972), 'The shore platforms of the Vale of Glamorgan, Wales', *Trans. Inst. Br. Geogr.* **56**, 127–44.
—— (1974a), 'The morphology and classification of shore platforms in England and Wales', *Geogr. Annlr* **56 A**, 103–10.
—— (1974b), 'The geometry of shore platforms in England and Wales', *Trans. Inst. Br. Geogr.* **62**, 129–42.
Wilkinson, H. R., de Boer, G. and Thunder, A. (1973), 'A cartographic analysis of the changing bed of the Humber', (Dept. of Geography, Univ. of Hull), *Misc. Ser.* **14**, 70 pp.
Willis, D. H. and Price, W. A. (1975), 'Trends in the application of research to solve coastal engineering problems' in J. Hails and A. Carr (eds.), *Nearshore sediment dynamics and sedimentation* (Wiley, London), 110–122.
Wood, A. (1968), 'Beach platforms in the Chalk of Kent, England', *Z. Geomorph.* **12**, 106–13.
Wright, L. W. (1967), 'Some characteristics of the shore platforms of the English

Channel and the northern part of the North Island, New Zealand', *Z. Geomorph.* **11**, 36–46.
—— (1970), 'Variation in the level of the cliff/shore platform junction along the south coast of Great Britain', *Mar. Geol.* **9**, 347–53.
Yalin, M. S. and Price, W. A. (1974), 'Formation of dunes by tidal flows', *Proc. 14th Conf. Coastal Engng* (Copenhagen).

15

APPLIED GEOMORPHOLOGY: A BRITISH VIEW

D. BRUNSDEN
(*King's College, London*)
J. C. DOORNKAMP
(*University of Nottingham*)
D. K. C. JONES
(*London School of Economics*)

Applied geomorphology is the application of geomorphological techniques and analysis to the solution of a planning, environmental management, engineering or similar problem. A global view of such a diffuse subject would be lengthy, difficult and beyond the scope of this paper. The following discussion therefore concentrates on the character, demands and philosophy of applied studies in Britain, and considers only those projects that have produced information that has actually been used. It is clear that many geomorphological studies are capable of providing information of value to land managers, planners and engineers (R. U. Cooke & Doornkamp, 1974; Cooke, 1976; I. Douglas, 1976). Unfortunately, it appears that much of this potential is not being realised, and hence many studies of an applied character must be excluded from further consideration because they fail to satisfy this definition. This is mainly due to a general lack of awareness, on the part of decision-makers, concerning the nature and scope of geomorphology as compared with the traditionally-used overlapping fields of geology and hydrology. Any change in the present situation will require both the establishment of better links between geomorphologists, essentially academic and geographically based, and other interested parties, and the development of better communications for the dissemination of information. Some success has already been achieved both by the British Geomorphological Research Group and by the Engineering Group of the Geological Society of London. However, this is but a beginning; ignorance as to the nature of contemporary geomorphology is still widespread.

THE CURRENT (1975) STATUS OF APPLIED GEOMORPHOLOGY IN BRITAIN

As a first approximation to the activities of British geomorphologists in applied geomorphology, we may turn to the *Register of current research in geomorphology* for 1975 (*British Geomorphological Research Group*, 1975). Only sixty-five (15·5 per cent) of the 420 entries indicated any interest in applied geomorphology. Of these, probably fewer than twenty would claim to practise as applied geomorphologists (i.e. less than 5 per cent). These figures give a falsely pessimistic impression and fail to indicate the growing strength of applied geomorphology in Britain. First, the register under-estimates the number of full-time, professional geomorphologists now working in such organisations as the Institute of Hydrology, Regional Water Authorities, the Nature Conservancy Council, the Institute of Oceanographic Sciences, the Soil Survey of England and Wales, Gas Boards, the Transport and Road Research Laboratory, the Land Resources Division of the Ministry of Overseas Development, Ove Arup and Partners, and Rendel, Palmer & Tritton (currently as post-graduate students). Secondly, these statistics conceal the range of work being carried out, which includes:

Application of remote-sensing techniques to highway engineering;
Geomorphological mapping and highway engineering design;
Development of geomorphological design-parameters for highway engineering;
Land classification (land-systems mapping and terrain analysis);
Geomorphological mapping for resource evaluation in developing countries;
Geomorphological evaluations with a view to water-use for irrigation;
Application of geomorphology to the classification and analysis of desert soils;
Land evaluation for rural planning;
Geomorphological contributions to statements on environmental impact;
Evaluation for planning legislation of sites of geomorphological importance;
Landslide analysis and mapping;
Identifying, classifying, analysing and mapping flood, landslide, avalanche or other hazard areas;
Analysis of contemporary weathering, fluvial, slope or coastal processes;
Study of jökulhlaups—water bursts from glaciers;
Management and conservation of coastal zones;
The evaluation of water resources;
The evaluation of aggregate resources;
Water-quality studies.

Thirdly, examination of the register fails to indicate the increasing emphasis on applied geomorphology in both undergraduate and post-graduate training in Britain. Undergraduate courses containing at least some applied geomorphology are taught in most Departments of Geography in British Universities. Post-graduate courses that contain a significant element of applied geomorphology include an M.Sc. course at Sheffield, the post-graduate course in Integrated Surveys at Reading, and the M.Sc. course in Conservation Studies at University College London.

Finally the register conceals the rapid growth in awareness and interest regarding the application of geomorphological information to decision-making. On an individual basis, British geomorphologists hold positions as consultants to, or have been consulted by, such bodies and organisations as UNESCO, Fairey Surveys Ltd., Rendel Palmer & Tritton (Consultant Engineers), Sir William Halcrow & Partners, G. Wimpey Ltd., Mobil Oil (North Sea) Ltd., several local government and water authorities and the Government of Bahrain. The full significance of this can be appreciated only if it is realised that it has happened in the last 10 years.

THE NATURE OF APPLIED GEOMORPHOLOGY

These wide-ranging applications of geomorphology are encouraging in their diversity, for they indicate both the possibilities open to established academic practitioners as well as the potential career opportunities for future students. It must be noted, however, that applied work can be divided into two groups, (1) 'problem solving' and 'data analysis' and (2) 'problem identification and data collection with regard to a particular development'. While both types of investigation have to be carried out in a professional and competent way, they require quite different approaches from the geomorphologists involved.

Studies falling within the first category are closely related to academic research, and can be considered intellectually satisfying since they usually involve the application of the scientific method and quantitative analysis. The main problems tend to be associated with lack of both data and technical ability (see page 258).

The second type of study poses more severe problems for the geomorphologist who will normally be required to work within tight time-schedules laid down within a contractual framework. Data are often lacking, and constraints of time and cost may preclude the establishment of a well-controlled monitoring programme. In such circumstances, value-judgements frequently have to be made regarding the origin and distribution of landforms, deposits, processes and hazards. Under these conditions, a broad geomorphological background, good space perception and an ability quickly to 'read' the landscape are of the utmost importance. Thus the academic instincts of the geomorphologist have to be curbed owing to the constraints imposed by the practical situation and he must perforce rely heavily upon his intuition and experience.

At present, applied studies largely use traditional expertise, rather than research programmes involving the development and application of sophisticated new techniques. Ninety per cent of the work will normally consist of routine geomorphological surveys using established methods or evaluating information from a geomorphological standpoint. Only some 10 per cent of time is likely to be spent on innovative research and then only when a problem occurs for which there are no existing techniques of measurement, recording and analysis. Hence, for most of the time, geomorphologists will find themselves trying to do a professional job using standard methods. The lesson is that they must be technically competent and capable of working to the best standard available in the profession (i.e. they must not be negligent).

As yet, no philosophy of applied geomorphology has emerged except to provide the client with the most accurate information and guidance as quickly as possible. Much of this work is prone to criticism by other geomorphologists on the grounds of its being unstimulating, repetitive or intellectually barren (see R. J. Chorley, in this volume, p. 11). An example of this is the use of morphological mapping as a basis for choosing a road alignment and providing information prior to the costly site-investigation phase [Brunsden *et al.* 1975a and 1975b]. While these criticisms may have some validity, it must be noted that sophistication and intellectual stimulation are not necessarily prerequisites for an expertise to be of value to management bodies and commercial organisations. The nature of applied work will be determined by what clients need and not by the fads and fashions of an academic discipline. In this respect the general appreciation of landforms and their use as indicators of process distributions, hazards and materials, is an area of expertise unique to the geomorphologist which should be highly valued.

THE BASIS OF APPLIED GEOMORPHOLOGY IN BRITAIN

One possible approach to applied work by geomorphologists trained in Britain might be that of the 'Geographical geomorphologist' (R. J. Russell, 1949). In this we may differ from our colleagues in the United States who are often trained in departments of geology, but less so from those in other countries. Most geomorphologists in Britain continue to be initially trained within the structure and course-content of geography. There are at least five abilities which arise from this training:

(1) **Ability to think in spatial terms.** Geomorphologists have been trained to think about distributions and generalisations, not only of one phenomenon at a time, but of several at a time. This shows most clearly in their familiarity with maps, in terms of their construction, use and analysis.

(2) **Ability to detect spatial correlations.** A natural step for any geographer is to accept that patterns betray associations. There is no statistical test of spatial correlation which can replace experience in the study of the relationships between mapped distributions.

(3) **Ability to change our scale of thinking in accordance with the nature of the problem.** The time-honoured distinction between *site* and *situation* has a long geographical antecedence. It is particularly important in assessing the influence of factors, from outside a particular site, on the conditions or behaviour of the site itself. Flood hazard, landsliding or avalanching from outside the boundaries of a site investigation area are obvious examples. The reverse is also true, in that major changes at a specific site often lead to widespread alterations in environmental conditions which, in time, can deleteriously affect conditions at the original site or elsewhere (see, for example, Brunsden and Jones, 1972). This forms one of the bases of statements on environmental impact.

(4) **Ability to comprehend the significance of the time-dimension.** In applied geomorphological work it is critical that we should be able both to recognise and to classify features of the land surface into those which are, for the most part, a legacy of past geomorphological events, and those which are currently undergoing significant change. Furthermore, we need to be able to comprehend and predict:

(i) Recurrence intervals of natural events and potential hazards (e.g. the frequency and magnitude of floods)
(ii) Rates of change under current processes (e.g. rate of erosion)
(iii) Prediction of the possible reactivation of a past process by unwise human interference (e.g. the solifluction lobes disturbed during construction of the Sevenoaks By-pass [A. W. Skempton and J. N. Hutchinson, 1969]).

(5) **Ability to use plan documents.** One traditional strength of geomorphology lies in the sophistication and skill with which we use the plan document—whether it be plans, maps or aerial photographs. We have acquired, along with other

disciplines, increasing skills of field instrumentation, model building and testing, advanced computational techniques and data analysis; and yet many of these still need to be related to the spatial context and to recognisable landforms in order to be seen as being meaningful. In the past, the two most-used plan documents in applied geomorphology have been maps using the land-systems survey approach, generally for small-scale analysis of large areas (C. A. Christian and G. A. Stewart, 1964; J. W. F. Dowling and P. J. Beavan, 1969), and geomorphological maps for more specific investigations of sites at a large scale (J. Demek, 1972). By these means it is possible to record the basic elements of land-form materials and processes, and their development through time. The geomorphological map also provides the necessary basis for further collection of data. It facilitates production of derivative or summary statements. It allows information to be abstracted for design or planning decisions and thus provides basic planning documents.

A study may, of course, include the use of a full range of techniques including remote sensing (e.g. interpretation of air-photo and satellite images), field monitoring, statistical and laboratory analysis. Training in these techniques is of paramount importance.

PROFESSIONALISM IN GEOMORPHOLOGY

Geomorphologists undertaking work of an applied nature need to be professional in two senses: first, they need to be seen to belong to a world of professionals; secondly, they must have something to offer which makes them professionally distinct. The first is an area of discussion too large to enter into here, except to note that there are recognised professional standards, codes of conduct and rules of discipline with which geomorphologists should acquaint themselves more closely and follow, even in these early days of a growing profession (*Report of the Geological Society of London*, 1974; F. F. Sherrell, 1975). The need to be professionally distinct is a matter for our immediate concern. The nature of our distinctiveness arises directly from our training and the abilities listed earlier. There are many geomorphologists who would hesitate at this point and say that these are surely not enough to provide the basis for professional practice. Response can only be based on experience which is that these abilities have been found useful (Brunsden, Doornkamp, P. G. Fookes, Jones and J. M. H. Kelly, 1975a, 1975b) and it has recently been predicted by engineering geologists that geomorphology will be one of the growth areas for site-investigation procedure in the next 10 years (W. R. Dearman and Fookes, 1974). Work already carried out (Brunsden *et al.* 1975a, 1975b) has provided unique data, by rapid means, in answer to specific problems in a manner readily absorbed and appreciated by a client. However, future expansion of applied geomorphology will depend on creating and exploiting opportunities and on educating personnel capable of providing the expertise required.

THE FUTURE FOR APPLIED GEOMORPHOLOGY IN BRITAIN

Opportunity

The opportunities to develop applications of geomorphology can be improved by publishing appropriate papers outside those journals that have a circulation largely restricted to an audience of geomorphologists. Publication should be sought, therefore, in journals reaching engineers, engineering geologists, ecologists, regional planners and landscape architects. In addition, geomorphologists should make a greater contribution at conferences organised by engineers, hydrologists, soil scientists and engineering geologists (e.g. Brunsden, Doornkamp, L. W. Hinch and Jones, 1975). The great value of such occasions often lies in the development of personal contacts, the informal discussions and the creation of a rapport between the geomorphologists and members of other disciplines. At the same time it should be possible to learn more about these subjects and their problems, and hence become aware of the areas in which geomorphological contributions are both pertinent and welcome.

In addition attempts should be made to utilise those academic studies that have potential applications. The most recent *Register of current research in geomorphology* (*B.G.R.G.*, 1975) lists many areas of geomorphological research which, although being pursued mainly out of academic interest, are capable of being brought more extensively into applied work. These include studies of:

Wind-erosion processes—the interaction of wind turbulence, soil erodibility and erosion rates;
Landforms as indicators of sand and gravel resources;
Physical, structural and weathering properties of glacial tills;
Surface and ground-water movements within limestone;
Measurement and modelling of soil erosion;
Slope processes revealed by variations in soil properties;
Controls on rates of sheetwash;
Factors controlling stream discharge, sediment and solute production;
Sand movement in coastal dunes;
Runoff and drainage-basin modelling for transient flows.

Further details of the application of these and other studies are summarised in Cooke and Doornkamp (1974). In addition there are several studies being carried out which concentrate on the human impact upon environment. These are largely concerned with the effects of management (including planning policies) and mismanagement (either with or without related planning policies) of the earth's surface. Insofar as these are studies motivated mainly by curiosity about the effect of man, they cannot be classified as forming a part of applied geomorphology; nevertheless, they are building a collection of case studies which in later years will be used in applied geomorphology as yardsticks against which the relative wisdom of future planning policies may be assessed. The list of such studies includes:

The impact of planning policies on coastlines;
Patterns and causes of accelerated soil erosion;
Variations in fluvial processes, discharge and water-quality properties in catchments affected by urbanisation;
Effect of management and recreational pressures on 'delicate' land;
Effect of afforestation on erosion and sediment transport;
Effect of bridge-building on estuarine sedimentation;
Effect of reservoir construction on river channels;
Human response to hazards (e.g. floods, landslides, etc.).

The further development and use of such studies within a framework of planning and decision-making will depend upon,

(1) A process of self-education on the part of geomorphologists, to help in anticipating what the decision-maker would most welcome from them;
(2) The education, in turn, of the decision-maker, so that he becomes more aware of what geomorphology, *as practised by geomorphologists*, has to offer;
(3) The establishment of a body of case studies of high technical merit *in applied geomorphology*; and
(4) Finding the right methods for communicating geomorphological information to the decision-makers. Publication in 'professional' and 'trade' journals and the development of specialised mapping techniques which show engineering interpretations are among the obvious first steps.

The potential obviously exists for the future growth of applied geomorphology. Much of the recent work on processes, monitoring and mapping has obvious applications in a wide range of management situations. Thus applied geomorphology in Britain could well grow quickly to attain a status comparable to that already achieved in North America, Eastern Europe and the Soviet Union. It is important to realise, however, that this will not be achieved without effort on the part of both individual members and the corporate body of geomorphologists.

The practising geomorphologist. Future growth in applied geomorphology is likely to result in a demand for geomorphologists at two distinct levels: that of (1) the consultant, and that of (2) the full-time employee in a variety of organisations. Both of these developments are likely to influence changes within the academic discipline and will create problems for the subject as a whole.

Consultancy. Consultants will almost certainly be required to advise during the design and construction of engineering schemes, to undertake research into geomorphological problems, and to work as members of teams making surveys of natural resources. In the first instance, such consultants will continue to be drawn from the ranks of academic geomorphologists. This has been the practice for the past decade, and although no figures are available regarding the number of geomorphologists involved in such work, it is safe to assume that it is increasing. Nevertheless, several years are likely to elapse before the first full-time consultant geomorphologist joins the growing ranks of professional engineering geologists and hydrologists.

From the outset, the practising geomorphologist must follow accepted codes of professional conduct. Only in this way will he maintain the goodwill of colleagues in adjacent disciplines and ultimately also of his clients. He must be seen to be 'professional' in the best sense of the word and be prepared to develop and stand by a professional code of practice. The practising geomorphologist must also work within a clearly recognised framework. His role as a geomorphologist, and the information he can supply, must be clearly understood by himself, his colleagues, members of other professions and his clients.

The continued use of academics may result in a conflict between academic consultants, on the one hand, and full-time practitioners on the other, both competing for the same areas of work. This problem has two distinct aspects. On a moral and financial level, there is disagreement as to whether salaried academics should compete with consultants whose livelihood wholly depends on this type of work. This conflict has already been identified within geology, where university personnel have undercut full-time consultants by virtue of the fact that their overheads are smaller. There would appear to be no reason why both forms of consultancy should not co-exist harmoniously to their mutual benefit, so long as academics are only employed to provide a level of expertise that is not otherwise available.

The second aspect is rather more disturbing, and concerns the generally-held view that all matters pertaining to the surface of the Earth lie within the realm of geology, while all studies involving water are covered by hydrology. Therefore, most aspects of geomorphology *appear*, in the professional world, to be covered by these overlapping and competing disciplines, both of which already have well-developed consultancy frameworks. This loss of identity and uniqueness is one of the fundamental problems facing the future of geomorphology as a discipline, and will be considered more fully later.

Full-time employment. The second main area of demand will be for geomorphology graduates as full-time employees on the staff of engineering firms, especially as design engineers, and on other bodies, such as the Countryside Commission, concerned with the planned use of the earth's surface. Employment of this type will go initially to qualified M.Sc. and Ph.D. students, whose role will be that of professionally trained staff capable of making planning decisions. Later, or even concurrently, graduates may find employment in site investigation and design teams. Their probable roles will be to operate as (1) field technicians concerned with data recording and primary mapping, (2) as office staff providing support in the interpretation of air-photographs and in library search services, (3) as primary data analysts (e.g. hydrological records) and (4) as cartographers converting field maps to fair documents. This is a potentially significant area of growth for the subject and it is important that the technical content of University and Polytechnic degree courses be modified and improved so that training programmes will produce the right kinds of personnel to meet this opportunity when it arrives.

In this respect, geomorphology is at a disadvantage when compared with other science subjects in general and geology in particular. The position of geomorphology in Britain, as an element of geography, means that geomorphology and related technical aspects of the subject only form part, and at times a relatively small part, of undergraduate degree structures. While this may be desirable in terms of the continued production of geographers in the classic mould, who are both generalists and synthesisers, it can also be argued that it may have the effect of constraining the further development of geomorphology as a discipline.

If geomorphologists are to acquire the competence required of a full-time professional applied geomorphologist, then their standard of technical training will have to be raised to a higher level than is current in most University courses. In other words there may be valid grounds for increasing the geomorphology content in many degree courses and also for introducing some student places for specialist degrees in geomorphology. The combination of subjects offered in such courses would, of course, depend upon the type of applied geomorphologist that the course is intended to produce. Thus in the case of 'engineering geomorphology' the emphasis would be placed on various aspects of geology, soil mechanics and hydrology, in addition to geomorphology, including air-photo analysis, geomorphological mapping, laboratory, mathematical, statistical and computational techniques.

In addition the single most important, most urgently required, and, we believe, professionally most desirable development, as regards the educational element of applied geomorphology is the establishment of a Post-Graduate Institute of Geomorphology. Such an Institute could provide one- and two-year post-graduate training courses in those aspects of the subject which do not at present form a part of most undergraduates' training and its programme would be geared to the training of scientists for the professional opportunities that have been recognised in this paper. However, the task of such an Institute should not be restricted solely to the training of students. Ideally, it would also need to foster research, particularly research with a potential bearing on the applied aspects of the subject, and to involve its staff as consultants undertaking the type of work listed earlier. Such consultancy work would serve to provide a sound body of experience and case studies, in addition to establishing potential-job contacts for the graduates of the Institute. More important, however, is the fact that such an Institute would provide a tangible, widely-recognised, focal point for the future development of a major area of growth in geomorphology.

Geomorphology as a profession

As with any other science subject, the future of geomorphology will be determined increasingly by its value to the community. While the role of geomorphology in the general training of geographers has long been recognised, its potential applications have tended to be either neglected or adopted by practitioners in adjacent disciplines. It must be stressed that in a period of limited financial resources, the strength of a scientific subject is often a function of its uses. The production of academics and teachers is not enough in itself; geomorphology must have a healthy applied wing if it is to expand.

Earlier in this paper, certain strengths of British geomorphology were indicated as resulting from its position within geography. This historical affiliation, recently reinforced by the incorporation of the British Geomorphological Research Group as a study group of the Institute of British Geographers, has provided the present distinctiveness of geomorphology as a subject. However, if it is accepted that the future development of geomorphology will depend upon higher standards of technical competence and stronger links with other Earth Sciences, then the established links with geography might be seen to restrict the development of the subject. The 'human-physical' dichotomy apparent within geography not only shows signs of increasing, but also greatly reduces the influence of geography, and therefore its satellite geomorphology, in policy-making. While geomorphology in Britain has never been healthier, if measured in terms of numbers, productivity and intellectual development, the subject must still be considered immature and powerless as regards the scientific community as a whole. It is time that the 'amateur status' of the discipline was dispelled and that it began to exert an influence in the worlds of science and management.

One reason why geomorphology has failed to fulfil this role is that it is not represented by either a professional organisation or qualification. Thus the role of geomorphology has been largely subsumed by geology and hydrology in professional circles. Geomorphology is considered part of geology by the Geological Society of London, is practised by the Engineering Group of that body, and falls within the sphere of influence dominated by the Institute of Geological Sciences. Thus in certain circles geomorphology is coming to be seen as *geomorphology practised by geologists.* This situation urgently requires rectification.

While the development of applied geomorphology can assist in this direction, it will still lack strength if it continues to be publicly represented as it is at present. There are strong grounds for arguing that geomorphology should be established as a profession and be represented by an Institute. A similar situation was recently identified in geology, where a Working Party on Professional Recognition (*Geological Society of London*, 1974) reported that 'All fields of engineering and all fields of science, other than geology, are now represented by professional bodies. The evidence. . .shows that these professions have been strengthened by the existence of a representative body and their public standing has been enhanced by professionalism'.

The main differences between a learned society and a professional institute are that while a learned society is concerned with extending, refining and disseminating knowledge, and may encourage practical application, a professional body exists primarily to maintain and protect standards of training, practical competence and ethics within a particular field. The five main functions of a professional body, as identified by the Geological Society of London Working Group are:

(1) The establishment and maintenance of a code of ethics,
(2) The laying-down of standards of education and training,
(3) The representation of interests and the exertion of influence at various levels including national government, local government, and the general public, and to assist in the formulation of legislation,
(4) To advise on employment and career structures, and
(5) To assist in the general dissemination of information.

It is obvious that the growth of applied geomorphology would be greatly assisted by the establishment of an Institute whose influence would considerably exceed that of individual members or small groups of applied geomorphologists.

However, the establishment and operation of such a body would probably prove difficult with the present number of geomorphologists. It is worth noting, therefore, that the Geological Society Working Group considered geomorphology as one interest to be included in a possible future professional body. We must await the establishment of the 'Institute of Professional Geologists' to see exactly how they intend to accommodate either geomorphology, or geomorphologists, in their ranks. It is a sad reflection on the present situation that consideration of such a body should be actively proceeding without the views, and in many cases, without the knowledge of currently practising geomorphologists.

Rather than await developments from outside their subject, most geomorphologists would prefer to have a say in its evolution and growth. We recognise that there are valid reasons why it is difficult, if not impossible, to make major changes to undergraduate courses in geography in order to accommodate the professional needs of one part of the subject. Similarly, we recognise that geomorphology is a discipline distinct from geology and do not find any long-term advantage to geomorphology in becoming subsumed by geology or lost under a general 'Earth Science' title. We have also discounted an imminent professional recognition for geomorphology, with its own professional body, because its strength, in terms of practitioners and its fullness as an applied discipline, is still not great enough (though we believe that the time will come for such a development).

As an interim step, therefore, the British Geomorphological Research Group, which represents most geomorphologists in Britain, should be encouraged and supported in its efforts to establish stronger links with those societies representing geology, engineering geology, soil mechanics and hydrology, while maintaining links with geography. In addition, it seems vital for the future growth of applied geomorphology and, we would argue, of geomorphology as a whole, that the group urgently considers its possible development as a professional body, with its own qualifications and the authority to speak for geomorphology in the professional world. If this is achieved, the identity and distinctiveness of the subject will be preserved and its scientific standing greatly increased.

REFERENCES

British Geomorphological Research Group (1975), *Current research in geomorphology*.

Brunsden, D., and Jones, D. K. C. (1972), 'The morphology of degraded landslide slopes in South West Dorset', *Q. J. Engng Geol.* **5**, 205–22.

——, Doornkamp, J. C., Fookes, P. G., Jones, D. K. C., and Kelly, J. M. H. (1975a), 'Geomorphological mapping techniques in highway engineering', *J. Instn Highw. Engrs* 22, 35–41.

—— (1975b), 'Large scale geomorphological mapping and highway engineering design', *Q. J. Engng Geol.* **8**, 227–53.

—— Doornkamp, J. C., Hinch, L. W., and Jones, D. K. C. (1975), 'Geomorphological mapping and highway design', *6th Reg. Conf. Africa Soil Mech.* (Durban), 3–9.

Christian, C. A., and Stewart, G. A. (1964), 'Methodology of integrated surveys', *Conference on Principles and Methods of Integrating Aerial Survey Studies of Natural Resources for Potential Development* (*U.N.E.S.C.O.*, Toulouse), 145 pp.

Cooke, R. U. (1976), 'Urban geomorphology', *Geogr. J.* **142**, 59–65.

—— and Doornkamp J. C. (1974), *Geomorphology in environmental management* (Oxford Univ. Press), 413 pp.

Dearman, W. R. and Fookes, P. G. (1974), 'Engineering geological mapping for civil engineering practice in the United Kingdom', *Q. J. Engng Geol.* **7**, 223–56.

Demek, J. (ed.), (1972), *Manual of detailed geomorphological mapping* (Academia, Prague), 368 pp.

Douglas, I. (1976), 'Urban hydrology', *Geogr. J.* **142**, 65–72.

Dowling, J. W. F., and Beavan, P. J. (1969), 'Terrain evaluation for road engineers in developing countries', *J. Instn Highw. Engrs* **16(6)**, 5–22.

Geological Society of London (1974), *Report of the Working Party on Professional Recognition,* 28 pp.

Russell, R. J. (1949), 'Geographical geomorphology', *Ann. Ass. Am. Geogr.* **39**, 1-11.

Sherrell, F. F. (1975), 'Professional liability and professional indemnity for consultants', *Br. Geol.*, 35-6.

Skempton, A. W., and Hutchinson, J. N. (1969), 'Stability of natural slopes and embankment foundations. State-of-the-art report', *Proc. 7th int. Conf. Soil Mech.* (Mexico), 291-335.

16

THE FUTURE OF GEOMORPHOLOGY

G. H. DURY

(*University of Wisconsin, Madison*)

Four years ago, in reviewing some current trends in geomorphology, mainly of the fluvial kind, I expressed strong hesitation about forecasting the shape of the discipline in the next decade (Dury, 1972). On the present occasion, I am in the rash position of having accepted an invitation to forecast, with no time limit specified. The ten-year span, however, appears to exercise a certain fascination. It was signalised by L. D. Stamp (1966) in a review of progress entitled 'Ten Years On', and is incorporated in the 'Next Ten Years' theme currently under exploration by the British Geomorphological Research Group (BGRG). Precisely 10 years have passed since R. J. Chorley (1966) made the first major written statement of the application of statistical methods to geomorphology; it is only just, however, to recall that this statement was composed as early as 1961, being delayed in publication by the tardiness of some other contributors to the volume where it appeared.

In a general and somewhat vague manner, we look back on the 1960s, not only as a period of liberal university funding and university expansion, but also as the decade of the quantitative revolution in geography, geomorphology included. Even though I. Burton (1963) was able to claim that the revolution had already succeeded in the opening years of the decade, and to identify geomorphology as leading the van, Stamp could still recognise a front-line struggle with invading quantifiers as late as 1966. By the end of the decade, Chorley (1971, p. 95) felt able to hint at a situation of Many Techniques In Search Of A Geographical Problem.

All this is now past history, and fast fading at that. As D. R. Stoddart (1967) has clearly shown, the citation-decay rate of our literature is rapid. Nevertheless, any attempt to forecast what is to come ought to make at least some reference to the present and recent state of the art. Although in the immediate context it might seem more useful to determine where we think we have got to conceptually, than to log recognisable practical achievements, these achievements deserve at least some passing notice.

I am not intending in this connection to usurp or replicate the functions of other contributors to the present symposium, nor to review recent reviews. It is enough to observe that the recent published record in fact includes reviews of endeavour in various subdivisions of geomorphology, for instance, fluvial, glacial, periglacial, arid, coastal, karstic and humid-tropical, plus pertinent items in *Progress in Geography*. Our discipline has lodged an overt claim to an encyclopaedic body of knowledge and investigation (R. W. Fairbridge, 1968). It has grown conscious of its history (Chorley, A. J. Dunn, and R. P. Beckinsale, 1964, 1973). It has produced specialised symposium collections, including those of the British Geomorphological Research Group, the Institute of British Geographers, the Supplementbände of the *Zeitschrift für Geomorphologie,* and the Binghamton series. Specialised texts dealing with particular subdivisions are now available: many of them have appeared during the last 10 years. Some contributions regarded as fundamental are being recycled, for example in compressed form in the *Geographical Readings* series published by Macmillan, and in selected to *extenso* form in the *Benchmark Papers in Geology* series of Dowden, Hutchinson and Ross.

The annual output of publications of direct geomorphological interest ran at a minimum of 60 000 items in 1970. Depending on the growth rate assumed, it seems capable of reaching 200 000 or 250 000 items by the end of the century. Foreseeable annual totals would certainly be increased, were allowance made for languages underrepresented, or not represented at all, in *Geomorphological Abstracts*. One suspects that publication in Japanese and Russian is underrepresented: and who can guess what allowance should be made for the potential one-quarter of the world's geomorphologists who publish in Chinese?

We have promoted a successful revolution, both technical and conceptual. We have developed professional self-consciousness. Our profession has produced, and continues increasingly to produce, an abundant literature of its very own. In natural science, needless to say, no actual achievement can ever be enough. But, this consideration apart, are we, within reason, gratified by our present status and perceived prospects? Some of us are certainly not.

The first half of the present decade has produced several expressions of grave unease. Chorley, having earlier identified the conceptual vacuum created by the breakdown of diastrophism and of the Davisian basis of geomorphology, later recognised an increasingly severe dilemma (Chorley, 1963, 1971). He heralded the danger that geomorphology (and climatology) might split into descriptive-teaching and dynamic-researching forms. Where geomorphology and climatology are taught mainly in departments of geography, the danger is that research may be mainly located elsewhere, for example, in geology, engineering or meteorology. J. N. Jennings (1973), declaiming against bandwagons and gimmickry, saw no salvation in mere fashion. K. M. Clayton (1973) recorded his disappointment with the geomorphological activities of the International Geographical Union which, despite the activities of the International Geological Union and of INQUA, he still took as the primary international forum for geomorphologists. He found that discussion of recent geomorphological advances fell flat, and concluded that philosophical debate at international congresses simply does not start. We must be duty bound to hope for something better from the international meetings organised by the BGRG, the present symposium included. In common with Chorley, E. H. Brown (1975) has concerned himself with the relationships and interrelationships of physical geography. He has identified both imbalance and a centrifugal tendency. The imbalance consists in the proportion of landscape (too much) to atmosphere (too little). Brown questioned whether human geography any longer needs physical geography, and regarded the operation of research funding as promoting centrifugal separation of the physical from the human aspects. The discussion of his address revealed a complete range of opinion, from deprecation of the physical/human split to complete lack of interest in the name under which research is conducted.

It seems highly ironical in retrospect that the bubbling self-confidence of geomorphologists in the 1950s should have been overtaken by a suspicion of being not really wanted. Human geographers in the 1950s were able to claim that, without a population, a given area of the earth's surface could have no geography. Their geomorphological colleagues, totally untroubled by considerations of population density, did not bother to point out that an unpeopled area merely represents an end-member condition. They were satisfied by the existence of a physical landscape. Now, however, the massive quantitative development of the geography of transport, cities, economic activity, industrial location and the rest enables human geographers to operate as if geomorphology did not exist. The hypothetical featureless plain, already being styled by some writers a plane, seems not only a simplification, a model or an abstraction, but also, in the most complete sense, an ideal.

PROJECTION IN THE SHORT-TERM

A short-term forecast, then, which in the fashion of the time we might describe as a linear model, would seem of necessity to allow for fission of the geographical discipline, wherever (as in the U.K.) geomorphology is largely contained within geography. For a time, the Geologists' Association were reluctant to accept geomorphological papers, and this might have occasioned some driving-back of geomorphology into geography and forced some geomorphologists, temporarily, at least, into the unenviable position of having to choose between geographical and other audiences.

A short-term forecast, or linear model, could attempt to predict whither geomorphology probably will go, say within the next 5 or 10 years. It could incorporate an index of preference, or an index of change in preference. Index values of preference for sundry variations of geomorphology can be derived from records of recent publication and of research in progress. For Britain, the B.G.R.G. *Register of Current Research* deals overtly with the popularity stakes (see, for instance, A. M. D. Gemmel, 1975). A more elaborate index could be derived from the listings in *Geomorphological Abstracts.* Although, as already suggested, there may be some linguistic bias in the latter one can easily recognise the national preoccupations mentioned on various occasions by Chorley and Clayton. These include process studies in North America, climatic geomorphology in France and Germany, glacial and periglacial morphology in Poland and applied geomorphology in the U.S.S.R. In the U.K. at the present time, applied geomorphology seems to be competing for attention against more traditional forms of study. Changes in national preoccupation ought surely to be amenable to measurement.

Mere counting and classification of recent publications and current research projects is not, however, enough. With respect to counting, D. S. Price (1975) makes much of Lotka's *Law of Scientific Productivity* which states that the number of scientists with just n publications is proportional to $1/n^2$. For a given group, about half the total output of papers will come from the highly productive élite whose number roughly equals the square root of the total number in the group. The latest (1975) B.G.R.G. *Register of Current Research* identifies 420 workers. The editor suspects that the total may be unduly low, on account of questionnaire fatigue (or possibly allergy?). Let us for the sake of argument assume a 35 per cent shortfall in returns. The actual number of workers then becomes about 640. According to the *Law of Scientific Productivity* and its corollaries, some twenty-five of these can be expected to produce one-half of the total research output of the group, and thus to define the overt direction of British geomorphology in the short (and possibly also the medium) term.

But impact is not necessarily proportional to volume. How many of the publications of R. E. Horton, aside from his 1945 paper, do we customarily cite? We must hope, too, that the volume/impact relationship is poor, in respect of those who publish much the same material, or even identical material, in more than one place. A measure of the impact of published research is derivable from the *Citation Index*. Anyone who wishes is welcome to promote two master's theses, on the relationship, if any, between impact as indicated by the *Citation Index* and volume of publication as indicated by the record. The first thesis would deal with the current impact/volume relationship, and would on this basis forecast the direction of geomorphology for the next 10 years. The second thesis, written 10 years later, would explain how and why the forecasts of the first thesis went wrong.

One drastic short-term problem, which any forecast must take into account, is that in some respects and in some countries, geomorphologists have been outflanked, as, indeed, have geographers in general. I heard an explicit statement of the claim that 'a good geographer can do anything' as late as 1962. The claim is of course nonsensical. On the other hand, the principle that a good geographer, or good geomorphologist, ought to be able to command respect in a second discipline, seems reasonable enough. I will pass over the current earnest attempts by geographers in Britain to secure recognition of their social usefulness. Moreover, I will freely admit to speaking from a position of privilege, if not indeed of prejudice, since my own employment has in the past been both full-time geography and full-time geology. It has now settled into a split. Be all this as it may: the ecological/environmental furore of the late 1960s and early 1970s has forced us to recognise that researchers under environmental names other than our own have, for a considerable time now, been doing some of the things which we take as falling within our own special province. Still more disconcertingly, in some ways they have been doing these things better than we have. The circumstance that ecology in the biosciences, after a generation or two in the outer fringes of scorn, has made a vigorous comeback, represents a gigantic straw in the environmentalist wind.

One possible reaction on the part of physical geographers is to strike alliances with workers outside. Brown (1975) has drawn explicit attention to the dangers of looking elsewhere for intellectual respectability. A clearly directed question is, if you cannot beat them, and they will not let you join them, then what do you do?

Much in this connection seems to depend on university structure. At the high risk (say, in the 90 to 99 per cent range) of proving unpopular, and at the perhaps total risk of seeming invidious, I will now state that the typical U.S. university structure impresses me as being far better able than the typical British structure to accommodate disciplinary shifts in the short-term. Although many U.S. departments of geography appear formerly to have decided (perhaps during the 1950s?) that physical geography was not for them, or not for them very much, the geomorphological situation remains far from hopeless.

Thorough grounding for physical postgraduates can be ensured through physically-relevant courses taken in other departments. The heavy reliance on formal coursework for postgraduates in the U.S., both at the master's and at the doctor's level, has at least three important facets. On the negative side, coursework is necessary to compensate for low baccalaureate standards. First, on the positive side, formal instruction of post-graduates brings the products of the U.S. and U.K. systems into approximate parity at completion of the master's degree, but in the U.S. predisposes a candidate to accept more formal instruction still. Thus, secondly on the positive side, a doctoral candidate in the U.S. system is exposed to the operations of at least two university departments, and by no means rarely to the operations of five or six. The sampling involved here is not shallow sampling, but sampling in depth. A postgraduate geomorphologist wishing to deal in palynology will be formally trained in palynology for one whole year. Those U.S. departments of geography which have retained or reclaimed their physical interests can call, as of right, on outside assistance. Their postgraduates enrol, as of right, in outside courses.

At the same time, the combination of centrifugal push from inside with centrifugal pull from outside has its risks. These are more serious in some ways than the risks run where geography, surviving more or less intact, is threatened by hiving-off. The advances of fluvial morphology during the last generation owe very much, by demonstration, to the researches and research programmes of Horton, Strahler, Leopold, Langbein and their associates, advancing almost exclusively beneath the geological banner. It is probably fair to say that geomorphologists in the U.K., based mainly in geography and working in an environment of geological thought where every rock from Cambrian onwards can be regarded as drift, freely incline to publish in geographical serials. Geomorphologists in the U.S. are based considerably, if not indeed principally, in geology. Geomorphological research is published mainly in geological and hydrological serials. The split of audience, as observed by geographers, could be taken to imply an already split discipline.

Nothing said here is meant to derogate certain noteworthy efforts of cooperation by academics of contrasted sorts. The linkage between geomorphology (in geography) and civil engineering at Ottawa comes readily to mind as a happy illustration. But there does seem to exist an important structural difference between *ad hoc* collaboration among congenial colleagues with shared interests, and the *de jure* entry of postgraduates into more than one departmental programme. Without pursuing this particular theme into the attitudes and practices prevalent on the European mainland, I may perhaps recall an earlier discussion of the prospect that physical geographers and physical geologists may find themselves driven together, *nolens volens*, by external pressures (Dury, 1970). Desirable though the end-result might prove, it would be preferable if cooperation were from choice. We need something superior to the defensive reactions of the minor kings of early England, when invasion menaced. We might perhaps rephrase Haldane's definition of love, and say that academic union could represent a synthesis of commitment and liking, arising out of common interest.

PROJECTION IN THE MEDIUM TERM

A medium-term projection might attempt to deal with the next 10 to 25 years, that is, with the remainder of the present century. It could consider whither geomorphology might go. The required model is a diffusion/infusion model. It should predict the increasing improvement of geomorphology itself, as discoveries continue to be diffused through the discipline, and as their implications continue to be explored. It should also predict the continuing impact of other disciplines on geomorphology, and of geomorphology on other disciplines. It should allow for at least a partial breakdown of interdisciplinary barriers.

Basic research is by nature slow. A new fundamental idea may demand more than a generation for thorough scrutiny and development. In this connection we think immediately of Horton, who began stating his analytical ideas in the 1930s and arrived at his final statement in 1945. Stream network topology is, by demonstration, still being actively investigated today. Although the vogue of inventing new ordering techniques bears some resemblance to an earlier vogue of inventing new map projections (and I plead guilty myself to having invented a Mercator projection with two standard parallels, but claim credit for never having published it), there can be no doubt that the continuing study of stream nets is, at its best, a desirable investigation of the systematic statistical operations of nature.

Several other directions of research, already followed by some but likely to require lengthy exploration, are clearly evident. Because of the time likely to be needed, they are more properly mentioned for the medium term than for the short term. I will name two.

Tropical geomorphology promises great new things, chiefly with the aid of radar. The RADAM project in Brazil is producing laydowns and mosaics, plus maps of geology, soils, plant ecology and geomorphology. The undertaking is synthetic from its outset. At the very minimum, we shall gain a vast bulk of new information. The intensive character of tropical weathering must surely mean that the geomorphological results will represent some kind of climax situation.

The second direction is that of palaeomorphology in relation to palaeoclimatology. Although some geomorphologists deplore attention to the record of the past, palaeoclimatologists and historical geologists incline to regard the present as the point to which the past has so far extended. Past conditions of climate and of morphogenesis have differed by at least an order of magnitude from conditions observable today. Thus, in a sense, the past record provides us with the only large-scale experimental data that can be obtained for systems outside those which we now observe. Medium-term progress can be hoped to include increased investigation of the geomorphological effects of climatic shift, with a view to predicting future effects of assumed or predicted future climates. The complexity of the evidence that can be employed is such as to ensure a long-drawn-out undertaking.

In both of the directions indicated, the diffusion/infusion process is already working, even if in somewhat subsidiary form. It could become primary. Despite the alarm at tendencies of fission, we can still argue that it is good in principle for geomorphology to reach out beyond its existing bounds. Equally, we may not deprecate the injection of ideas and methods from outside. We might then expect that, in the medium term, differences among so-called related disciplines will become increasingly blurred. But cross-fertilisation, effected in the real world by accident, by goodwill or even by nervousness, is not enough in itself. It requires structural accommodation.

Universities such as Macquarie in Australia and East Anglia in England, by declining to adopt the traditional departmental structure, secured great flexibility for themselves. But even where traditional departments exist, increase of flexibility is by no means impossible to contrive.

I have mentioned a *rapprochement* between geography and geology under the short-term head, because it has already occurred in my own university. However, it provides an essential introduction to further likely developments in the medium term. The development about to be described could equally well have involved, say, geography and history, or geography and economics.

At Madison, as at many other places, geology and geography were originally one. They split apart as knowledge accumulated and as perceived interest diverged. Some reconvergence has now been effected, by provision for joint Ph.D. majors. Although such degrees are provided for by several other departments, Geography and Geology have not united in the direction of a double-major doctorate until the last 2 years. There is now one doctoral candidate well on the way to a Ph.D. in Geography/Geology. By the time that this paper appears, it is highly likely that there will be a second candidate, this time in Geology/Geography. It is scarcely necessary to add that both candidates are geomorphologists.

Joint or other interdepartmental direction of postgraduates is likely to be easiest where academics are cross-appointed to more than one department. Of the twenty geographers at Madison, eight hold additional appointments outside Geography, eleven additional appointments in all, involving eight outside units. Supra-departmental direction, although a different matter, can also be accommodated. Two chief means of accommodation are already in practice: the supersession of the former organisation and grouping of departments by some new kind of structure, and the provision for formal cooperation on a university-wide basis.

Acts of, and proposals for, supersession seem to direct a particularly severe general threat against geography, which is especially liable to dismemberment, and a particular threat against geomorphology, insofar as this has involved itself in environmental affairs. The powerful sense of territoriality displayed by a typical department is kept in check, and is protected against assault, by mechanisms designed to suppress undue overlap of claimed interest and actual activity. Most university structures betray minor strains, but even minor structural failure appears to be uncommon. However, proposals for drastic reorganisation in an existing departmentalised university are likely to meet determined opposition. Geographers fear complete obliteration. Geomorphologists look with apprehension on a potential, mere-handmaiden role. More generally, a drastic structural reorganisation may prove extremely difficult or even impossible to bring about. It could also be undesirable, if it meant the production of bachelors' degrees which, on account of unfamiliar names, had low values in the employment market. At the postgraduate level, a different situation prevails. It is already possible to integrate geomorphology–and, for that matter, geography in general–into postgraduate studies with ostensibly non-geographical names.

For the sake of British readers, I need here to digress into contrasts of typical university structure. In the U.S. sense, the faculty are the academics. A Faculty in the British sense (e.g. Arts, Science) is in the U.S. sense a College (e.g. Letters and Science, Engineering). Arrangements can easily be made in the U.S. for postgraduate training to cut across the College structure (in the U.S. sense). Thus the University of Wisconsin–Madison has recently developed a postgraduate programme in Land Resources, leading to the M.S. and Ph.D. degrees. It demands formal study in (a) natural and physical sciences, (b) social sciences and humanities, (c) methodology and (d) an individual area of programme focus. The first two Ph.Ds to complete will probably be two candidates who began as geographers. One is concentrating on environmental mediation, the other on a geographical data bank. Their formal coursework ranges from computer retrieval systems to environmental law.

Arrangements of this kind constitute a response to the (usually unstated) fact that a traditional department is too limited for what it needs to do. The supra-college (committee degree) programmes greatly broaden the scope of postgraduate study, while allowing departments to retain their identities. They also permit a given department to contribute to the training of postgraduates who would otherwise be far outside the departmental range. In addition, they bring highly disparate faculty together on the advising and examining committees. The Land Resources Programme is only one of several of its sort already in existence at Madison. The medium term promises further additions to the list of committee degrees. Geography in general, and geomorphology in particular, has an evident stake in developments of the kind indicated.

Coming back to the more general context, and concentrating on geomorphology, we might attempt to refine the predictions of our diffusion/infusion model by at least two means. We could scrutinise the orientation of the current crop of postgraduate geomorphologists, and could speculate on the likely medium-term impact of project funding.

Postgraduate training amounts, at least in part, to trial by ordeal. It also presumably means commitment to some central idea. And, for the Ph.D. candidate at least, it means living closely with this idea for some time. On the principle that a Ph.D. thesis is worth little if it does not raise at least as many questions as it answers, we might suppose that the general direction of a thesis will point the general research direction of its author during the short term of the first post-doctoral decade, and that a drastic switch of subsequent direction is on the whole unlikely. I have very little hard information to offer in this connection. Perhaps a third master's candidate could be induced to make the necessary investigation? But if the supposition is in any way correct, then the medium-term direction of geomorphology is already predictable, even if only in somewhat linear terms.

It is probably needless to stress that research possibilities today depend very largely on money. R. F. Swindel and T. O. Perry (1975) have directed attention to an important corollary of the Golden Rule of Arts and Sciences (GRASS:

discussed further below), namely, Whoever Has The Gold Makes The Rules. I do not propose to discuss the extent to which projected research in geomorphology falls ambivalently between the social sciences (to which geography is thought to belong) and the physical sciences (whose adherents mainly fail to comprehend what geography is, and who are usually suspicious of geomorphology located outside geology). Plenty of loud expressions of recent dissatisfaction with the whole funding process, and especially with its methods of project selection, are on record on both sides of the Atlantic.

It seems inevitable that, as research expenses increase, and as central control over funding becomes increasingly firm, proposers will be increasingly inclined to play it safe. Especially is this so where projects funded on an annual basis appear to demand, or do in fact explicitly demand, concrete results by the year's end. Hence, the very writing, not to speak of the adjudication, of research proposals could well reflect the kind of influence identified by Swindel and Perry (1975) as operating on the writing and adjudication of research papers. Although the discussion of these two writers is meant as satire, it is telling enough to be painful. Beginning with the phenomenon that the better a manuscript submitted for publication, the worse the reaction from the referees, they arrive at the Golden Rule of Arts and Sciences: the probability that a manuscript will be accepted by referees is related to the quality of the manuscript by a Gaussian acceptance curve. Proofs include the fact that in a particular field there is a small but dependable trickle of unbelievably poor publications, and that mankind is hardly more diligent in anything than the suppression of master craftsmen and masterpieces. Tail probabilities are therefore small. A corollary is that if one's research is well received, it is mediocre. It merely characterises the discipline without advancing it. In-house editors, attempting to maximise the chance that a manuscript will be published, improve the poor papers and botch the good ones. The general result, according to Swindel and Perry, follows from the circumstance that referees are recognised leaders in their professional fields, having written the books and papers of a few years ago. They are still active and thus concerned about their own careers. They are unlikely to approve work far better than their own, whereby they would diminish their own life's work.

Here, then, is another practical means of refining our medium-term projection. Let us attempt to identify the likely editors and referees of geomorphology in the 1980s and 1990s.

Having now lived and worked through a medium-term span as here defined, I propose to end this section on a distinctly sour note. Simply because the medium-term spans a generation or so, geomorphologists developing their researches near the end of the term seem capable of overlooking what happened near the beginning. The known citation-decay rate implies a serious risk that former research results will be neglected. As P. E. James (1967) has said, 'One major source of error that impedes geographic scholarship is the failure of too many geographers to read what other geographers, past and present, have written'.

Just as every new lamb imagines that it has made an original discovery of spring, so in geomorphology, the passage of a generation seems enough to permit certain advanced ideas, or the advanced treatment of certain themes, to be forgotten and then discovered anew. Just as needless repetition is by its nature wasteful, so is the overlooking of pertinent previous work both careless and offensive. The rapid progress of our discipline ought not to obscure the fact that some early works, even those more than 20 years old, are still worth reading and still worth citing. Although each generation may continue to write its own history, it would clearly be economic if each generation would read, as opposed to partly rediscovering, its own geomorphology.

PROJECTION IN THE LONG TERM

It will have been clear that, in suggesting means for predicting the future of geomorphology in the medium-term, I have not attempted to forecast whither it ought to go. Doubtless in common with many readers, I regret the opulent days of the 1960s, when generous research funding could be provided in the hope, by no means always unjustified, that something would happen, and when Luna Leopold could happily say of fluvial morphologists, 'We're just a bunch of goddam mavericks!' Those days, alas, seem gone for ever. The prospective increasing regimentation of research offers a grim prospect for the long-term, for which we might look to forecast whither geomorphology could go.

Or does it? I most earnestly hope that it does not. Swindel and Perry (1975) maintain that each of us tends to produce manuscripts of similar quality throughout our professional careers, on the principle that wild variation of quality on the part of a given individual is highly improbable. Some of us may feel inclined to argue this principle, not only in the context of the developmental stages examined by Price (1975), but also on the basis of built-in obsolescence. But, if the principle is at all valid, it suggests that the linear and diffusion/infusion models offered for forecasts in the short- and medium-terms could be reliable. We seem to me bound to hope that they are anything but reliable, and that the long-term effect will in fact be step-functional.

We are accustomed to paying tribute to ideas whose time has come. It is easy enough to look back through the history of science, including the history of geomorphology, and to identify starts which were false, merely because a particular thinker was ahead of his time. Regrettable though the fact may be, discontinuity means a new start. Cadastral survey was a distinct potential in Roman days, and was indeed practised in ancient Egypt. Modern cadastral survey in the western world seems to owe nothing, either to Egypt or to Rome. The form of the earth needed to be rediscovered, centuries after the day of Eratosthenes.

For our own day, we can readily enough identify ideas and research themes whose time has already come, quantitative geomorphology, data-processing, computer modelling, systems description and analysis, plus major topics on which enormous labour remains to be expended. Whether or not the required labour will be expended on the assimilation of the Seventh Approximation, to my mind, a wholly necessary exercise in linguistics, time has yet to prove. If the proof is negative, then the Approximation's concept is likely to seem ahead of its time. To the time-demanding topics already mentioned may be added climatic geomorphology and the implications of plate tectonics for geomorphology. But is there to be nothing further? Think of chemistry, which a generation or more ago could be regarded as having successfully defined all its basic terms.

Apparent stagnation has been succeeded by vigorous and productive revolution. Mathematics has entirely reshaped itself in our own lifetime. Physics, however troubled by having to deal with particles which seem to be in two places at once, or with substances which possess negative mass or negative gravity, has also changed out of recognition. Geology has been revolutionised by plate tectonics.

If we, as geomorphologists, conclude that our presently available conceptual designs are the ultimate best, we shall surely stagnate. We shall be doing more of the same in the short term, and more of the same but different in the medium term. Let us ask, with Rupert Brooke's fish, 'Can This Be All?'–and trust that the answer is to be a resounding *No*! Let us hope that, 25 years from now, those of us who survive will be doing something very different from what we are doing today.

Discussing scientific revolutions, T. S. Kuhn (1970: pp. 92, 97, 149, 150–151) cites Max Planck as remarking, sadly, that scientific truth does not triumph by convincing its opponents, but because these opponents eventually die. Kuhn also cites Karl Popper to the effect that ultimate verification is not possible, whereas falsification, the test with negative outcome, demands the rejection of established theory. Recognised and stubborn anomalies, which cannot be assimilated into existing paradigms, lead eventually to new theories. The necessary revolutions are inaugurated by small groups, instructed by established paradigms but impressed by a sense of malfunction. Here, then, is the revolutionary situation. Rather than seeking to identify those who cannot wait to inherit the land, we should seek out those who are, with reason, dissatisfied.

There seems to be no other way of identifying the vector of the hoped-for leap. Especially is this so, since several fundamental scientific discoveries were purely matters of luck. I have myself long been an advocate of brilliant strokes of luck in geomorphological research. Unless our collective luck has finally run out, we must hope for further jumps in our direction and our thinking. If we could readily foresee the next new revolution in advance, it would come as no surprise. If no surprises are in prospect, I consider us lost. We are, I suggest, obliged to hope that surprises are in fact in prospect, and that a forecast of surprise in the long term will not be falsified by events.

REFERENCES

Brown, E. H. (1975), 'The content and relationships of physical geography', *Geogr. J.* **141**, 35–40.

Burton, I. (1963), 'The quantitative revolution and theoretical geography', *Can. Geogr.* 7, 151–62.

Chorley, R. J. (1963), 'Diastrophic background to twentieth-century geomorphological thought', *Bull. geol. Soc. Am.* **74**, 953–70.

—— (1966), 'The application of statistical methods to geomorphology', in G. H. Dury (ed.), *Essays in geomorphology* (Heinemann, London), 257–387.

—— (1971), 'The role and relations of physical geography', *Progr. Geogr.* 2, 87–109.

——, Dunn, A. J., and Beckinsale, R. P. (1964), *The history of the study of landforms, vol. 1: Geomorphology before Davis* (Methuen, London), 678 pp.

——, Dunn, A. J., and Beckinsale, R. P. (1973), *The history of the study of landforms, vol. 2: Life and work of William Morris Davis* (Methuen, London), 874 pp.

Clayton, K. M. (1973), '22nd International Geographical Congress, Montreal, August 1972', *Z. Geomorph.* N.F. **17**, 246–9.
Dury, G. H. (1970), 'Merely from nervousness', *Area* **4**, 29–32.
—— (1972), 'Some current trends in geomorphology', *Earth Sci. Rev.* **8**, 45–72.
Fairbridge, R. W. (ed.) (1968), *The encyclopedia of geomorphology* (Reinhold, New York), 1295 pp.
Gemmel, A. M. D. (ed.) (1975), *Current research in geomorphology* (Geo-Abstracts, Norwich), 75 pp.
James, P. E. (1967), 'On the origin and persistence of error in geography', *Ann. Ass. Am. Geogr.* **57**, 1–25.
Jennings, J. N. (1973), ' "Any milleniums today, lady?"—the geomorphic band-waggon parade', *Aust. geogr. Stud.* **11**, 115–33.
Kuhn, T. S. (1970), *The structure of scientific revolutions* (Univ. of Chicago Press), 210 pp.
Price, D. de S. (1975), 'The productivity of research scientists', *1975 Year-book of Science and the Future* (Encyclopaedia Britannica, Inc., Chicago), 409–21.
Stamp, L. D. (1966), 'Ten Years On', *Trans. Inst. Br. Geogr.* **40**, 11–20.
Stoddart, D. R. (1967), 'Growth and structure of geography', *Trans. Inst. Br. Geogr.* **41**, 1–19.
Swindel, R. F., and Perry, T. O. (1975), 'A previously unannounced form of the Gaussian distribution: the golden rule of arts and sciences', *J. Irreproduc. Results* **31**, 8–9.

INDEX

Abney level, 77
aggradation, 25–29, 32, 34–36, 38
aggregate resources assessment, 217–218, 252, 256
aggregate exploitation, impact of, 230–231, 234–236, 242
air-photographs, 85, 150, 229, 242
  use in mapping coastal features, 229, 242
  use in mapping mass movement features, 85
  use in mapping permafrost, 150
algae, 213
algae reefs, 213
alas, 142–144
Alleröd soil, 157
alluvium, 25–26, 32–33
animals, geomorphological impact of, 78, 212
anthropogenic influences, 33–34, 37, 49–50, 56, 63–65, 79, 171–173, 207, 213, 219–220, 230–231, 234–236, 242
applied geomorphology, 74, 76, 79, 84–87, 149–151, 172–173, 189, 217–218, 220, 224, 230–233, 235–236, 251–261
arènes, 194–195
arenisation, 194–195
arroyo, 25–39
asymmetrical valleys, 157
automatic monitoring, 98, 100–102
  of landslides, 101
  of mudslides, 98, 100
  of pore-water pressure, 102
avalanching, 165, 254

bankfull capacity, 57
bankfull discharge, 56
barrages, 232
barrier bar, 204
base level, changes in, 34, 37–39, 41
basins, glacially over-deepened, 108, 110, 113–115, 119
baydarakh, 143
beach change studies, 228
beaches, 207, 210, 228, 230–231, 234–236, 242
beach erosion,
  due to hurricanes, 210
  due to man, 234–236
beach materials,
  long-shore grading, 236
  studies of, 207, 235–236, 242
beach replenishment, 230–231, 236
bedrock meanders, 57
Bingham rheological model, 103
bio-erosion, 212–213
biochemical planation, 195
biological deposition, 207, 212–213
biological erosion, see bio-erosion
biostasy-rhexistasy concept, 171, 189, 192
bog-bursts, 90, 93
borehole investigations, 233, 236, 241
boundary roughness, 208
British Geomorphological Research Group, 1, 43, 45, 73, 75, 132–133, 136, 243, 251, 259, 261, 263–264
buried channels, 57, 241

cartographic analysis of slopes, 77
catchments, 48–55, 167–168, 189–190
  experimental, 48–55
  instrumentation of, 48
  instrumented, 167–168, 189–190
  limestone, 48, 53, 60
cavitation, 125, 127
cementing agents, bio-chemical, 207, 212
Chalk landforms, 80, 157–158, 190, 194–195
channel capacity changes, 63–65
channel changes, 61–63
channel cross-sections, 28–30, 35, 56, 61
channel deposition, 25–29, 32, 34–36, 38, 56
channel erosion, 32, 55, 61
channel geometry, 20, 56, 61, 63
channel gradient, 26, 28, 30–32, 34, 36–37, 61
channel networks, 20, 57, 63
channel patterns, 57, 61, 63
channel size and climate, 174
channel stability, 164
channel width variations, 34, 37–38, 56, 64
channels,
  anastomosing, 169
  braided, 56
  meandering, 56, 61
chemical denudation, 60
chemograph, 54–55
cirque, 118
Citation Index, 266
citation-decay rate, 263, 271–272
Clay-with-Flints, 189, 194
cliffs, 76, 158, 204, 206, 237, 239–240, 242
  and sea-level change, 80
  erosion of, 237, 240, 242
  frost-riven, 158

climate and landforms, 163–167, 169–171, 174, 176, 186–189
climatic geomorphology, 6–7, 272
coastal defence, 231–237, 239
coastal erosion, 237–238
  see also beach erosion
coastal processes, 207–214, 225–229, 242
coastal terraces, 206
coastal types, 215
computer block-diagram, 79
conduits, 48
consultancy, 257–259
continental shelf types, 207
control systems, 45
control theory, 21
convexities, formation of, 78
coombe rock, 154, 157
coral reefs, 213
cryergic morphogenesis, 155
cryogenic phenomena, 139–151
  alas, 142–144
  baydjarakh, 143
  cryopediment, 140, 149–150, 157–158
  dell, 148–149
  duyoda, 143
  head, 139, 154, 156–159
  ice wedges, 141–143, 147–148, 155
  ice wedge polygons, 141, 143, 155
  khonu, 143–144
  pingo, 141, 143–144, 155–156
  rampart, 142
  relict, 147, 155–159
  solifluction, 143, 147–149, 154, 254
  talik, 141, 144, 147–148, 150
  thermocirque, 142
  thermokarst, 142–144, 147, 150, 155–156
  influence of lithology, 158
cryogenic texture, 144–147
cryonivial morphogenesis, 157–159
cryopediment, 140, 149, 150, 157–158
cryopedimentation, 140, 149
cryoplanation terrace, 157–159
cusps, 227
cycle of erosion, 4–5, 40–41, 44, 185–186

dams, effect of, 33–34, 37, 63, 65, 150–151, 207, 219–220, 256
dead-ice features, 121–122
debris slide/debris flow, 90, 92, 103
dell, 148–149
deltas,
  coastal, 207-208, 220
  inland, 162
  lagoonal, 204
deltaic coast, 204
denudation chronology, 5, 8, 40–41, 80, 155–159, 164, 188–189, 192–196, 240–241, 268
denudation rates, 78–79, 163, 169–170, 190–193
diagenesis, in coastal deposits, 211–212
differential erosion, 193, 206
discharge, 47 ff.
  determination, 48
  impact of urbanisation on, 50, 63–64, 256
  values, major drainage systems, 207
  temporal variations, 26, 50
discharge and sediment supply, 206–207
drainage basin,
  as an ecosystem, 167
  characteristics, 45, 60
  erosion studies, 51–56, 60, 167–168, 189–190
  morphometry, 45–47, 59
drainage density, 169
drainage evolution, 5, 164
drainage network, 45–47
  variable definition of, 46
  variation over time, 46
drainage systems of the world, 207
dry valley, 45, 66, 148, 157
duricrusts/duricrusting, 187–188, 190–191
duyoda, 143

earthflow/earthslide, 89, 165
earthquakes, impact of, 218–219
ecology/ecological conditions, 166–168, 173–174
ecosystem, tropical, 166–168, 190
edge-waves, 227–228, 235
effective stress, 89–91, 93, 95
employment potential, 136, 252, 258–259
engineering and geomorphology, 86–87, 172, 231–234, 252, 256–258
englacial drainage, 119
entropy modelling, 17
erosion rates, 51–53, 60, 166
  see also denudation rates
erosion surfaces, 80, 159, 164, 169, 174, 176, 241
escarpment retreat, 194
esker, 120–121
etch-planation, 186–187, 193–195
evapotranspiration, 58, 167
exfoliation, 163
experimental catchments, 48–55

fan, 30–31, 37
fjord, 110, 119
flocculation, 207
flood-plain changes, 62
flood-plain morphology, 62–63
flood-produced landforms, 41–42
flood/floods,
  affects on flood-plain morphology, 63
  hazard mapping, 254

impact of, 41–42
influence of urbanisation, 50
mean annual, 47
variation downstream, 56
flow slides, 91
fluvial landforms, 61–65
fluvioglacial
deposition, 119–125
deposits, 123
erosion, 119–125
forest nutrient cycling, 167–168
Fourier analysis, 227
frost action, 118, 139–151, 154–159
relative importance of, 166
frost creep, 149, 156, 158

gauging stations, 43–44, 54, 56
gelifluction deposits, 156–158
gelifraction, 158
geography and geomorphology, 252–255, 258–259, 264–267, 269–271
Geological Society of London, 251, 260
geomorphological mapping/maps, 85–87, 150, 164, 172, 253
geomorphologists, employment potential, 136, 252, 258–259
geomorphology, relationship with other disciplines, 258–261, 265–267, 269–271
geomorphology, teaching of, 1, 73, 252, 254–255, 258–259, 267, 269–270
geomorphology and geography, 252–255, 258–259, 264–267, 269–271
geomorphology training, 254–255
gibbsite, 170
glacial deposition, 119–125
glacial deposits,
volume of, 119
glacial drainage, 119–123
englacial, 119
supraglacial, 119, 121
subglacial, 120–121, 123
glacial erosion, 108–119, 125–129, 195–196
influence of jointing, 108, 110, 116
influence of lithology, 108–109, 115
influence of pre-glacial weathering, 195–196
glacial geomorphology,
analysis of research, 132–133
glacial landforms, 108 ff.
basins, 108, 110, 113–115, 119
cirques, 118
fjords, 110, 119
hanging valleys, 113–114, 116–117
hill-and-basin forms, 108, 195–196
slopes, 108–110
strandflat, 118
valley-side benches, 110, 116, 121–124
valleys, 109–117
glacial scouring, 125–128
glacially over-deepened basins, 108, 110, 113–115, 119
glaciated valleys, 109–117
cross profiles, 110–116
hanging, 113–114, 116–117
longitudinal profiles, 110, 112–113
steps in, 110, 112–113
Golden Rule of Arts and Sciences, 270–271
grade, 37–38
granite/granite landforms, 158–159, 163, 169, 187, 195
growan, 194
groynes, 231
gully/gullies, 26 ff.
bed material, 29, 32–33
cross-sections, 28–30, 35
development, 57, 64, 148, 173
discontinuous, 28–32, 34, 37–38
erosion rates, 57
gradient, 26, 28, 30–32, 34, 36–37
headcut retreat, 29
profile, 30, 33–35, 37

hanging valleys, 113–114, 116–117
hazard mapping/maps, 86, 172–173, 254
head, 139, 154, 156–159
Hortonian runoff, 78
humid tropics, 162, 165–168, 170
hurricanes, coastal effects, 210, 213
hydraulic geometry, 36–37, 56, 61, 171
hydrogen budget, 167
hydrogeomorphology, 43, 57–65
hydrograph,
changes due to urbanisation, 50, 63–64
changes due to damming, 63
sediment, 54–55
solute, 54–55
storm, 50
hydrology, 43 ff.
hydrological assessment, 58
hydrothermal alteration, 192

ice plucking, 110, 113
ice wedges, 141–143, 147–148, 155
ice-margin lakes, 123
ice-wedge polygons, 141, 143, 155
icings, 148
inclinometer, 99
insects, geomorphological effects, 78, 169
inselberg/inselberg landscape, 162, 164–165, 168–169, 186, 190, 193
Institute of British Geographers, 259, 264
Institute of Geological Sciences, 137, 260
Institute of Hydrology, 44, 46

instrumented catchments, 48–55, 167–168, 189–190
International Biological Programme, 166
International Geographical Union, 73, 264
International Hydrological Decade, 166
involutions, 154, 155
isostatic movements, 241

jointing, influence of, 108, 110, 116

kame/kame-terrace, 123
kaolin/kaolinite, 170, 192
karst morphology, tropical, 164, 169, 171, 175, 187
khonu, 143–144
kinematic analysis, 103
'knoch-and-lochan' landscapes, 108, 195–196

laboratory simulation,
  coastal processes, 232
  landslide processes, 103
laminar flow, 127
land evaluation, 172
land-system mapping/maps, 172, 255
landform evolution
  see denudation chronology
landslides, 76, 79, 84–104, 165, 240, 254
  classification, 87–88
  monitoring, 96–103
  morphological measurements, 87
Late Glacial Zone III, 156
laterite, 162, 164, 168–169, 175, 185–186, 191–192
leaching, 80, 165
leaf litter, 173
limestone, solution of, 53–54
limestone catchments,
  determination of, 48
  erosion in, 53
  water movements in, 60
limnological studies, 168
lithology, effect on landforms, 108–109, 115, 158, 169–170, 174, 193, 205–206, 240
loess, 147
longitudinal profiles,
  glaciated valleys, 110, 112–113
  gullies, 30, 33–35, 37
  mudslides, 100–101
  steps in, 110, 112–113
long-shore currents, 209–210, 228
long-shore sediment movement, 228, 236–238, 240, 242
Lotka's Law, 265–266

mangroves, 213
mapping,
  channel change, 61–62, 64
  drainage basin morphometry, 59
  drainage density, 59
  flood hazard, 86, 172–173, 254
  flood-plains, 62
  geomorphological, 85–87, 150, 164, 172, 253
  hydrological, 58
  land-systems, 172, 255
  morphological, 11, 77, 253
  runoff patterns, 58–59
  slopes, 76–77
  slope instability, 85–87
  soils, 164
  solutes, 59
  streamflow patterns, 58
  water quality, 59
maps, use of, 254–255
meandering valleys, 57
meanders, 57, 61–62, 64
mega-ripples, 208
meltwater channels, 156–157
mineral assessment, using geomorphology, 150, 217–218, 252, 256
modelling,
  coastal processes, 229, 232
  landform change, 66
  mudflows, 96, 99
  runoff, 65–66
  slopes, 76, 79–80
  water flows, 60
morphological mapping, 11, 77, 253
morphometry, 45–47, 59
morphometric analysis, 57, 171
muddy coasts, 207, 209
mudflat, 204
mudflow,
  process-response system/model, 96, 99
mudflow simulation, 103
mudflow surges, 100–101
mudflows, 57, 87–90, 96, 99–101, 103, 151, 169
mudslide, automatic monitoring, 98, 100
mudslide surges, 98, 100, 102
mudslides, 90, 98, 100, 102
Munro water-level recorder, 101

Natural Environment Research Council, 134–137
neotectonics, 169
nesses, 237–238
nitric acid in rain, 163
nivation, 158
nivation hollows, 149
nutrient cycling in forests, 167–168

oyster reefs, 213

palaeomorphology
  see denudation chronology
palaeopedology, 191

'paleic' surface/landforms, 109–111, 116, 118
paleosols, 191
paleo-temperatures,
  marine 191
parallel roads, 121–124
pediment, 76–78, 186–187, 189, 193, 195
pediplanation, 186–187, 194–195
pedological studies/surveys, 164, 172, 188
Peekel micro-strain recorder, 101
percolation, 48
percolines, 41–42
periglacial phenomena
  see cryogenic phenomena
periglacial zone, 140
permafrost, 139–151
  aggradation, 141–144
  degradation, 141–144, 147, 150–151
  dynamics, 141
  epigenetic, 141, 144, 146–148
  syngenetic, 141, 143–148
  thickness, 139, 141
photography, time-lapse, 97
physical model, Thames estuary, 232–233
piezometer, 101–102
pingo, 141, 143–144, 155–156
pipes/piping, 47, 55
plastic scouring forms, 125–129
plate tectonics, 174, 214–215, 241, 272–273
plunge pool, 32
point bar, 56, 62
polygenetic landscapes, 109–110, 155–159, 169, 192–196
pore-water pressure, 89–90, 94–96, 100–102
pore-water pressure monitoring, 95–96
pre-failure creep, 101, 103
process-response system/model
  mudflow, 96, 99
professionalism, 255, 257, 260
profile analysis, 76–77, 80

quantitative landform analysis, 171
quantitative revolution, 263
quick clay, 91

radar, application of, 268
rainsplash, 78
reclamation, 233–234, 239
reefs, 213
resource inventories, 172
resource surveys, 172, 217–218
Richter slope, 77
Ring Shear test, 93–94
ripples, 208–209
river course changes, 42
river profiles, 41
river terrace, 21, 26, 36, 41, 63
roches moutonnées, 127
runoff, 58, 78
  patterns, 58
  plots, 75

salt marshes, 224, 238–239
salt spray, 214
sand waves, 229
sandur plain, 120
sand-dunes, 238–239, 256
saprolite, 185–186, 190–191, 194–195
screes, 76–77
  characteristic angle of, 77
scientific theory, 15
sea-ice, effects of, 211
sea-level change, impact of, 168, 174, 216–219, 225, 235, 240–242
sediment transport in coastal zone, 208–211, 227–229, 232, 235–239, 242
sedimentation studies, 191–193
  shallow marine, 191–192
  continental, 192–193
  paleoenvironments, 192
sediment rating curves, 54
sediment rating loops, 55
shallow planar failure model, 18
shear strength, 75, 89
shear stress, 32
shearing tests, 93–94
shore platforms, 240–241
shorefast ice, 211
shoreline length, changes over time, 214–219
sichelwannen, 125–127
silica,
  removal, 166, 170, 190
  in stream water, 166, 170
simulation, 20
slope as a series, 77
slope evolution, 78, 165, 168–169, 188
  in tropics 165, 168–169, 188
slope form, 76
slope instrumentation, 76, 78
slope mapping, 76–77
slope models, 76, 79–80
slope monitoring, 75, 78–79
slope pantometer, 77
slope profile recorder, 77
slope profiles, 77
slope profiling, 75, 77, 80
slope retreat, 78–79
slope stability concepts, 89–91
slope theory, 76
slope-angle frequency, 76–77
slope-angle mapping, 77
slope-unit analysis, 77
slopes, cartographic analysis of, 77
slopes, future research, 79–81
slopes, glaciated, 108–110
slopes, in differing environments, 76–77

slopes, long-term stability, 80
slopes, analysis of publications, 73–76
small catchment studies, 48–55, 167–168, 189–190
soil catena, 80
soil creep, 78, 79, 165, 169
soil geochemistry, 191
soil mechanics, 74, 79, 80, 89–96
soil erosion, 66, 74–75, 79, 164, 172–173, 189, 256
  accelerated, 79, 172–173, 256
solifluction, 143, 147, 149, 154, 254
solutes, 53–56, 59–61, 190
  load estimation, 59–60
  proportion of, 190
solution, 53–54, 78, 79–80, 165–166, 192, 194–195, 212
solutional lowering of landscapes, 190–192, 194–195
spring catchments,
  determination of, 48
stability analysis, 89–91
stacks, 206
stone-lines, 169
storm breach channels, 210
storm surges, 210, 224, 239
strandflat, 118
stream gradient, 26, 28, 31–32, 34, 36–37, 61
stream head extension, 42
stream meanders, 61–62
  changes due to man, 64
stream networks, 45–47
  variable definition of, 46
  variation over time, 46
stream ordering, 268
strength and stress concepts, 89
stress, effective, 89–91, 93, 95
striations, 127
structure and landforms, 169–171, 174, 193, 205–206
sturzstroms, 103
subglacial drainage, 120–121, 123, 125, 127
subglacial processes, 107, 120–123, 125–129
subglacial scouring forms, 125–129
subsidence, 220
subtalus slope, 115
suffosion, see thermokarst
supraglacial drainage, 119, 121
surface runoff, 78
surf-zone dynamics, 209, 227–228
suspended sediment, 54–56, 190, 206–207
systems analysis, 8
systems theory, 17

taliks, 141, 144, 147–148, 150
talus, 115
teaching of geomorphology, 1, 73, 252, 254–255, 258–259, 267, 269–270
tectonic activity and landforms, 193, 218–219, 241
termites, 78, 169
Tertiary, 109, 187, 192–195, 240–241
theory in geomorphology, 1 ff.
  Conventionalist, 2, 10–11
  Functional base of, 2, 7–9
  Historical base of, 2, 4–6
  Immanent base of, 2–4
  Realist base of, 2, 9–10
  Taxonomic bases of, 2, 6–7
  Teleological base of, 2–3
thermocirque, 142
thermokarst, 142–144, 147, 150, 155–156
thermo-abrasion, 142, 144
thermo-erosion by streams, 142, 148
thermo-planation, 142
throughflow, 50, 78
  troughs, 79
tidal currents/streams, 208, 229, 235, 238, 242
  measurement techniques, 229
tidal-creeks, 239
tidal morphology, 229, 238, 242
tidal range, 208, 240
tidal surges, 232
tide-gauge records, 241
tides, 208, 224, 238, 242
time-series analysis, 228
tors, 157–158, 187, 189, 192–193
tracer experiments, 228, 232
translational slides, 87
tropical ecosystem, 166–168, 190
tropical geomorphology, growth of, 162–164, 185–189
tropical rainforest, 162, 165–168
tropical soils, investigations of, 191
tsunamis, 208
tunnel valleys, 120

U-shaped valleys, 113, 115–116
  description by mathematical functions, 115
underfit streams, 41–42, 57
underground catchments,
  determination of, 48
  erosion in, 53
  water movements, in, 60
undrained loading, 100–101
urbanisation,
  effect on channel capacity, 63–65
  effect on discharge, 50, 63–64, 256
  effect on floods, 50
  effect on sediment concentrations, 56

valleys, glaciated, 109–117

valley-side benches, 110, 116, 121–124
vegetation, influence of, 33–34, 79, 142, 213, 239
vegetation removal, impact of, 79, 142
Vigil Network, 44–45, 54

Water balance,
 determinations, 51
 modelling, 50
water quality, 59
waterfalls, 165
watershed breaching, 113
wave climate, 210
wave environments, global pattern, 209
waves, 118, 208–211, 224–228, 232–235, 237–240, 242
wave refraction, 227–228
wave-measurement techniques, 209, 227
wave-produced landforms,
 see waves
weathered land surface, 195
weathering,
 frost, 118, 139–151, 154–159, 166
 of granite, 169–170, 187, 194–195
 pre-glacial, influence of, 195–196
 rates, 170
 solution, 53–54
 tropical, 163–166, 169–170, 175, 186–188, 190–195, 268
weathering front, 165, 195
wind transport, 210
wind-water interaction, 209–210
Woodhead sea-bed drifter, 229